International Conference on Sustainability in Architectural Design and Urbanism (ICSADU 2017)

IOP Conference Series: Earth and Environmental Science
Volume 99

Semarang, Indonesia
9 – 10 August 2017

ISBN: 978-1-5108-5528-1
ISSN: 1755-1307

Printed from e-media with permission by:

Curran Associates, Inc.
57 Morehouse Lane
Red Hook, NY 12571

CURRAN ASSOCIATES INC. proceedings.com

Some format issues inherent in the e-media version may also appear in this print version.

Printed by Curran Associates, Inc. (2018)

For permission requests, please contact the Institute of Physics
at the address below.

Institute of Physics
Dirac House, Temple Back
Bristol BS1 6BE UK

Phone: 44 1 17 929 7481
Fax: 44 1 17 920 0979

techtracking@iop.org

Additional copies of this publication are available from:

Curran Associates, Inc.
57 Morehouse Lane
Red Hook, NY 12571 USA
Phone: 845-758-0400
Fax: 845-758-2633
Email: curran@proceedings.com
Web: www.proceedings.com

Table of contents

Volume 99

International Conference on Sustainability in Architectural Design and Urbanism 2017

9–10 August 2017, Semarang, Indonesia

Accepted papers received: 5 December 2017
Published online: 5 January 2018

Preface

Papers

PREFACE

The Conference on Sustainability in Architectural Design and Urbanism (ICSADU 2017) is an international scientific forum for discussing and sharing ideas, thoughts, practices from around the world in the field of architecture and urbanism. The first ICSADU was held by Department of Architecture, Faculty of Engineering, Universitas Diponegoro in Aston Hotel Semarang, Indonesia as part of their annual international program. We highly appreciated the authors and audience contribution for the conference program and have made their way to attend the conference.

Resiliency is the issue in 21th century regarding urbanism & architecture. Climate change, environmental degradation, inaccurate and risky in urban planning process, and the emergence of the hazards have become major problems in human habitation on earth. A few things can be done to anticipate them to occur so human existence can be sustained. Some communities have been launched plethora of design innovations as one of the solutions that pay attention to the environment. However, if those problems emerge, the risk is unlikely inevitable.

Based on the afore mentioned facts, it is necessary to remind designer and planner to think carefully and eager in providing solutions in the form of a design that can make a building and its surroundings more comfortable, safer and more sustainable when exposed to incoming risks.

The objective of this conference are to be an international forum for the academia, researchers, practicioners , governments, students,and other parties in communing their research and development results on fundamental and applicable of architecture and urban planning, to extensively share and exchange ideas, and thoughts related to the themes, to facilitate network among participants for improving the quality and benefits of the researches. The Conference theme was 'Resilient City and Resilient Design toward a Better Living Space'. The range of topics includes: (1) management of policy, (2) infrastructures, (3) environment, (4) communities & societies, (5) housing, also (6) building technologies.

Putting together ICSADU 2017 was a team effort. We first thank the authors for their contributions to the program. We must also thank the organizing and scientific committee for their invaluable efforts in reviewing papers and constructive feedback to the authors. We are very grateful for Prof. Tomohiko Yoshida from Ritsumeikan University, Prof. Gagoek Hardiman from Diponegoro University, Dr. Saehoon Kim from Seoul National University, and Thana Chirapiwat, Ph.D from Silpakorn University for their contribution as keynote speakers in this conference; Journal of Architecture and Urbanism, especcialy for editor in chief - Prof. Almantas Samalavicius; IOP Conference Series Earth and Environmental Sciences; and finally we would like to thank Department of Architecture, Faculty of Engineering, Universitas Diponegoro for the support in putting together this first ICSADU.

We hope that this conference will foster the exchange of new ideas and promote new contacts between researchers on the field of urbanism and architecture.

Sincerely,
On behalf of the conference committee,
Dr. Atiek Suprapti
Chair of ICSADU 2017

Steering committee

Dr. Ir. Atiek Suprapti, MT
Edward Pandelaki, ST, MT, PhD
Satriya W Firmandhani, ST, MT
Chely N Bramiana, ST, BBE, MSc
Ratih Widyastuti, ST, MT
Dr. Eng. Bangun I R H, ST, MT
Anang W Sejati, ST, MT
Hana Faza S R, ST
Mustika Wardani, ST
Rizqi Jamaludin, ST
Aldilla H Latifa
Irfan F Zuhair
Hira Kemala
Farelio Artha

Scientific committee

Prof. Tomohiko Yoshida (Ritsumeikan University, Japan)
Prof. Gagoek Hardiman (Universitas Diponegoro, Indonesia)
Assoc. Prof. Sae-hoon Kim (Seoul National University, South Korea)
Thana Chirapiwat, Ph.D (Silpakorn University, Thailand)
Prof. Alamantas Samalavicius (Vilnius Gediminas Technical University, Lithuania)
Dr. Ir. Bram Entrop (University of Twente, The Netherlands)
Prof. Dr. Jun-Ichiro G. Tsutsumi (University of the Ryukyus, Japan)
Dr. Atiek Suprapti (Universitas Diponegoro)
Edward E Pandelaki, Ph.D (Universitas Diponegoro)
Prof. Bambang Setioko (Universitas Diponegoro)
Dr. Erni Setyowati (Universitas Diponegoro)
Dr. Agung Budi Sardjono (Universitas Diponegoro)
Dr. R. Siti Rukayah (Universitas Diponegoro)
Dr. Suzzana Ratih Sari (Universitas Diponegoro)
Dr. Djoko Indrosaptono (Universitas Diponegoro)
Dr. Sunarti (Universitas Diponegoro)

This conference is sponsored by:

Department of Architecture
Faculty of Engineering
Universitas Diponegoro

Lembaga Penelitian dan Pengabdian Masyarakat
Universitas Diponegoro Research Centre

and Resilient City Toward A Better Living Space

IOP Conf. Series: Earth and Environmental Science **99** (2017) 011002 doi:10.1088/1755-1315/99/1/011002

IOP Conf. Series: Earth and Environmental Science **99** (2017) 011002 doi:10.1088/1755-1315/99/1/011002

IOP Conf. Series: Earth and Environmental Science **99** (2017) 011002 doi:10.1088/1755-1315/99/1/011002

IOP Conf. Series: Earth and Environmental Science **99** (2017) 011002 doi:10.1088/1755-1315/99/1/011002

 doi:10.1088/1755-1315/99/1/011002

IOP Conf. Series: Earth and Environmental Science **99** (2017) 011002 doi:10.1088/1755-1315/99/1/011002

 doi:10.1088/1755-1315/99/1/011002

IOP Conf. Series: Earth and Environmental Science **99** (2017) 011002 doi:10.1088/1755-1315/99/1/011002

IOP Conf. Series: Earth and Environmental Science **99** (2017) 011002 doi:10.1088/1755-1315/99/1/011002

IOP Conf. Series: Earth and Environmental Science **99** (2017) 011002 doi:10.1088/1755-1315/99/1/011002

IOP Conf. Series: Earth and Environmental Science **99** (2017) 011002 doi:10.1088/1755-1315/99/1/011002

IOP Conf. Series: Earth and Environmental Science **99** (2017) 011002 doi:10.1088/1755-1315/99/1/011002

IOP Conf. Series: Earth and Environmental Science **99** (2017) 011003 doi:10.1088/1755-1315/99/1/011003

Peer review statement

All papers published in this volume of *IOP Conference Series: Earth and Environmental Science* have been peer reviewed through processes administered by the proceedings Editors. Reviews were conducted by expert referees to the professional and scientific standards expected of a proceedings journal published by IOP Publishing.

The Role of Stakeholders Related to the Management of Ecological Function of Urban Green Open Space. Case Study: City of Depok, Indonesia

Andy Anton Mangopa Malik
Doctoral Student in School of Architecture, Planning, and Policy Development Bandung Institute of Technology, West Java, Indonesia

Correspondinng e-mail: andymalik_80@yahoo.com

Abstract. Urban green open space is one of the assets that provide substantial benefits to the urban community. One important function of urban green open space is a function of ecology. This study will provide initial explanation on the various studies related to the ecological function of urban green open space. The study of urban space management approach related to ecological function will explain the extent of the role of stakeholders in the urban areas that will further strengthen the importance of the existence of green open space, especially in city of Depok. With so many problems related to the supply and use of green open space in the city of Depok. This approach was originally applied by the private sector and many applications made a great contribution, so it began to be used by the government in managing public assets there. This study will use descriptive method, at the beginning of the study will explain the existence of the reality of urban green open space as part of the urban space by viewing it from theoretical overview of space, function and role of the various problems that occur in it. The results of this study indicate there are six problems in the management of green open spaces in city of Depok. Using the stages in asset management will provide space for participation of existing stakeholders in the management of green open spaces in city of Depok.

Keywords: management, space, urban open space, city of Depok

1. Introduction

Urban environment has become a major habitat of the community at this time. It also means that the number of persons affected by the conditions of the urban environment continues to increase (van Poll, 1997) [1]. The urban population is growing rapidly, in 1975, there are only 1.54 billion people living in urban areas, in 2000, there were 2.92 billion projected world population living in urban areas. This figure will be doubled in 2025, of which approximately 5.07 billion people will live in urban areas (UNFPA, 2007)[2]. In terms of area, the urban area has an area of only 3% of the Earth's surface, but more than half the world's population, has become the center of social cultural transformation, the engine of economic growth, as well as places of innovation and knowledge creation (Wu, 2009) [3].

Some environmental experts have realized that every urban problems, particularly social problems, have root causes related to the environment (Byrne and Wolch, 2009)[4]. This is the background why urban communities still need a harmonious closeness to the natural environment. Urban park actually is an open space which can integrate between the environment, society, and health in the urban environment by promoting an ecological approach to health and well-being that is based on contact with nature. In addition, the urban park is also useful in environmental, aesthetic, recreational, psychological, social, and economical for urban communities. But in fact there is still a urban park problems that reduce the presence and functionality.

This essay discusses the asset management as an approach in the management of the urban park, which includes the role of government, the public and the private sector. In an effort to understand the problems of managing the urban park, this essay comprehensively will also discuss in detail the existence of an urban park as part of urban space, function and problems.

1.1 The History of Green Open Space

The role of green open space is also gradually changed since the mid-19th century when it was first introduced as a refuge from the urban centers increasingly crowded as described in the previous section regarding the "garden city" and "radiant city". A study of the history identifies four types of open green spaces based on the characteristics, social objectives and user base by Roberts (2014) [5]:

(1) "Pleasure garden" (1850- 1900), where the park is regulated environment to reduce the impact of the industrial city;
(2) "Reform parks" (1900-1930) that gave birth to the concept of the park for the children of the working class and immigrant families;
(3) "Recreation facility" (1930-1965), he went on to extend this fantastic concept to the suburbs and the urban with the entry of the stadium, a swimming pool and indoor facilities; and
(4) "Open space system" (1965-1990) where the trend of green open spaces such as streets, plazas and waterfronts are used for recreational purposes.

Harnik (2006)[6] suggests that the desire for green open space utilization declined after World War II due to the development of more focused on the suburbs. As a result, plenty of green open spaces deteriorate due to lack of funds for maintenance and operation. Interest in the development of open green space refurbishment in the early 1990's where the urban population began to grow. In the last two decades of social and environmental issues regarding the provision and discussion of sustainable green open spaces began to be discussed (Cranz and Boland, 2004)[7]. Green open spaces are an important component of the social structure and ecological landscape of urban centers. In addition to providing aesthetic value and a place for recreation, urban green open space plays an important role in promoting economic development, improving public health, providing employment opportunities, and become self-sufficient in the use of existing resources (Cranz and Boland, 2004; Walker, 2004) [8]. Of the various previous studies have shown that green open spaces have an important role related to physical activity and social citizens who created the "sense of community" and can improve the quality of life of the urban community (CABE, 2005; McCormack., et.al, 2010; Roberts N , 2014) [9]. Green open space also has a role in improving the economy of the community (Cranz and Boland, 2004; CABE, 2005; Rabare, .et.al, 2009) [10].

1.2 Green Open Space Standards Review

According to Indonesian Minister of Home Affairs Regulation No. 1 of 2007 on the green open space arrangement of urban areas, the definition of Green Open Space is part of an open space of an urban area filled with plants and plants to support ecological, social, cultural, economic and aesthetic benefits. Green open space itself is divided into two types, namely Green Open space of Urban Green Open Space Public and Private Urban Area. Green Open Space is a Public Urban Area Urban Area of green open space provision and maintenance are the responsibility of the District / City Government.

Based on the type, urban green open space consists of parks, nature, recreation parks, neighborhood parks housing and settlement, neighborhood parks office and commercial buildings, forest parks, city forests, protected forests, landscapes such as mountains, hills, slopes and valleys, nature reserves, botanical gardens, zoos, public cemeteries, sports fields, the parade ground, open parking, land urban agriculture, the path under high voltage, the river banks, beaches, buildings, lake and marsh, the path of road safety, the median road, railways, gas pipelines and pedestrian, the area and the green belt buffer zone (buffer zone) airfield and park roof (roof garden).

In determining the land required to meet the public demand for parks and open space, National Recreation and Parks Association (NRPA) has developed national standards for the level of service and community environment park ranges from 9.25 to 14.5 acres (1 acre = 4046.86 m^2) per 1,000 people within a mile (1 mile = 1.60934 km), which is used as a criterion to evaluate the level of service standards of green open space) [12] (See table 1). Provision of green open spaces of both standardization lies in providing orientation. The provision of urban green open space based on minimum population and service range in each type.

Table 1. Association National Recreation and Parks Association (NRPA) Standard for Green Open Space

Type	Acres (1 acre=4046.86 m^2) per 1,000 persons	Service Area (1 mile=1,60934 km)	Accessibility
Community Park	8.0-10.0	1/2- 3	15 minutes driving
Neighborhood Park	1.0-2.0	1/4-1/2	5-10 minutes walking
Mini-Park or Pocket Parks	0.25-0.50	<1/4	5-10 minutes walking
Open Space / Open Lands	1-2	Entire Community	45 minutes driving
Total:	**9.25-14.5**		

Source: National Recreation and Parks Association (2014)

In Indonesia, to determine the extent of green open space is based on the number of residents performed by multiplying the number of population served by the standard of green open space according to the applicable regulations. While on the provision standard issued by National Recreation and Parks Association using the extent approach, service coverage areas and accessibility to existing green open spaces .

Table 2. Provision of Green Open Space Based Population in Indonesia

No	Service Unit	Green Open Space Type	Minimum Area/unit (m2)	Minimum Area/household (m2)	Location
1	250 inhabitant	Small Neighborhood park	250	1,0	Center of Neighborhood resident
2	2500 inhabitant	Neighborhood park	1.250	0,5	Center of Neighborhood resident
3	30.000 inhabitant	Sub District park	9.000	0,3	grouped by school / village center
4	120.000 inhabitant	District park	24.000	0,2	grouped with the school / center subdistrict
		Cemetery	adapted	1,2	spread

Source : Regulation of the Indonesian Minister of Home Affairs 5, 2008 on Guidelines for the Provision and Utilization of Green Open Space in Urban Area

The second similarity lies in the provision of this standard that focuses on the human user. Widespread use minimal capita in use by Regulation 5, 2008 that differ based on the type of need attention. In contrast to the Indonesian Minister of Home Affairs 5, 2008[13] About Procedures for Environmental Planning Housing Urban wear range of services in the supply area of existing green space (See table 2).

The orientation of provision of some existing standardization indicates that the paradigm of "people centered development" thinking is very strong in the provision of green open space. Attention to environmental conditions, the carrying capacity of the land can only be seen within the reach of services as well as the timing of achievement in the provision standards issued by the NRPA. It is also not enough to see the conditions and environmental characteristics in the provision of green space represented.

1.3. Ecological Functions Of Green Open Space in Urban Area

Green spaces that feature good connectivity and act as "wildlife corridors" or serve as "urban forests", can sustain populations of existing species in urban areas (Haq, 2011; Byrne and Sipe, 2010) [14]. Local green space is based on the protection and optimization of natural ecological systems and refers to suburban green spaces on a large scale not only enhancing the ecological environment of the entire city area and also providing important support from urban environmental improvements. Furthermore, the introduction of green space in the suburbs to the city also acts as the basis of ecological balance (Wuqiang et al., 2012) [15].

Supply of urban green space ecosystem services ranging from the maintenance of biodiversity through the urban climate regulator. Compare with the rural areas, the difference in solar input, the pattern of rainfall, solar radiation, air temperature, wind speed and relative humidity varies significantly because of the built environment in urban areas (Heidt and Neef, 2008) [16]. Therefore, the availability of green open space, vegetation around the houses, to the management of water bodies by the authorities can help to reduce the improvement of temperature situation.

The presence of motor vehicles produce pollutants that harm children, the elderly and people with respiratory problems (Sorensen et.al., 1997) [17]. Greening activities in urban areas can reduce air pollution directly when the dust and smoke of particles are trapped by vegetation. The effect of "urban heat" caused by large heat-absorbing surface, the combination of high energy use causes increased urban temperatures by 5 ° C. Research has shown that on average 85% of the air pollution that got into the park can be filtered (Bolund and Sven, 1999) [18]. Noise pollution from traffic and other sources can cause stress and create health problems for people in urban areas. The overall cost to reduce noise pollution is estimated to be in the range of 0.2% - 2% of EU gross domestic product. In the presence of urban green space most of the cities in the European Union to reduce the noise level depends on the quantity, quality and distance from the source of noise pollution. In contemporary studies on urban green space required consideration of complex urban ecosystem, urban green space conservation to maintain natural ecological network for environmental sustainability in the city.

For cities with high urbanization and economic growth such as China must consider the dynamic form and expand urban areas in the management of green spaces of urban effective that will be able to contribute in reducing overall CO2 to maintain or even improve the uptake of CO2 by natural ecosystems (Huang et al., 2009) [19].

Reduced pollution vary in each city depends on the amount of tree cover, with an increase in tree cover leads to a reduction of greater pollution, but also for example the length of leaves each season as well as a variety of meteorological variables that affect transpiration tree and speed deposition (Paoletti et al ., 2011) [20]. Trees can also reduce the temperature of the air through transpiration, affecting photochemical ozone by reducing the production of ozone. Cavanagh et al, (2009) [21] explain in more detail about the specific role of urban trees in reducing air pollution which mentions the function of vegetation in terms of atmospheric particles intercept and absorb various gaseous pollutants (Yin et al, 2011) [22].

1.4 Problems in the Provision of Green Open Space in Urban Area

In the theory of urban planning, Melville (1996) [23] said that city planning is the result of two groups which both have strengths, namely the government and non-government. The government is representative and administrative structure in the form of a formal organization and legal power to regulate and organize the life of the city. Non-government consists of various forces that are not part of the official government, but nevertheless have an influence or determine the planning of the city. The linkage between the two forces in order to determine the shape of urban planning, it is very different with inter-community, inter-political system and inter-period history. However, the city planning is the result of the various activities undertaken by organizations and individuals, both within, and outside the government.

According Purnomohadi (2006) [24] one of the factors is the lack of available urban green open space inconsistency of policy and strategy of the Urban Spatial Planning, lack of understanding and attention on the urgency of the existence of green space in urban areas unity. Strategic planning for development in the area of green space is not adequate, because it is considered as a public space (common property) that is not economically profitable so mutually irresponsible. Green open spaces are often regarded as dustbins, wild huts and disease-carrying vector nests, which tend to be more of a problem than benefits.

Population growth resulting in densification of population and settlements rapid and uncontrolled in parts of the city. This leads to the need for increased space to accommodate their interests. The increasing demand for space, especially for residential and undeveloped land increasingly impact the deterioration of environmental quality. Spatial Plan that has been made is not able to prevent urban land conversion so that the existence of green open space is threatened and the city even more uncomfortable to move.

In Indonesia according to Siahaan (2010) [25] that the likelihood of a decrease in the quantity of green space, particularly green space in the last 30 years is very significant. In big cities such as Jakarta, Surabaya, Medan and Bandung, green space area has been reduced from 35% in the early 1970's to 10% today. Green open spaces that exist largely been converted to urban infrastructure and new residential areas. The main problem is the reduction in the presence of green space green space because of limited land and inconsistent of applying spatial policy. Reduced of green open space caused by the conversion of land that is to switch function green open space for other space designation.

According Dwihatmojo (2015) [26], Indonesian Law No. 26 Year 2007 on Spatial Planning mandates that city spatial planning must contain a plan provision and utilization of comprehensive Green open space minimum of 30% of the total area of the city. Green open space in urban green open space consists of public and private green open space where the proportion of green open space in urban areas is at least 30%, consisting of 20% green public open space and 10% consists of green open space private. The proportion of 30% is the minimum size to ensure the balance of the ecosystem of the city, good balance and equilibrium microclimates hydrological system, as well as other ecological systems that can increase the availability of clean air that is needed by the community, as well as well as to improve the aesthetic value of the city. The target area of 30% of the area of the city can be achieved gradually through the allocation of urban land is typically (Regulation of the Indonesian Minister of Public Works No. 5 Year 2008 on Guidelines for the Provision and Utilization of Green Open Space in Urban Area).

But the facts that the existence of urban green open space in Indonesia is far from the ideal proportions, the dominant market forces change in land use so that the existence of green open space increasingly marginalized even ignored the functions and benefits. Spatial expected to accommodate not seem powerless to resist the market mechanism. Legal instruments set of spatial planning should be implemented properly by the decision maker. The government must be consistent in carrying out spatial planning. Provision of green open space should be adjusted to the appropriations specified in the spatial plan. Law on Spatial Planning that includes sanctions can be used as a legal basis to meet the needs of green open space. The government should be able to provide green open space for the community so as to provide comfort because of its quality environment. Identification of the availability of green open space needs to be done so that the government determine the availability of green space as one of the ingredients of evaluation in determining the direction of policy and the protection of open spaces. Green open space as a component of the space availability levels both in quality and quantity must always be taken into account in urban planning process in order to create environmentally sustainable cities (Dwihatmojo, 2015) [27].

Another cause of the limitations of green open space provision is limited urban land due to high land prices. For municipal governments to own land and manage it to be a green area for urban lungs, the government must buy land. Private parties are also beginning to understand the potential of this government does not necessarily give the price of normal, this condition occurs in several regions in Indonesia.

IOP Conf. Series: Earth and Environmental Science **99** (2017) 012001 doi:10.1088/1755-1315/99/1/012001

1.5 Methods

This study will use descriptive method, at the beginning of the study will explain the existence of the reality of urban green open space as part of the urban space by viewing it from theoretical overview of space, function and role of the various problems that occur in it. This study will provide guidance in the form of an asset management approach in view of the role that is more specific to the aspect of urban environment in city of Depok, Indonesia.

2. The Role of Stakeholders in Managing Ecological Function

In this section will explain the various problems that occur in the provision of green open space and the role of stakeholders in the provision of green open space in the city of Depok.

2.1 Case Study of Green Open Space Provision in Urban Area (city of Depok)

Based on the Long-Term Vision Development Plan of Depok City 2006-2025 that makes Depok as a religious and eco-friendly trading and service city. City of Depok is as one of the buffer city which belongs to Jabodetabek region. Jabodetabek area is the region with the largest urbanization process in Indonesia because of the tendency to become one of the centers of trade and industry and services that encourage residents to work and settle in this region. City of Depok as a buffer city is also directed to settlements, educational towns, trade service centers, tourism and water absorptions. This situation causes the mobility of activities of the population will be higher and competitive so that urbanization process will continue to increase which impact on the increasing area of settlement and conversion of agricultural land to non agriculture. What happens next is the very visible pace of development in this area. The city of Depok actually leads to a city of trade and education services. For it is not surprising that the housing construction is as fast as the construction of trade and education facilities.

Based on the Urban City Spatial Plan of Depok 2012-2032, the amount of green open space in city of Depok is far from what is said according to Law No. 26 of 2007 and regulation of Public Works No. 05 of 2008 on Spatial Planning. In the regulation made by the central government, the provision of green open space urban must meet 30% green open space from city of Depok area, with the composition of 20% public and 10% private. The number of green open spaces of Depok City currently up to 2011 is 3,110, 88 Ha or only 15.53% of the total area of Depok City with an area of 20,029 Ha.

There are several factors that affect the lack of availability of green open space in city of Depok. The first factor, beginning with development planning, is the more focused development planning focus that focuses on its development infrastructure in the commercial city. The green open space from the planning side has become one of the top priorities. City of Depok has begun to pay attention to the arrangement of space and environment. In 2011, city of Depok only had 9 priority scales which did not discuss about urban environment. On the priority scale of 2011, city of Depok did not focus its environmental priorities, but only mentioned the spatial layout on the seventh priority scale, as one of the priorities in the city of Depok APBD. In 2013, the city of Depok has begun to focus on the environment.

The second factor, is the problem in implementation of the planning that are created by the city government. The implementation in this case is related to the implementation of the agencies related to green open spaces, such as Bappeda, Spatial Planning and Settlement Service, Office of Sanitation and Gardening and Environment Agency. Implementation that is a core of a plan must be guided by the planning that usually always meet the barriers, will affect the success of the program. In the implementation of annual work programs created by each of these agencies found some problems that cause impeded implementation namely the dimensions of shifting space that resulted in the dimension of time also shifted and the dimensions of costs are also affected.

The third factor, budgetary constraint factor is also another factor in the availability of green open space in city of Depok. When viewed from the existing land use, Depok still looks green, but the ownership of the land that is green is not from the public element, but most private property that is not managed properly, and also can lost at any time because it was built by the owner. The city government cannot do anything because the land is private property, if the government wants to manage it means the city government must buy the land. Private parties who also have begun to understand the potential possessed by this does not necessarily give a normal price. This is where the difficulty is owned by the government of Depok City.

The fourth factor, is the weak of supervision. Supervision in the case of land provision in city of Depok is still limited related to the quantity and quality of the supervising human resources. Supervision is not effective, because there are areas that are not affordable, due to limited human resources. The fifth factor, is the limited land, the land for green open space in city of Depok is still limited due to the increasingly density of population every year. The last factor is the lack of public awareness.

2.2 The Role of Stakeholders in Urban Green Open Space Management

There are three approaches in Urban Management, namely: traditional; market-based; and participatory, (Mohamed, 2009) [28] (See table 3). Urban management approach seeks to fight injustice, community involvement informal settlements policy making in the urban space. However, to achieve this successful involvement of communities should be active in the policy-making process. The urban authorities should also have the right credentials that allow them to deal with issues that are important to the inhabitants of informal settlements such as the access to urban land, housing, basic services, education and health. In addition, the urban government should be open to and supportive of informal settlements, and have appropriate mechanisms for their participation in policy-making. On the other hand, informal settlement communities can benefit from the opportunities of participation if they have a strong social capital, both the structure of society that are connected to make their voices can be heard by the policy makers. In an open democratic system, participation continues to grow through the interaction between policy and legal provisions and elements of the micro context, with the facilitators (local officials and community organizations). For explanation can be seen in the table below:

Table 3. Urban Management Model Approach

Approaches to urban management	Traditional	Market-based	Participatory
Key objective	Maintenance of public order - provision of services	Managing urban growth	Social equality and expanding citizenship
Role of citizens	Subjects - voters	Customers/consu-mers	Partners
Role of urban authorities	Rulers - trustees	Managers	Partners
Type of relationship between citizens and authorities	Coercive - delegation	Responsiveness	Collaboration
Policy-making	Technical processes	Less emphasis on policy processes	Political and technical processes
Participation of disadvantaged population	No citizen participation	Citizen participation not desired	Participation of citizen, particularly the poor is encouraged

Source: (Mohamed, 2009)

This table shows that the participatory urban management approach offers a conducive environment for the participation of the poor, such as the informal settlement communities, and policy-making at the municipal level. Participatory model also shows the usefulness of the participation process with the government that enable the poor to take part, unlike the two other approaches (traditional and market-based). In the book "Managing The City" (Diamond. J, et.al, 2007) [29] states that a holistic approach is needed and also integrated in solving various problems in urban areas. All parties must make a real contribution to the planning so as not to bring up a new gap in terms of thought or practical terms. In such efforts it is necessary to explore whether to allow an agreement of the interests of each party.

In the current conditions according to Diamond et.al (2007) [30] this is difficult due to a conflict of interests such as the imbalance of community aspirations with the central government. This condition also often occurs in the provision of public facilities (in this case Green Open Space). One approach to solving the complexity of this urban green open space problem is "asset management". In Roos and Lukman (2010) [31], this concept was first spawned by the private industry, integrating various thoughts from Demming, Baldridge, and several other figures (FHWA, 2007) [32]. Asset management is a process for improving the understanding of asset conditions, improvements in operating costs, and performance, which assist in improvements in decision making (World Bank, 2000) in Siregar (2004) [33]. Asset management as a process of maintaining or maintaining and utilizing public capital. The focus of public asset management is on decision-making on the construction, use, maintenance, repair and cessation or replacement of roads, sewers, and other infrastructure. The goal is to achieve maximum total capital payback where the approach used is a participatory approach (Lemer, 2000 in Siregar, 2004) [34].

There are three main factors that prompted the government needs to do its management asset (Bertovic, et al., 2009 in Akbar, et al., 2010) [35], namely:

1. Decentralization, local government must provide a number of services continues to increase with limited financial resources.
2. Local government property real estate assets are often much more significant than annual revenues from related assets. Although many assets are not intended to generate revenue, there is still a possibility to increase the revenue. In addition, small expenditures may indicate that there is a need to increase expenditures for asset maintenance purposes.
3. Local governments usually hold more power and flexibility in managing their assets. Rather than raising taxes and charges that are politically process which is sensitive, it would be better if the assets managed by local government to be able to contribute to increase local revenue.

As shown in Table 4, illustrates the stages of asset management as well as the role of the various parties involved management of urban parks (Vogelsang-Coombs and Miller, 1999; Rakodi 2003, Siregar, 2004; Akbar and Lukman, 2010) [36].

Table 4. Stages of Asset Management And The Role of Local Government, Public and Private Parties

STAGES	DISCUSSION	THE ROLE		
		LOCAL GOVERNMENT	PRIVATE SECTOR	PUBLIC SECTOR
Inventory	Collecting the essential characteristics of the park so that it can provide relevant information in urban green open space management decisions.	The local government's role here is to give responsive attention to the needs of poor urban residents, including the provision of public (Rakodi, 2003)	Encourage the private sector to conduct an inventory of activities and events that will have an impact on the use of urban green open space.	Efforts inventory could also involve the community (community organization) in a community for the accuracy of the information about the presence of existing urban green open space.
Valuation	Efforts to resolve the legal issues of assets, in this case related to the status of	In general, local government role is determined in accordance with the urban management	Participates in the collaborative management and provision of urban green open space	Urban communities are entitled to contribute to urban assets

	control of park assets and strategies to solve various legal issues related to the acquisition or transfer of assets.	model adopted, the task generally make laws, policies and budget allocations as an authoritative decision of administrative (Vogelsang-Coombs and Miller, 1999)	with the government for the sake of improving the quality of urban areas (social, economic, ecological, aesthetic)	are critical to the local government through discussion forums or suggestions through the media information.
Legal Audit	Assessment of urban green open space assets refers to the four functions of the ecological value of green space, social, aesthetic, and economic.	Conducting the process of assessment of the existence of urban green open space that is based on a specific standard or guideline	Create a private community of caring and critical of the arrangement of urban green open space.	In some cases, community organizations can be a mediator who can bridge the desire of the urban population about the existence of the existing urban park.
Optimazion	Efforts to take advantage of opportunities that exist so urban green open space can fulfill its functions optimally	Responsive to inputs from various parties regarding the existence of public space by providing information media and discussion forums that will be useful in the preparation of budgetary provision and maintenance of public space (urban green open space)	Encourage local governments in the improvement and optimization of public space in the form of mutually beneficial cooperation in the form of cost sharing management and maintenance of urban green open space.	Efforts to optimize from the existing urban green open space that can be offered by community : maintaining and caring for the existing urban green open space assets through the establishment of community-based environmental / administrative.
Monitoring and Control	Setting up the right instruments to perform asset management area in a professional, transparent, accountable, efficient, and effective from	Running evaluation and control functions(such as local regulations) periodically using various methods of effective and efficient. This approach can be done by the provision of human resources	Cooperating with the government in providing information on the availability of urban green open space (related to accountability, transpartasi management) as	Efforts supervision and control of existing assets can be realized with the social role of internal controls its own citizens. It is helpful to

	planning, management / utilization, as well as supervision.	and the use of information technology in the management of urban green open space.	well as contribute to the monitoring of existing assets.	see where the asset itself in terms of utilization

Sources : Adaptation dan Interpretation (Vogelsang-Coombs and Miller, 1999; Rakodi 2003, Siregar, 2004; Akbar and Lukman, 2010)

The community can play a role by improving the quality of the environment in the settlements for example by planting crops and building absorption wells and waste management. Home yard is used optimally with various types of plants either planted directly or in intermediate media such as pots and other intermediary media. The private sector can also play the role of providing green open space in facilities built (malls, plazas, housing, and so on). The government places itself as a facilitator by placing society as the main actors. Society in this case in the form of individuals, groups, institutions, or in it including the private (Dwihatmojo, 2015) [37].

3. Conclusion

Based on the results of existing studies from the conclusions can be drawn, namely:

1. The various studies that exist ecological benefits of green open space include: as a controller of air and sound pollution, natural conservation, biodivesity, improving air quality, water absorption areas, reducing the effect of "urban heat".
2. The availability of urban green open space in city of Depok is influenced by several factors, namely: first factor, starting from development planning, development planning focus more focused on commercial city that focus on development infrastructure, second factor, is at the time of planning implementation there are problems that are closely related so that in the implementation of work program made by the city government, the third factor, budgetary constraint factor is also another factor in the availability of urban green open space in city of Depok, the fourth factor, is the lack of supervision, supervision in terms of land provision in city of Depok is still limited due to the quantity and quality of the supervising human resources, the fifth factor, is the limited land, the land for urban green open space in Depok City is still limited due to the increasingly density of population every year, the last factor is the lack of public awareness.
3. Asset management approach in urban space is one approach that plays an important role in the implementation stages of spatial planning (utilization and control). This approach is one of the most valuable approaches to urban green open space management. This concept together with the complexity of planning science is able to create a holistic understanding of assets so that better decisions can be made in optimizing assets. Its ability to identify important information related to park management functions and opportunities, has provided a new way to optimize urban green open space potentials. In addition, the implementation of asset management information systems, which further strengthens the asset management position as the right approach in park asset management.
4. An asset management process that includes asset inventory, legal audit, asset valuation, legal audit, asset optimization and asset management information system development. The process of attribute identification is the most important part of this research. This process is a representation of the application of an asset management approach that seeks to capture comprehensive information about park management. The information provided by these attributes is a complete picture. Through these attributes, the information obtained is not only about the existing conditions of green open space alone, but also can be obtained alternative opportunities for green open space management in the future.
5. The asset management approach is a participatory approach, in which the role of all parties must be seen in asset management. This approach also provides space for people who previously had difficulty providing criticism, information to the government in order to get the green open space that is in line with their expectations.

4. Acknowledgments

The authors are grateful for the support of the promoters Prof. Roos Akbar, M.Sc., Ph.D. Sri Maryati, ST., MIP in making this paper, also for support from the Indonesian ministry of higher education which has been assisting the author in financing the study and completion of the paper. Thanks also to the city of Depok Government who provided assistance in the form of information and data in supporting the completeness of the data for the analysis of this paper.

5. References

[1] van Poll,H.,1997, *The perceived quality of the urban residential environment: a multi-attribute evaluation*, http://dissertations.ub.rug.nl/FILES/faculties/science/1997/h.f.p.m.van.poll/thesis.pdf

[2] UNFPA, 2007, *State of world population.*

[3] Wu, J. (2009): *Urban sustainability: an inevitable goal of landscape research.* Landscape Ecol 25: 1–4. http://leml.asu.edu.

[4] Byrne, J. dan Wolch, J. , 2009, *Nature, race, and parks: past research and future directions for geographic research.* Progress in Human Geography 33(**6**): 743–765. http://phg.sagepub.com.

[5] Roberts,Nina., 2014, *The modern urban park : access and programming; where have we been... where shall we go?,* San Fransisco University.

[6] Harnik, P.,2006, *The excellent city park system: what makes it great and how to get there,*: Trust for Public Land, Washington, DC.

[7] Cranz, G., & Boland, M., 2004, *Defining the sustainable park: a fifth mdel for urban parks.* Landscape Journal, *23 (***2**), 102–120.

[8] Cranz, G., & Boland, M., 2004, *Defining the sustainable park: a fifth mdel for urban parks.* Landscape Journal, *23 (***2**), 102–120.
Walker, C., 2004, *The public value of urban parks*, Washington, DC: The Urban Institute.

[9] Commission for Architecture and the Built Environment.(CABE), 2005,*Start with the park : creating sustainable urban green spaces in areas of housing growth and renewal,* CABE, USA.
McCormack.G.R.,.2010, *Characteristics of urban parks associated with park use and physical activity: a review of qualitative research*, Health and Place Journal, Elsevier.
Roberts, Nina., 2014, *The modern urban park : access and programming; where have we been... where shall we go?,* San Fransisco University.

[10] Cranz, G., & Boland, M., 2004, *Defining the sustainable park: a fifth model for urban parks.* Landscape Journal, *23* (**2**), 102–120.
Commission for Architecture and the Built Environment.(CABE), 2005, *Start with the park : creating sustainable urban green spaces in areas of housing growth and renewal,* CABE, USA.
Rabare E.,et.al.,2009,*The role of urban parks and socio-economic development, case study of kisumu Kenya,* Theoritical and Empirical Research in Urban Management Number 3(**12**),CCASP Terum.

[11] Indonesian Minister of Internal Affairs Regulation 1 of 2007, 2007,*The arrangement of open green space of urban areas.*

[12] Association National Recreation and Parks Association (NRPA), 2004, *National standards for the level of service and community environment park.*

[13] Minister of Home Affairs 5, 2008, *Guidelines for the Provision and Utilization of Green Open Space in Urban Area*

[14] Haq., 2011, *Urban green spaces and an integrative approach to sustainable environment*, Journal of Environmental Protection, 2(**5**): 601-608.
Byrne J, Sipe N.,2010, *Green and open space planning for urban consolidation – A review of the literature and best practice*, Urban Research Program, ISBN 978-1-921291-96-8.

[15] Wuqiang L, Song S, Wei L.,2012, *Urban spatial patterns based on the urban green space system: A strategic plan for Wuhan City* P. R. China Shi Song.

[16] Heidt V, Neef M.,2008, *Benefits of urban space for improving urban climate, ecology, planning and management of urban forests,* International Perspective.

[17] Sorensen M, Smit J, Barzetti V, Williams J., 1997, *Good Practices for Urban Greening*, Inter–American Development Bank, http://www.iadb.org/sds/doc/ENV109KKeipiE.pdf.

[18] Bolund P, Sven H.,1999, *Ecological services in urban areas*, Elsevier Sciences: Ecological Economics, 29: 293-301. doi:10.1016/S0921-8009(99)00013-0.
[19] Huang D, Lu CC, Wang G. 2009,: *Integrated management of urban green space: the case in Guangzhou China*, 45th ISOCARP Congress.
[20] Paoletti E, Bardelli T, Giovannini G, Pecchioli L. 2011, *Air quality impact of an urban park over time,* Procedi a Environmental Sciences 2011(**4**), 10-16.
[21] Cavanagh, J.A.E, Zawar Reza P, W ilson J. 2009, *Spatial attenuation of ambient particulate matter air pollution within an urbanis ed native forest patch,* Urban Forestry & Urban Greening 8(**1**), 21-30.
[22] Yin S, Shen Z, Zhou P, Zou X, Che S, W ang W, 2011, *Quantifying air pollution attenuation within urban parks: An experimental approach in Shanghai, China*, Environmental Pollution 159(8-9), 2155-2163
[23] Branch, Melville., 1996, *Perencanaan Kota Komprehensif - Pengantar dan Penjelasan.* Gadjah Mada University Press,Yogyakarta.
[24] Purnomohadi, Ning., 2006, *Ruang terbuka hijau sebagai unsur utama tata ruang kota.*, Direktorat Jendral Penataan Ruang Departemen Pekerjaan Umum, Jakarta.
[25] Siahaan, James.,2010, *Ruang Publik : Antara Harapan dan Kenyataan,* Buletin Tata Ruang, Badan Koordinasi Penataan Ruang Nasional, Edisi IV (Juni-Juli 2010), p.11-16.
[26] Dwihatmojo Roswidyatmoko. 2015, *Pemetaan neraca sumberdaya air kabupaten sabu raijua, nusa tenggara timur, indonesia,*Geoplanning Jurnal, Diponegoro University, Semarang.
[27] Dwihatmojo roswidyatmoko. 2015, *Pemetaan neraca sumberdaya air kabupaten sabu raijua, nusa tenggara timur, indonesia,*Geoplanning Jurnal, Diponegoro University, Semarang.
[28] Mohamed S.S, 2009, *The participation of informal settlement communities in city-level policy-making processes in Johannesburg,* Johannesburg.
[29] Diamond.J,et.al,(ed).2007, *Managing the city*, Routledge, New York.
[30] Diamond.J,et.al,(ed).2007, *Managing the city*, Routledge, New York.
[31] Akbar, Roos, Lukman Azhari ,2010, *Manajemen taman milik pemerintah kota Bandung berbasiskan pendekatan manajemen aset*, jurnal teoritis dan terapan bidang rekayasa sipil, Institut Teknologi Bandung.
[32] Federal Highway Administration (FHWA), 2007, *Asset Management Overview*, US Department of Transportation.
http://unpan1.un.org/intradoc/groups/public/ocuments/UNTC/UNPAN013655.pdf.
[33] Siregar, D.D.,2004, *Manajemen aset: strategi penataan konsep pembangunan berkelanjutan secara nasional dalam konteks kepala daerah sebagai ceo's pada era globalisasi & otonomi daerah.* PT Gramedia Pustaka Utama, Jakarta.
[34] Siregar, D.D.,2004, *Manajemen aset: strategi penataan konsep pembangunan berkelanjutan secara nasional dalam konteks kepala daerah sebagai ceo's pada era globalisasi & otonomi daerah.* PT Gramedia Pustaka Utama, Jakarta.
[35] Akbar,Roos, Lukman Azhari, 2010, *Manajemen taman milik pemerintah kota Bandung berbasiskan pendekatan manajemen aset*, jurnal teoritis dan terapan bidang rekayasa sipil, Institut Teknologi Bandung.
[36] Vogelsang ,Vera D. Coombs , Miller M.. *Developing the Governance Capacity of Local Elected Officials Public Administration Review* Vol. **53** Iss. 3.
Rakodi, C., 2003, *Politics and performance: the implications of emerging governance arrangements for urban management approaches and information systems*, Habitat International, 27(**4**), 523–547.
Siregar, D.D.,2004, *Manajemen aset: strategi penataan konsep pembangunan berkelanjutan secara nasional dalam konteks kepala daerah sebagai ceo's pada era globalisasi & otonomi daerah.* PT Gramedia Pustaka Utama, Jakarta.
Akbar,Roos, Lukman Azhari, 2010, *Manajemen taman milik pemerintah kota Bandung berbasiskan pendekatan manajemen aset*, jurnal teoritis dan terapan bidang rekayasa sipil, Institut Teknologi Bandung.
[37] Dwihatmojo roswidyatmoko. 2015, *Pemetaan neraca sumberdaya air kabupaten sabu raijua, nusa tenggara timur, indonesia,*Geoplanning Jurnal, Diponegoro University, Semarang.

IOP Conf. Series: Earth and Environmental Science **99** (2017) 012002 doi:10.1088/1755-1315/99/1/012002

Passive fire protection in high density village (case study, Bustaman Semarang)

Sukawi Sukawi, Satriya Wahyu Firmandhani*, Gagoek Hardiman
Department of Architecture Universitas Diponegoro, Semarang, Indonesia

*Corresponding e-mail: satriya_firmandhani@yahoo.com

Abstract. Fire hazard is the disaster that always has an unpredictable process of coming. When it comes, its level scope and the magnitude of the effects cannot be predicted. Dense settlements especially in big cities, among others Bustaman Kampong Semarang never escape from physical problems such as flooding and wildfire. If both are compared in dense settlements scope, so that, wild fire is the most potentially catastrophic. It is necessary to do a research on passive fire protection in a village of high density city such as Bustaman. Qualitative research was conducted using descriptive method to conduct observations and interviews in the Bustaman.

Bustaman as a high density village, with narrow roads and dense rows of houses. The terraced buildings are also encountered, and found many buildings use combustible material. That environmental conditions can facilitate the propagation of flames in case of fire.

To improve the established Bustaman's environment, in terms of the application of passive fire protection systems, it is recommended to utilize the road as the dividing buildings. Need to build the separation wall fireproof in every each series in several units of too long buildings and attempted open space procurement that separates rows of buildings that are too long, and also the replacement of combustible material with a material that is more incombustible.

Keywords: passive fire protection, dense settlement

1. Introduction

Fire hazard is the risk that needs to be watched, especially in dense settlements. Firewild will always be harmful. Flame and firewild occurred involves several elements. Therefore fires can be prevented by controlling the causing elements of the fire. Fire-fighting effort usually done with active and passive protection systems.Active protection system is a system of fire protection from the inside of buildings and the environment itself, and pursued by the owners themselves, such as installing utilities prevention of fire hazards. Furthermore, in the prevention and management of fire, the building design and the environment must be considered passive fire protection, such as flame-retardant material selection, compartmentalization, the spacing of the room / building, design footprint of the building which is accessible to fire engines [1].

Fire danger in dense settlements which has been described above it is important to be studied, in order to avoid casualties and great losses in case of fire. Not only within the scope of one building, the dense settlement also closely involving the environment around the building units. So, we need a thorough review within the scope of the building and the environment.

There are some items that should be considered in planning efforts secure building against fire: (1) Prevention of sources of ignition and spread of wider fire. (2) Fire protection to the building structure. (3) Smoke detection and path planning of emergency. It can be used as background knowledge in the conduct of research [2].

Building and environment structures as a condition of physical / spatial built-environment which are in that area found the facts forms-physical / spatial manmade form of buildings (facilities and environmental infrastructure), directly adjacent to the facts physical form / natural spatial [3].

Based on Minister of Public Works Regulation No. 26/PRT/M/2008 on Technical Requirements for Fire Protection System in Buildings and Environment. That any required environmental areas have the safeguards against fire, either passive or active.

Bustaman is densely populated settlements in the core city of Semarang. A village covering an area of 0.6 hectares consisting of 2 RT (neighborhood association), those are RT 04 and RT 05. Villages with a population of 366 people or 114 households. Bustaman is categorized as a very high population density because it reaches 610 inhabitants / ha [4].

Besides to the condition of a dense population, with the condition that only the residential area of 0.6 hectares, also found that the neighborhood Bustaman only have a narrow neighborhood street, with lots of terraced houses, especially the installation of biogas potential sources of ignition. The purpose of fire protection systems, especially passive system, among others, (1) Protecting the building from collapsing simultaneously due to fire, (2) Minimizing the intensity of the fire in case (that flashover does not occur), (3) Giving time for occupants to save themselves, (4) to ensure continuity of the function of the building, but still safe, (5) Protecting the safety of firefighters when fire-fight rescue operation occurred [5]. These things support the importance of passive fire protection assessment in Bustaman so can reach the objectives mentioned above.

2. Methods

Assessment of passive fire protection in Bustaman using qualitative descriptive method by describing the data in the field and analyze the corresponding variables passive fire protection qualitatively with reference to rules / regulations applied. After conducting a qualitative analysis, conclusions are drawn regarding the application of passive fire protection system in Bustaman, and formulated recommendations in terms of the implementation of passive protection system.

3. Discussion

Bustaman widely known as a producer of processed goats meat such as Bustaman goat curry. Bustaman location as a locus of research can be seen in Figure 1.

3.1. Fires Pottential

Bustaman never experienced a fire in the era of the 60s. This event was caused by citizens cooking activity, who at that time was located at gedong 10. From the description of Mrs. Siti Haliman (67), it happened because of tinder burning in the kitchen processing curry goat. Tinder wood shavings is a powder that is used to prefix burn firewood. The fire quickly spread because the building material are made of wooden planks. Looking at the Bustaman's history which has ever experienced a fire, it is fitting to be more vigilant in the next period, one of them is implementing a system of passive fire in Bustaman.

Before studying the fire protection variables, keep in mind the potential causes of the fire in Bustaman are the source of fire and electrical networks.

3.1.1. Fire Source

Some fire sources in Bustaman which potentially cause a fire:

- Citizens' kitchen are located at the front of the house which is directly adjacent to the road can be seen in figure 2. The presence of the kitchen at the front of the room in the house due to the house has narrow rooms and it is not possible to put a kitchen in it. In terms of fire protection, kitchen is outside the house has a positive side that easily to reach the source of the fire in case of fire from the kitchen, fire-fights efforts easier to do. But there is also a downside, namely in case of fire arising from the kitchen outside the house, and did not immediately extinguished, the fire would easily spread to the other house or building around them.

- Kitchen stalls and goat curry processing.

The existence of the kitchen at the stalls and goat curry processing are also the sources of fire to watch. The existence of stalls outside the home is also possible to allow the fire to spread into surrounding buildings. In contrast to the kitchen stalls, goat curry processing that is in the house has a tighter guard due to be inside the building and usually use incombustible wall (incombustible material).

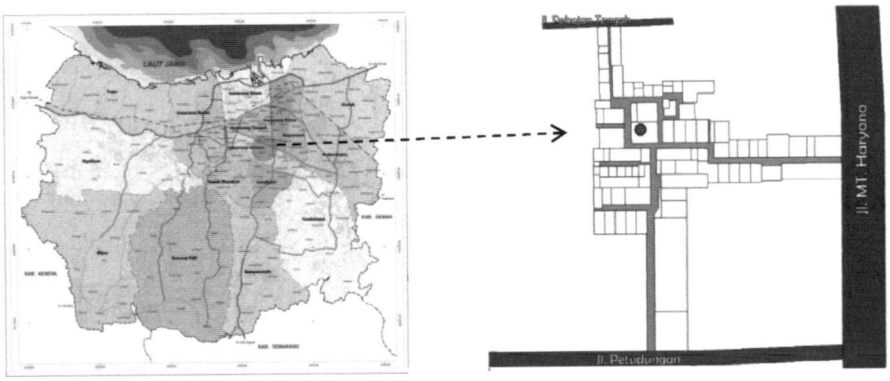

The map of Semarang City
Source: Spatial Plans of Semarang City 2011-2031

Figure 1. Location of Bustaman
Source: author, 2016

Figure 2. Kitchen Located in front of the house
Source: author, 2016

- The house and the installation of biogas
  One of the Bustaman's advantages compared to other settlements is the installation of biogas from sewage / human waste.The collected organic waste from public toilets processed into biogas. MCK (toilet) using anaerobic treatment system consisting of bio register as a tank, septic tank as a sedimentation basin, the baffle reactor to reduce Chemical Oxygen Demand (COD) by 80%, and anaerobic filter as a residence and the development of anaerobic bacteria [6]. Houses and biogas installations are also a source of fire that really needs to be watched out. Biogas is a fuel that is very effective, so the presence of the biogas installation can always be monitored. Biogas home and MCK (toilet) as storage of waste can be seen in Figure 3.

Figure 3. Biogas stove and Public Toilets
Source:Writer, 2016

3.1.2. Electric network/ electric grid.

Electrical installations in Bustaman include electrical fuse to prevent short circuiting. However, prevention of surge is not enough just to use a fuse, it is observed there is still electrical wiring in some homes that do not use the services of certified Electrical Engineering Bureau (BTL). In fact, some are undertaken solely by home owners. The type and quality of existing electrical wiring, not all in accordance with the capacity of electric current through the cable. It is very vulnerable because it can lead to short-circuits. For regions that are very dense, very important to have an electrical network certificate that is always monitored by PLN (State Electricity Company). The conditions and the situation of electric network system environment also need to be recked, the compliance community to equip his home with electrical installations using BTL (Electrical Engineering Bureau) legitimate services which is equipped with electrical installation plan made by BTL.

3.2. Variable of Passive Fire Protection

Preventive variable for passive fire hazards that exist among others: the road conditions, the building layout arrangement, buildings rows, and material use. It also includes attention to sources of ignition such as building and environment's electrical networks which has been described previously.

3.2.1. The width of the road

The road environment in Bustaman have widths from 1 to 2.5 meters, it does not meet SNI 03-1736-2000 on procedures for planning passive protection system for fire protection in houses and buildings which is requiring that a minimum road width is 6 meters [7]. With a width of that roadway, Bustaman difficult to be accessed by emergency vehicles such as fire trucks.The narrowness of the road, coupled with the dense buildings in Bustaman, will hamper the circulation and evacuation in case of fire in Bustaman. A fire truck can only be stopped by the roadside MT. Haryono, and burnout done with hoses that are required to have a range of up fires. With this kind of cramped conditions, efforts outages can also be done from within Bustaman is by spraying water from a nearby water source.Existing condition in the road representing Bustaman can be seen in Figure 4.

3.2.2. The Height of Building

The height element of the building is also a passive fire protection variable of a settlement. The higher a building, fire evacuation efforts would be more difficult because the location is more difficult to reach. Bustaman existing condition in the village consists of several terraced buildings / has two floors. With a narrow road conditions, the presence of 2-storey building needs to be watched out, because in case of fire on the upper floors, allowing for burning building debris collapsed onto the road so it can propagate the fire. The existence tritisan (sun shading) almost connect between buildings also makes a fire easily spread. Therefore this kind of situation should be changed by eliminating the building's tritisan. Conditions which is representing building height can be seen in Figure 4.

IOP Conf. Series: Earth and Environmental Science **99** (2017) 012002 doi:10.1088/1755-1315/99/1/012002

3.2.3. The distance between the Building (dilatation)

Dilation or the distance between the buildings also affects the passive fire protection in a dense settlement. It is also regulated in the classification of buildings SNI 03-1736-2000 Type C is required a separated-fire wall or walls within which limits a single dwelling unit or separate units adjacent. The conditions in Bustaman, not all houses have dilatation, but already all have walls in each. Some houses which have dilatation can be used as lengkong(side entrance) or an alternative access to enter the house when the main access cannot be bypassed (eg. being held an event).

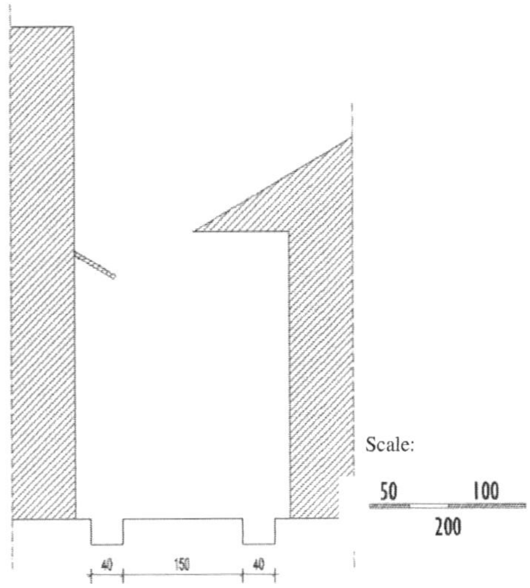

Figure 4. Bustaman road cross section
Source:Author, 2016

3.2.4. Building materials

The building materials used in houses in villages largely use permanent material that is reinforced concrete frame construction with brick masonry wall barrier. The use of these materials can inhibit the propagation of flames in case of fire. But still there are buildings that use semi-permanent materials such as walls and columns of wood. This material is not fireproof material, making it easily spread fire in case of fire (combustible material). Semi-permanent building can be seen in Figure 5.

3.2.5. Water sources

In Bustaman, water sources are found in the form of wells located in toilets or public toilets. These wells are managed by the citizen for the common good. There are also private citizens' wells and the water is piped into the bathroom or to the front of citizens' house. However, it is commonly used for fire protection is jointly owned wells located at MCK / Common toilet, so the operation be more accessible and not disturb the privacy of people's wells. Wells that are ready to be used to extinguish fires are at the center and in the east of Bustaman. Location of the wells can be seen in Figure 6.

Figure 5. Semi-Permanent houses
Source:Author, 2016

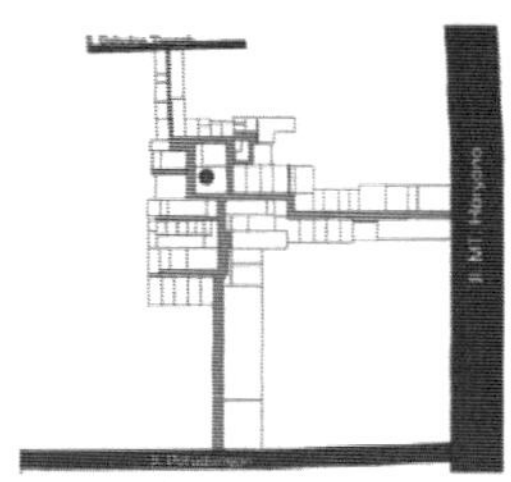

Figure 6. Well Location
Source:Author, 2016

Given the narrowness of the access in the village Bustaman, the distribution wells are assessed as insufficient. Wells should be increased again in the southern area of Bustaman towards Jl. Petudungan.

3.2.6. Society Anticipation

Passive fire protection here is not only in the context of the physical environment, but also required the participation of society as a preventive. Public awareness about the importance of fire prevention is necessary because a man with all the instruments (technology and organization) determines a good or bad environment [8]. Humans as a social component in the use of technology / organizing things will be a major impact on the natural components and cause symptoms / consequences such as fire. Some efforts should be made public to prevent fire hazards, among others:

- Passive fire protection, in the first line is determined by the detection system symptoms of the fire, which is carried out by people day and night could be in the form of patrolling.
- Flammable materials around the center of the symptoms of the fire should be broken / removed so that flame propagation is hindered. Initial extinguishing actions can be done by citizens before the fire tank car come.
- Extinguish the fire starting point in a simple way: for example, flush with water from nearby sources (wells, etc.). So that the procurement wells evenly in Bustaman environment is also a fire hazard prevention protection efforts passively.

The passive fire protection efforts will give a positive impact on the physical environment. The physical condition of a good environment will contribute to the increasing of social economic aspects [9]. So that the improvement of environmental conditions will provide benefits back to humans.

4. Conclusion

4.1. Conclusion

After analyzing the data in Bustaman, it can be concluded that Bustaman is still not implementing passive fire protection systems. This was concluded because:

a. The layout of the Bustaman's majority building is directly coincide. The layout of the building is a risky factor for the propagation of fire if any one of the buildings in the area on fire. The separation walls between the buildings are very easily carry on the flame by conduction on the building side, in addition, the temperature transmitting by convection can also spread fire to the surrounding buildings.
b. If the two-storey buildings on the dense settlements as Bustaman are burnt, the combustible material will easily fall into another building because of the narrow dividing road.
c. There are still contiguous buildings made of a-high-value combustible material or very easily-flammable.
d. The narrow neighborhood streets are also made the fire truck difficult to get in. Up to think that to make all buildings are covered by a jet of water from a hose of the fire extinguisher tank, the position of the car is on a highway.
e. Public awareness to prevent and treat symptoms of passive fire is still lacking. Proven with many temporary buildings in front of the house causes the function of road as a path separator is not optimal because every each buildings are connected by temporary equipment/ buildings.

4.2. Suggestions

It has been concluded that the Bustamanis still very lack of attention in the passive fire protection systems, so that, it needs adjustments again to prevent a fire hazard. Some suggestions / recommendations from this research results include:

a. Utilization of the road as the dividing of building must be implemented properly. Separation between buildings are aligned to contribute towards the prevention of fire propagation to surrounding buildings in case of fire.
b. Separation of the building also needs to be done into one side. The use of the building dilatation as lengkong (side entrance) of the building, be an alternative to inhibit the propagation of flame laterally. If not possible, can be applied fireproof walls between the buildings.
c. It also sought the procurement of open space that separates rows of buildings that are too long. Grouping of buildings in a certain amount by dividing a road, open space, the dividing wall is very useful to prevent the propagation of fire to surrounding homes until the arrival of firefighters.
d. Water resources construction in the southern area of Bustaman as an effort to provide the firefighters which is built in with the neighborhood.
e. Considering that the street in Bustaman is very narrow, road corridors should not be used to place the equipment and built/ installed the shade like tritisan (tent) that are allowing to spread fire into front buildings.
f. Replacement of combustible material with an incombustible material, which is one of them by rejuvenating house buildings with wood / bamboo materials.

5. Acknowledgments
The authors thank the parties that support this article writing and publication, and prior to the village residents and the staffs of Bustaman Purwodinatan village who have given permission for data collection both primary and secondary.

6. References

[1] Basuki, Ahmad. 2002. *Mencermati Standar Pengamanan Gedung untuk Antisipasi Bahaya Kebakaran*. Sebelas Maret University: Surakarta.
[2] Suzuki, Hiroaki. 2003, "*Fire Safety Desain in Building*" Lecture and discussion of Technical Systems Against Fire Safety and Rescue Building, Public Works Department and the Japan International Corporation Agency, Bandung
[3] Pangarso, budiwidodo, 2013. *Pemahaman Pedoman RTBL untuk Desain Arsitektural Gedung & Penataan Lingkungan Binaan*. Graduate School of Architecture, Parahyangan University: Bandung.
[4] Sukmawati, Annisa M; Yuliastuty, Nany. 2016. Eksistensi Kampung Lama Melalui Kearifan Lokal di Kampung Bustaman Semarang. *Tata loka Volume 18 no. 2, mei 2016, 108-117: Urban Planning Publisher Berau Diponegoro* University: Semarang.
[5] Suprapto, 2007. Sistem Proteksi Kebakaran Pasif Kaitannya dengan Aspek keselamatan Jiwa. *Jurnal Permukiman Vol. 2 No. 2* .Settlement Research and Development Center
[6] Permana, Achiara. 2006. MCK Plus Kampung Bustaman. Suara Merdeka: Semarang
[7] SNI 03-1736-2000 ,Tata cara perencanaan system protekasi pasif untuk pencegahan bahaya kebakaran pada bangunan rumah dan gedung.
[8] Hadi, Sudharto. 2009. *Manusia dan Lingkungan*. Publisher Agency Diponegoro University Semarang.
[9] Grigg, Neil, 1988. *Infrastructure Engineering And Management*. John Wiley and Sons

The evaluation the magnitude radiation exposure dose rate in digital radiography room design

Agung Dwiyanto[1*], Wahyu Setia Budi[2] , Gagoek Hardiman[1]
[1] Department of Architecture, Faculty of Engineering, Universitas Diponegoro, Semarang, Indonesia.
[2] Department of Physics, Faculty of Sciences and Mathematics, Universitas Diponegoro, Semarang, Indonesia.

Corresponding e-mail: agungdwie@gmail.com

Abstract. This study discusses the dose rate in digital radiography room, buit according to meet the provisions of KEMENKES No.1014 / Menkes / SK / XI / 2008 and Regulation of BAPETEN No. 8 / 2011. The provisions primary concern of radiation safety, not comfort, by considering the space design. There are five aspects to consider in designing the space: functionality, comfort, security, movement activities and aesthetics. However provisions only met three aspects of the design, which are a function, security and movement activity. Therefore, it is necessary to evaluate digital radiography room in terms of its ability to control external radiation exposure to be safe and comfortable
The dose rate is measured by the range of primary and secondary radiation in the observation points by using Surveymeter. All data are obtained by the preliminary survey prior to the study. Furthermore, the review of digital radiography room is done based on architectural design theory. The dose rate for recommended improvement room is recalculated using the same method as the actual room with the help of computer modeling.
The result of dose rate calculation at the inner and outer part of digital radiography observation room shows that in-room dose for a week at each measuring point exceeds the allowable dose limit both for staff and public. During a week of observation, the outdoor dose at some measuring points exceeds the dose limit set by the KEMENKES No.1014 / Menkes / SK / XI / 2008 and Regulation BEPETEN No 8/2011. Meanwhile, the result of dose rate calculation in the inner and outer part of the improved digital radiography room can meet the applicable regulations better.

Keywords: dose rate, range of primary and secondary radiation

1. Introduction

In the hospital architectural design, radiology ward is a central point of interest needed by other wards to support their functions. Almost every medical treatment in a hospital is done based on test and observation result of radiology ward. This ward uses the lowest exposure radioactive radiation from x-ray ionizing equipment. Radiology ward consists of three main parts: diagnostics containing ultra-sonic application, fluoroscopic, *etc.*; radiation therapy; and nuclear treatment [9].
Diagnostic radiology needs special and wide building with suitable shape and construction for its usage. Since diagnostic radiology uses ionize radiation, the ward design should consider the safety against radiation.

In designing diagnostic radiology ward, current design must refer to the ministry of health regulation number 1014/Menkes/SK/XI/2008 and the regulation of head of BAPETEN number 8 year 2011. Those regulations emphasize the safety aspect but do not consider other aspects of architectural spatial design theory. It is important to design a safe and comfortable digital radiography ward.

This study will evaluate the magnitude of dose rate in an existed ward built based on current regulations and compare the result with the suggested ward design built based on architectural aspect design. The research object is digital radiography ward in government hospital Semarang.

1.1. Theoretical background

1.1.1. Radiation and Protection

Radiation is the emission of energy from a source to its surrounding without certain medium. Radioactive radiation is not always dangerous for human and environment; it can be used for human advantages if used properly. The negative effect of radioactive radiation when hitting human body will cause genetic, somatic, teratogenic, stochastic, and non-stochastic effect. Those effects lead to congenital disease such as radio dermatitis, cataract, infertility, and acute radiation syndrome. The danger of radioactive radiation cannot be felt and seen directly and is able to penetrate many materials.

The danger of external radiation depends on its penetration power through a medium; the higher the power, the bigger the radiation threat. Radiations with high penetration power are gamma, x-ray, and neutron radiation, while alpha and beta radiation have lower penetration power.

The protection to external radiation can be done through these steps:

- Time factor
  Radiation dose is the amount of radiation absorbed by certain objects including human body. The value of dose is equal to the accumulation of radiation intensity in certain period of time. Thus, the longer a person exposed to radiation, the bigger the amount of radiation absorbed by the person.
- Distance factor
  The source of radiation emits radiation to every direction forming ball shape. So, the further a person from the sources (the core of the ball), the less radiation absorbed.

NCRP stated those x-ray shields are in two categories, source shield and structural shield. The source shield is usually provided by x-ray tube manufacturer in the form of ***Pb*** shield which is also tube housing. The structural shield is designed to dispel the effect of x-ray. Calculating the regulated thickness of the shield can be done physically.

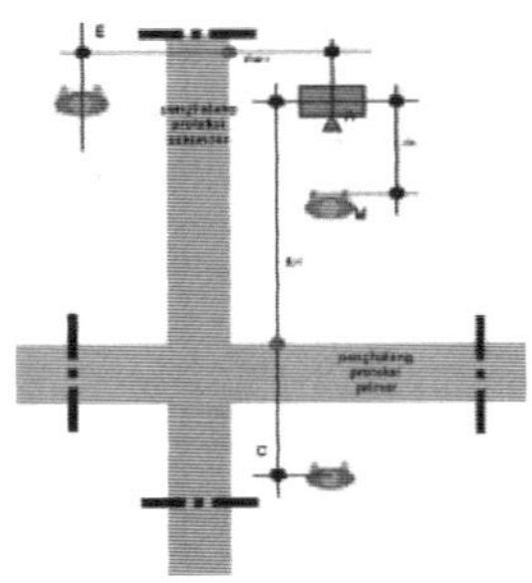

Figure 1. Rendening the Rate of Radiation Dose
Source: *NCRP, Report* No.49 Bethesda, 1976.

The x-ray instrument emits x-ray pulse (primary pulse) directed to patient (M) in the distance dsca. That pulse penetrates the patient then weakened by the primary shield. The radiation leakage and primary pulse that hit the patient cause radiation scattering. X-ray tube leakage radiation and scattered radiation will be weakened by secondary shield. Clearer picture of radiation can be seen in figure:

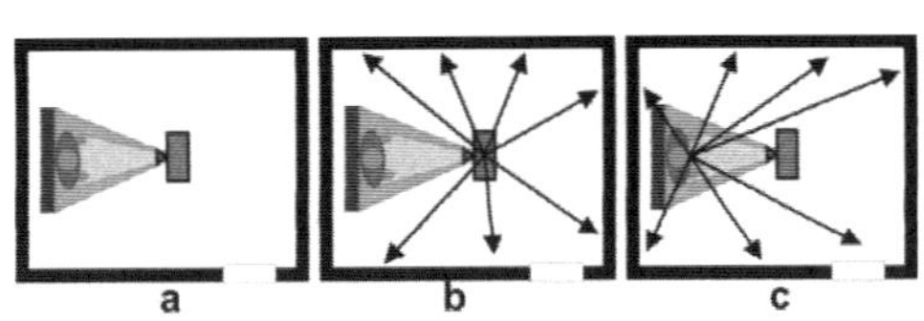

Figure 2. Exposure in x-ray ward

(a) Primary radiation from focal spot.
(b) Leakage radiation from x-ray tube.
(c) Scattered radiation from the patient.

1.2. Attenuation
Attenuation is the disappearance of foton from x-ray pulse after penetrating the matter caused by absorbance and scattering on primary foton [2]. The process of x-ray radiation weakening in a certain shield material is exponential based on equation (1).

I is radiation intencity after penetration, ***I*o** is the intencity before penetration, μ is coefficient of linear absorbent of material (cm^{-1} or m^{-1}), and x is the thickness of material (cm or m) [3].

Kara,U in his journal *A Study on* **Radiation** *in Operating* **Room** *in Suleyman Demirel University* (2016) stated that the minimum radiation safety is the most important thing considering the long term effect of radiation.

Rostamzadeh in his journal *Evaluation of the Level of Protection in Radiology Departments of Kermanshah, Iran* (2015) stated that the awareness of medical staff in using equipment and applying radiation protection procedure is very low.

1.3. Spatial Design in Architecture
Many experts in spatial and building design argued that there are no fixed rules in terms of designing since everything depends on the types, functions, and kinds of space. However, there are five fixed aspects that should present and be considered in designing a space or a building. These five aspects are connected to each other to make optimal result as seen in table:

Table 1. Basic aspects in space and building design

ESTETIC	SAFETY	COMFORT	FUNCTION	TRAFFIC
Harmony	power/strong	Sense	Detail	Enclosure
Scale	rules	View	displey element	achievment
color		Light&shade	truth	Efficien
balance		shelter	pressure	
layout		lighting	Vitality	
Simetri			Sense of outline	
Euritmik			volume	
similarity			character	
continue			Identity	
unity				
Focalpoint				
contras				
performer				

Table 2. The relationship among basic aspects of design

	Safety	
Estetic		comfort
function		traffic
	INTEGRATED	

2. Methods
2.1. Time and Place for Data Collection
Data are collected from June to August, 2015 in government hospital Semarang at digital radiography room of radiology ward.

IOP Conf. Series: Earth and Environmental Science **99** (2017) 012003 doi:10.1088/1755-1315/99/1/012003

2.2. Tool and Material
The researcher uses *surveymetergamma* tool to detect rate of radiation dose in digital radiography room and uses measuring band to measure the range of primary and secondary radiation of focal spot (0.3 meters outside the shield).
The material of the shield is the composition of wall and door of the room as shown in the figure.

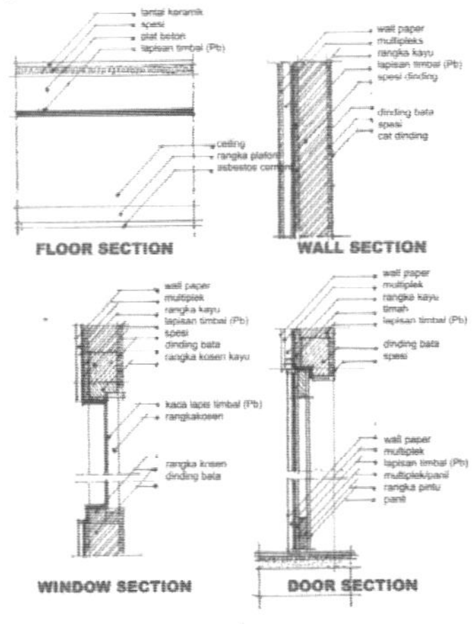

Figure 3. The composition of wall and door of the room

2.3. Research Procedure
Generally, this research consists of four steps. First, the initial survey conducted before the research. Second, the measurement of dose rate measured inside and outside the room. Third, complete data collection which is shown in the research diagram.

3. Discussion
3.1. The Measurement of Radiation Exposure Dose Rate in Existing Digital Radiography Room
The position of measurement of dose rate inside and outside the digital radiography room is shown on figure

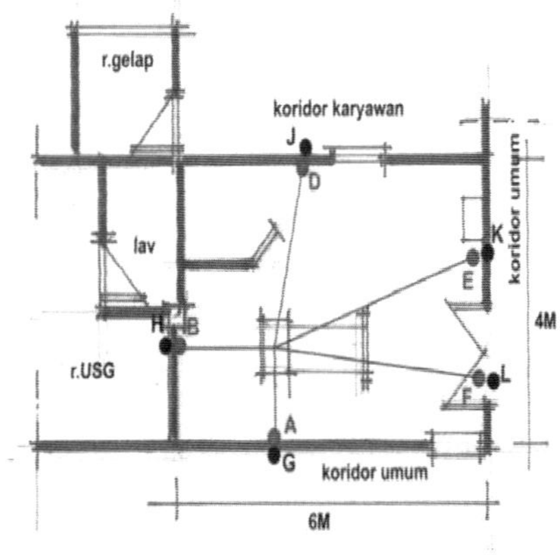

Figure 4. Measuring point positions for dose rate measurement in digital radiography room government Hospital

POSITION	MEASUREMENT POINT	DISTANCE (M)	EXPOSURE DOSE RATE (mSv)/week
INDOOR	A	1,45	9,47
	B	2	8,39
	C	1,71	8,41
	D	3,2	2,97
	E	5,9	1,03
	F	4,7	2,08
OUTDOOR	G	1,6	0,0991
	H	2,15	0,0418
	I	1,77	0,056
	J	3,35	0,0305
	K	6,05	0,0259
	L	4,74	0,0328

Table 3. The Calculation of Exposure Dose Rate

3.2. Evaluation

The evaluation of NBD exposure dose rate for staff is effectively 20 mSv/year or equal to 0,4 mSv/week. Based on the Ministry of Health regulation number 1014/Menkes/SK/XI/2008, NBD for public is effectively 1 mSv/year (0,02 mSv/week).

The rate of radiation dose in the digital radiography room fulfills the principle of distance arrangement of radiation safety in which the rate of radiation dose inversely proportional to quadrate of distance. So, the bigger the range of radiation to a certain measurement point, the lower the measured dose rate will be. In terms of measurement, the rate of dose is different in point D and point F; point F has lower rate. This is because the radiation on measuring point F has been weakened when penetrating the patient and pre-shielding. Thus, the detection rate of the radiation dose on that measuring point is very low.

The measuring of dose rate outside the area of digital radiography room shows that point G has the highest rate because the rate of the dose before penetrating the shield on point A is also the highest one. Thus, the highest outdoor radiation is on point G. Then, the lowest rate of dose is found in point K since the rate of the dose before penetrating the shield on point E is the lowest. However, the increase of dose rate occurs in point L caused by the leakage through unlocked door.

A week dose in the room on each measuring point exceeds the dose tolerance for staff and public. However, the outdoor dose for staff on point I and J is under the dose tolerance regulated by Perka BAPETEN. Moreover, the dose on point G, K, and L also exceed the dose tolerance regulated by the Ministry of Health.

Considering the fact that patients, their families and relatives do not stay for a long time in measuring point G, and H, the rate of dose received by them on point G,H,K, and L is still under the dose tolerance.

Moreover, based on the dose received in every measuring point, point G as the only spot full of patients and their guardians has the dose above 0,02 mSv/week, while point K and L has the dose under 0,02 mSv/week. The point I and J, where the staff works, has the dose above 0,4 mSv/week. Point H, where the staff and public commonly interact, has the dose above 0,4 mSv/week.

*3.3. Digital Radiography Room Based on Review (after completed based on **iso-exposure** and architectural spatial design aspects)*

The measurement spots to calculate dose rate inside and outside the digital radiography room are planned similar to the existing measuring points with improved design. The design can be seen on figure.

IOP Conf. Series: Earth and Environmental Science **99** (2017) 012003 doi:10.1088/1755-1315/99/1/012003

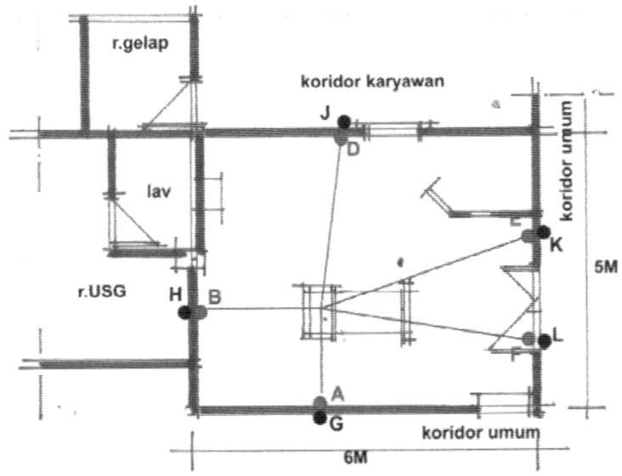

Figure 5. Revised groundplan

The improvement/ the review result on existing room is based on iso-exposure as portrayed on the result of measurement of the dose rate on the initial research. The initial research result shows that measuring point G (USG room) and point H (public corridor) do not fulfill the regulated weekly tolerance level of exposure for staff and public. Both of the room also do not fulfill the aspect of spatial design in architecture which are the traffic aspect for patient and medical staff and the positioning aspect of radiography equipment inside the room.

By adding the width of the room and changing the position of the equipment, the dimension of the room becomes 6mx5m.

This way, the measurement of dose rate of radiation exposure on digital radiography room is improved. The improvement is in terms of the decrease of the exposure dose rate inside and outside the room. However, the weekly rate of dose inside the room is still above tolerance level which is 0,4 mSv/week. The weekly rate of dose on the outside is also still above tolerance level which is 0,02 mSv/week.

Eventhough the measurement of the radiation dose rate is relatively the same, the improved design of the digital radiograpy room has minimized the radiation exposure. The measurement in this room shows that its rate of radiation exposure dose is still under tolerance level considering the real condition on the field and the position of medical staff and public on exposed site. This can be seen on the table below:

POSISI	MEASUREMENT POINT	eksisting DISTANCE (M)	EXPOSURE DOSE RATE (mSv)/week	revision DISTANCE (M)	EXPOSURE DOSE RATE (mSv)/week
INDOOR	A	1,450	9,47	2,000	8,39
	B	2,000	8,39	1,600	9,308
	C	1,710	8,41	1,500	9,4648
	D	3,200	2,97	3,000	6,8324
	E	5,900	1,03	6,000	0,715
	F	4,700	2,08	4,800	1,975
OUTDOOR	G	1,600	0,0991	2,200	0,05518
	H	2,150	0,0418	1,800	0,08446
	I	1,770	0,056	1,700	0,09178
	J	3,350	0,0305	3,200	0,029935
	K	6,050	0,0259	6,200	0,024106
	L	4,740	0,0328	5,000	0,031006

Table 4. The Result of Radiation Exposure Dose Rate Measurement in Recommended

4. Conclusion
Based on the result and explanation of this research, the conclusions are:

- The design of digital radiography room with five aspects of architectural design gives protection to the exposure of radiation and adds comfortability to the room.
- The position of medical supporting room should be parallel with the position of digital radiography equipment. It is minimal 2.3m behind the equipment.
- The position of the main door should be parallel with the radiation absorber panel.
- The corridor and waiting room should be positioned on the horizontal side toward digital radiography equipment.

5. References
[1] Akhadi Mukhlis, (2000). Dasar-dasar Proteksi Radasi, Rineka Cipta, Jogyakarta.
[2] Badan Tenaga Atom Nasional, Pedoman Proteksi Radiasi di Rumah Sakit, 1995
[3] Perka Bapeten nomor 8 tahun 2011 tentang, KESELAMATAN RADIASI DALAM PENGGUNAAN PESAWAT SINAR-X RADIOLOGI DIAGNOSTIK DAN INTERVENSIONAL
[4] Cember,H., dan E.J. Thomas, 2009, *Introduction to Health Physics,* Fourth Edition, The McGraw-Hill Companies, Inc., New York.
[5] Depkes RI, (1999), *Standar Pelayanan Radiologi Rumah Sakit Umum Klas B,C dan D.*
[6] Departemen Kesehatan RI, (1997) Pedoman Pemasangan Dinding Ruang Radiasi,
[7] Djoko Maryanto, (2008), *Analisis Keselamatan Kerja Radiasi Pesawat Sinar-X di Unit Radiologi RSU Kota Jogyakarta*, makalah Seminar Nasional IV SDM Teknologi Nuklir, Jogyakarta,
[8] Francis DK Ching,(2000), Arsitektur-Bentuk, Ruang dan Tatanan, Erlangga.
[9] Friedr,V, (1982), Standart Rumah Sakit-Bagian Radiologi, Braunschweig, Germany.
[10] Hendee,W.R., dan E.R. Ritnour, 2002, *Medical Imaging, Fourth Edition,* A JOHN WILEY & SONS, INC., New York.
[11] Kara,U (2016), *'A Study on* **Radiation** *in Operating* **Room** *in Suleyman Demirel University',* Acta Physica Polonica, A.. 2016, Vol. 130 Issue 1, p401-403. 3p
[12] Kementran Kesehatan RI, Pedoman Teknis Bangunan dan Prasarana Rumah Sakit 2013
[13] Permenkes RI No.780 tahun 2008 tentang *Penyelenggaraan Pelayanan Radiologi di Rumah Sakit.*
[14] Permenkes RI No.1014 tahun 2008 tentang *Standar Pelayanan Radiologi Diagnostik di Sarana Pelayanan Kesehatan.*
[15] Perpu No. 33 tahun 2007 tentang *Keselamatan Radiasi Pengion dan Keamanan Sumber Radioaktif.*
[16] Rostenberg Bill, FAIA, FACHA,(2006), *The Architecture of Medical Imaging-Designing Healtcare Facilities for Advance Radiological Diagnostic and Therapeutic Techniques,*John Wiley, New Jersey,Canada.
[17] Rostamzadeh, Ayyob (2015),' *Evaluation of the Level of* **Protection** *in Radiology Departments of Kermanshah, Iran,* Iranian Journal of Medical Physics. Summer2015, Vol. 12 Issue 3, p200-208. 9p
[18] Sukada, B. (2006). *Kegagalan sebuah Karya Arsitektur; dapatkah diukur?* Jakarta: Universitas Pelita Harapan-Lippo Karawaci

The effect of earthquake on architecture geometry with non-parallel system irregularity configuration

Livian Teddy[1*], **Gagoek Hardiman**[2], **Nuroji**[3], **Sri Tudjono**[4]
[1] Doctoral Program of Architecture and Urban, Diponegoro University, Indonesia,
[2] Departement of Architecture, Faculty of Engineering , Diponegoro University, Indonesia,
[3,4] Doctoral Program of Civil Engineering, Diponegoro University, Indonesia

*Corresponding e-mail: livianteddy@gmail.com

Abstract. Indonesia is an area prone to earthquake that may cause casualties and damage to buildings. The fatalities or the injured are not largely caused by the earthquake, but by building collapse. The collapse of the building is resulted from the building behaviour against the earthquake, and it depends on many factors, such as architectural design, geometry configuration of structural elements in horizontal and vertical plans, earthquake zone, geographical location (distance to earthquake center), soil type, material quality, and construction quality. One of the geometry configurations that may lead to the collapse of the building is irregular configuration of non-parallel system. In accordance with FEMA-451B, irregular configuration in non-parallel system is defined to have existed if the vertical lateral force-retaining elements are neither parallel nor symmetric with main orthogonal axes of the earthquake-retaining axis system. Such configuration may lead to torque, diagonal translation and local damage to buildings. It does not mean that non-parallel irregular configuration should not be formed on architectural design; however the designer must know the consequence of earthquake behaviour against buildings with irregular configuration of non-parallel system. The present research has the objective to identify earthquake behaviour in architectural geometry with irregular configuration of non-parallel system. The present research was quantitative with simulation experimental method. It consisted of 5 models, where architectural data and model structure data were inputted and analyzed using the software SAP2000 in order to find out its performance, and ETAB2015 to determine the eccentricity occurred. The output of the software analysis was tabulated, graphed, compared and analyzed with relevant theories. For areas of strong earthquake zones, avoid designing buildings which wholly form irregular configuration of non-parallel system. If it is inevitable to design a building with building parts containing irregular configuration of non-parallel system, make it more rigid by forming a triangle module, and use the formula.A good collaboration is needed between architects and structural experts in creating earthquake architecture.

Keywords: earthquake, non-parallel system irregularity configuration

1. Introduction

Indonesia is an area prone to earthquake that may cause casualties and damage to buildings. The fatalities or the injured are not largely caused by the earthquake, but by building collapse. The collapse of the building is resulted from the building behaviour against the earthquake, and it depends on many factors, such as architectural design, geometry configuration of structural elements in horizontal and vertical plans, earthquake zone, geographical location (distance to earthquake center), soil type, material quality, and construction quality [1,2]. One geometric configuration that may result in building collapse is irregular configuration of non-parallel system (figure 1). According to FEMA-451B [3],

IOP Conf. Series: Earth and Environmental Science **99** (2017) 012004 doi:10.1088/1755-1315/99/1/012004

irregular configuration of non-parallel system is defined to have existed if the vertical lateral force-retaining elements are neither parallel nor symmetric with main orthogonal axes of the earthquake-retaining axis system.

Such configuration may happen on: 1). The building is regular, however as the walls of the room are tilted, the wall slides (figure 1a) or the beam arranged is also tilted (figure 1b), 2). The building is regular, but column arrangement is not in one axis, making the beam connecting the column tilted (figure 1b), 3). The building is irregular, for adjusting to the shape of the site, or the building is deliberately tilted for architect's aesthetic consideration (picture 1c). The irregular configuration of parallel system may lead to torsion, instability and local damage to buildings [4]. It does not mean that non-parallel irregular configuration should not be formed on architectural design; however the designer must know the consequence of earthquake behaviour against buildings with irregular configuration of non-parallel system. The present research has the objective to identify earthquake behaviour in architectural geometry with irregular configuration of non-parallel system.

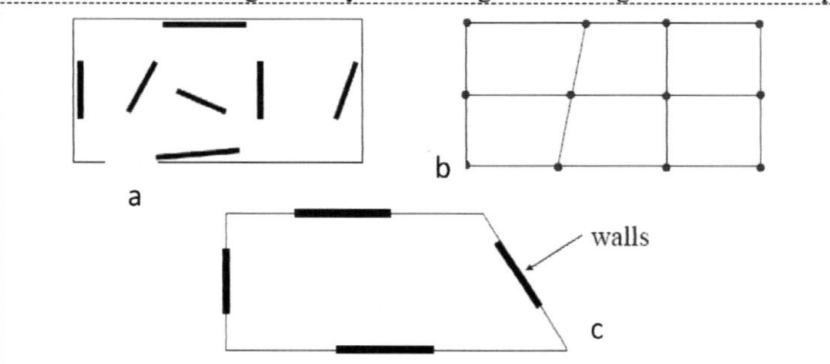

Figure 1. Non parallel system irregularity configuration

2. Methods

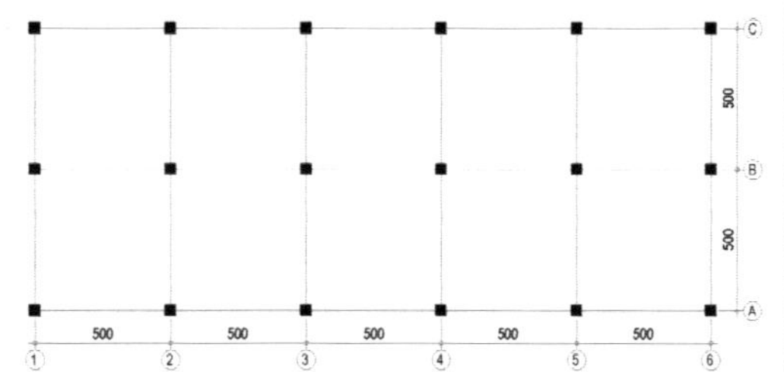

Figure 2. Model a

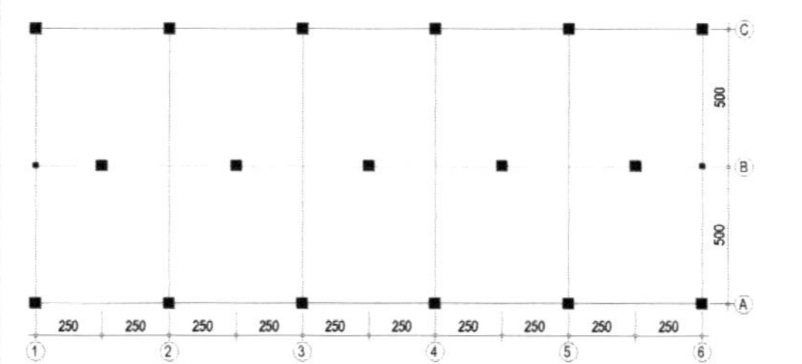

Figure 3. Model b

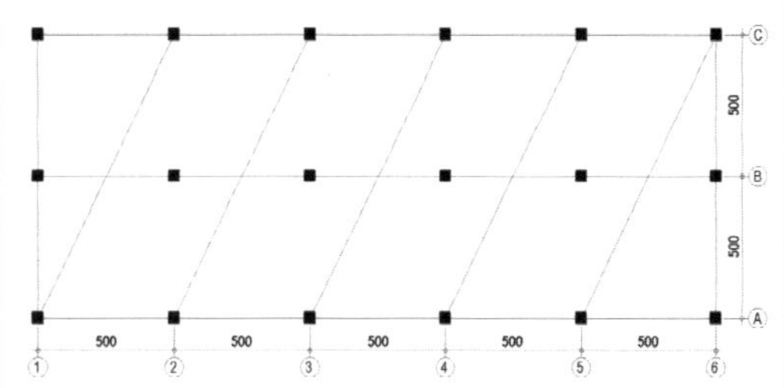

Figure 4. Model c

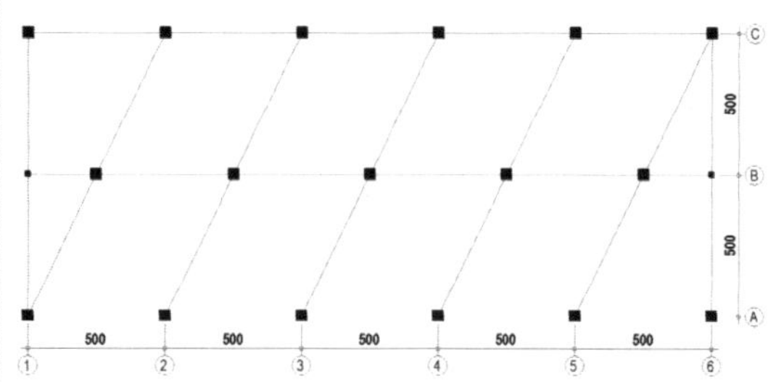

Figure 5. Model d

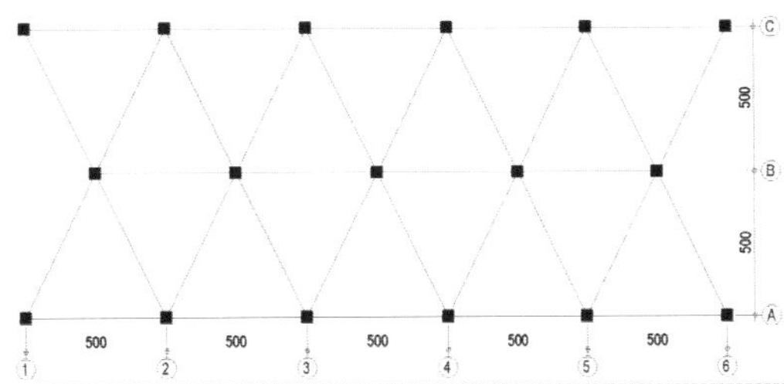

Gambar 6. Model e

Table 1. The property of structure model a-e

Model	Number of floors (height-m)	Dimension beam (cm)	Dimension column (cm)	Thickness of floor plate (cm)	Dimension building (m)
a	4 (16 m)	25X40	40X40	12	10X25
b	4 (16 m)	25X40	20X20, 40X40	12	10X25
c	4 (16 m)	25X40	40X40	12	10X25
d	4 (16 m)	25X40	20X20, 40X40	12	10X25
e	4 (16 m)	25X40	40X40	12	10X25

Table 2. Grade of structure model a-e

Model	Grade		
	Concrete (Kg/cm2)	Reinforcement (Kg/cm2)	Stirrup (Kg/cm2)
a	300	3000	2400
b	300	3000	2400
c	300	3000	2400
d	300	3000	2400
e	300	3000	2400

It is a simulated experimental study by using pushover analysis, and center of mass and rigidity analysis. In order to conduct both analyses above using the geometry model of the buildings (figure 2 to 6) and the structural properties (table 1 and 2), those were inputted to SAP2000/ETABS softwares, and were then analyzed by static pushover earthquake analysis method to be identified for its performance level and to be analyzed for its eccentricity to find out the potential torque occurred. The numerical outputs of the analyses were tabulated and compared between models and graphs. It is assumed that the models are in a high earthquake zone with spectral value of $S_S = 0.97$ g and $S_1 = 0.328$ g with medium soil (D) condition and office function.

3. Discussion

3.1. Target displacement

Pushover analysis is a static non-linear analysis in which the effect of earthquake plans on the structure of a building is considered as a static load capturing at the center of mass of each floor, which value is gradually increased to exceed the loading that causes the first yield (plastic joint) in the structure of the building, then through further load increase, it has a changing major post-elastic shape until it reaches the expected target displacement or until it reaches the plastic condition [3].

Table 3. Target displacement model a-e

Model	Target displacement	
	X (m)	Y (m)
a	0.187	0.197
b	0.201	0.246
c	0.183	0.235
d	0.158	0.208
e	0.163	0.178

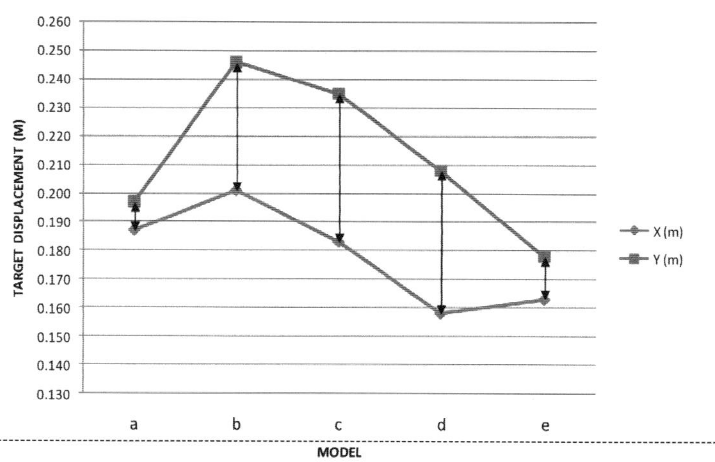

Figure 7. Target displacement model a-e

Reviewed from the results of pushover analysis in table 3 and figure 7, it can be concluded as follows:

- For the –X axis, all columns in one axis and its rigidity are dominated by the models with tilted beams (model c, d and e) rather than those that do not have tilted beams (model a and b). It means that there is an angle formation on the beams which can give additional rigidity to the structure of the building.
- For the –Y axis, model a and b of all beams are in one axis, while model c, d and e have their beams not in one axis. For columns in one axis are model a and c, while model b, d and e the columns are not in one axis. On the –Y axis, model a and e have the largest rigidity compared to other models, whereas the weakest rigidity is in model b. It means that the ideal building structure is when the beams and columns are in one axis but if it is inevitable, connect the columns with the triangle module beams so that it will have more rigidity. In addition, the columns which are not in one axis have weaker rigidity than the beams which are not in one axis.
- For regular buildings, the target displacement for the –X and –Y axes are generally relatively similar (model a and d), and the more irregular the building geometry the larger the target displacement of both axes, –X and –Y. It should be avoided, since both axes should have the same ability in facing the seismic loads.

It is similar with Shopping Center in Ercis District's-Turkey which has an irregular geometry (see figure 8). In order to reduce the irregularity, the building is blocked and separated by dilatation, but from these three blocks A to C, block C have greater non-parallel irregular configurations than others. When the earthquake stroke Ercis District's -Turkey in 2011, from the three blocks, block C was more severely damaged than blocks A and B. Based on the research results, the non parallel irregular configuration in the two-way axes -X and –Y, block C compared to blocks A and B which only on the X-axis, gives a significant contribution to the severe damage of block C in this shopping center building [6].

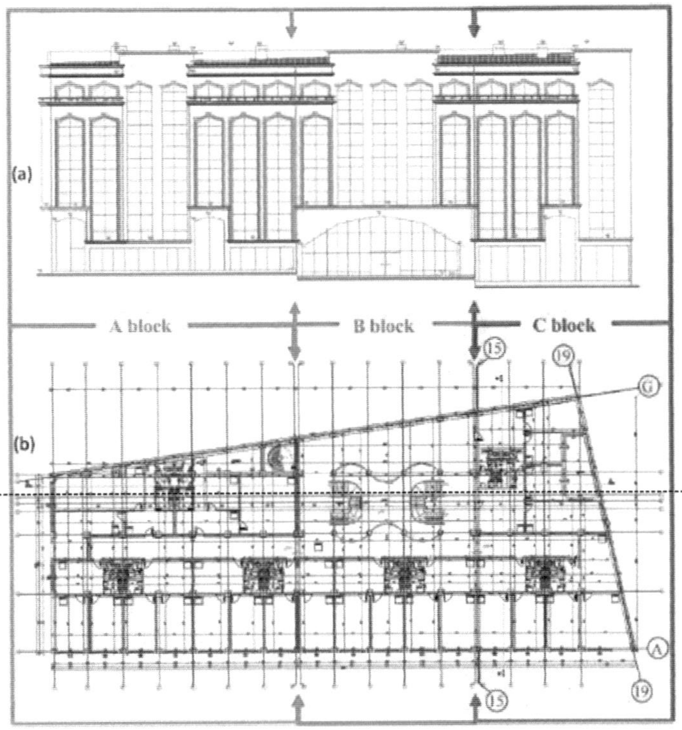

Figure 8. Shopping Center in Ercis-Turki District's: (a). Fasad architectural, (b). Building plan (source : [6])

In order to avoid the occurrence of the above, simple formula is proposed to evaluate the potential for the formation of non-parallel irregular configuration in the building that is still in the process of architectural design:

$$j_r = \sum j / \sum j_t \quad (1)$$

Table 4. Ratios quality level of the number of beams, columns and shear walls elements (j_r) irregular configuration of non-parallel system

Source	Quality Level		
	Good	Moderate	Poor
Recommendation	j_r<30%	30%≤j_r≤50%	j_r>50%

Where, j_r= ratios quality level of the number of beams, columns and shear walls elements, $\sum j$=the number of beams, columns and shear walls elements that do not follow the orthogonal axis and $\sum j_t$=the total number of beams, columns and shear walls elements at the level being reviewed . The result of j_r is then compared with table 4 so as to find out whether the formation of non-parallel system irregular configuration is still in tolerance level (good), moderate or poor.

3.2. Eccentricity
The torque force formed inside the building is caused by the lack of balance between the location of the retaining elements and the mass structure of the building. It is the eccentricity between the center

of mass and rigidity which makes the building experience a twisting ground motion around the center of rigidity, which results in torque - twisting effects in the building plan. This effect is undesirable and allows the danger of stress concentration [2].

Table 5. Eccentricity model a-e

Floor	Center of mass (m)		Center of rigidity (m)		Eccentricity (e)	
	Xm	Ym	Xr	Yr	X=Xm-Xr	Y=Ym-Yr
1st Floor	12.5	5	12.5	5	0	0
2nd Floor	12.5	5	12.5	5	0	0
3rd Floor	12.5	5	12.5	5	0	0
Roof Floor	12.5	5	12.5	5	0	0

According to table 5, the models a-e does not form eccentricity or in other words, the center of mass and the center of rigidity are overlapped, making its eccentricity= 0. It also negates the potential for torsion in the building model. Therefore, the main eccentricity occurs by building mass geometry form, and the irregularity effect of the beam/column arrangement is relatively minimal but it can cause another irregularity, i.e. non-parallel irregular configuration. Eccentricity can cause torsion in the building, the causes include uneven loading, rigidity and strength, and others [7] (see figure 9).

For avoiding excessive torque, then a simple formula is propose to evaluate the potential for the formation of torsional irregularity configuration in buildings that are still in the process of architectural design :

$$e_r = e/w \qquad (2)$$

Where, e_r= the ratio between the length of eccentricity (e) and the width of the building being reviewed (w).

Table 6. The eccentricity ratio quality level (e_r) of torque irregularity configuration

Source	Quality Level		
	Good	Moderate	Poor
JBDPA [8]	$e_r \leq 0.1$	$0.1 < e_r < 0.3$	$e_r \geq 0.3$

The result of e_r is then compared with table 5, so it is known whether the configuration of torque irregularity being formed is included in tolerance level (good), moderate or poor.

Figure 9. Building damage from torque eccentricity (source : [2]).

4. Conclusion
From the explanations above, there are some conclusions that could be useful for architects in designing the building, as follows:

- The ideal building structure is when the beams and columns are in one axis but if it is inevitable, connect the columns with the triangle module beams so that it will have more rigidity.
- Columns which are not in one axis have weaker rigidity than the beams which are not in one axis.
- The random seismic motion requires both axes to have equal ability in facing the seismic loads.
- Primarily, eccentricity occurs by the geometric shape of the building mass and the effect of irregularity of the beam/column arrangement is relatively minimal but can cause another irregularity, the non-parallel irregularity configuration.
- Avoiding the excessive formation of torque irregular configuration and non-parallel irregular configuration in the building design can be done by evaluation using formula 1 and 2 above.

For future research can be investigated the effect of earthquake on geometry architecture with non-parallel system irregularity configuration in irregular form.

5. Acknowledgments
The authors thank to the Faculty of Engineering of Sriwijaya University which has funded the publication of the part research of the authors' dissertation, especially for the staffs of finance in the Faculty of Engineering, Hermawan Yuliansyah, SE dan Herman, SE, who had helped the authors in processing the administration of funding distribution.

6. References
[1] Inan T, Korkmaz K. Evaluation of structural irregularities based on architectural design considerations in Turkey. Struct Surv [Internet]. 2011;29(4):303–19. Available from: http://dx.doi.org/10.1108/02630801111162378
[2] Özmen C, Ünay AI. Commonly encountered seismic design faults due to the architectural design of residential buildings in Turkey. *Build Environ.* 2007;42(3):1406–16.
[3] FEMA. NEHRP Recommended Provisions for New Buildings and Other Structures: Training and Instructional Materials-FEMA 451B. Washington DC: Federal Emergency Management Agency (FEMA); 2007.
[4] Arnold C. Seismic Issues In Architectural Design. In: Designing For Earthquakes A Manual For Architects - FEMA 454. California: Engineering Research Institute (EERI); 2006.
[5] Pranata YA, Wijaya PK. Kajian Daktilitas Struktur Gedung Beton Bertulang dengan Analisis Riwayat Waktu dan Analisis Beban Dorong. *J Tek Sipil Univ Atmajaya* [Internet]. 2008;8(3):250–63. Available from: http://cpanel.petra.ac.id/ ejournal/index.php/uaj/article/ viewArticle/17538
[6] Bikçe M, Çelik TB. Failure analysis of newly constructed RC buildings designed according to 2007 Turkish Seismic Code during the October 23, 2011 Van earthquake. Eng Fail Anal [Internet]. 2016;64:67–84. Available from: http://dx.doi.org/10.1016/j.engfailanal.2016.03. 008
[7] Crisafulli F, Reboredo A, Torrisi G. Consideration of Torsional Effects in the displacement control of ductile buildings. 13th World Conf Earthq Eng. 2004;(1111).
[8] Okada T, Murakami M, Kabeyasawa T, Katsumata H, Nakano Y, editors. Guidelines for Seismic Retrofit of Existing Reinforced Concrete Buildings. Tokyo: The Japan Building Disaster Prevention Association (JBDPA); 2005

The uniqueness and complexity of kampung city Bustaman Semarang Indonesia

Budi Sudarwanto[1]*, Gagoek Hardiman[2], Atiek Suprapti[3], Agung B Sarjono[4]
[1,2,3,4] Architecture Department, Engineering Faculty Universitas Diponegoro, Semarang, Indonesia,

*Corresponding e-mail: budisud@gmail.com

Abstract. The purpose of this study is to describe the uniqueness and complexity of one village town Bustaman Semarang Indonesia. This research is in the domain of qualitative research paradigm with ethnography research strategy. The critical ethnography technique is chosen with the consideration of the purpose of this research, that is besides describing also ~~the~~ analysing urban spatial condition over its complexity. Bustaman town village has ~~artifacts~~ outdated and local economic activities namely culinary processed goat meat. Kampung Bustaman shows the common character as a city hometown, which is chaotic and shabby. Some artifacts and historical value of Bustaman village as well as community culture become potential to be developed and conserved. But for that the quality of Bustaman village hometown should be improved. Bustaman's urban halls are under distress by changes in urban modernization, including the pattern of life of Bustaman villagers. However, the existing phenomenon that the potential of local and cultural communities are still able to survive and exist, within the limitations of space and resources available. The process of culinary production of goat meat, the limitation of the kampong space, the social awareness of the villagers, the role of community leaders, and the necessities of daily life are the drivers of the complexity of Bustaman village space. The village road corridor and the house terrace become the victim room for all the activities of the town. Clutter seems clear can be found in every corner of Bustaman village space. Kampung Bustaman can be categorized as a spontaneous and compact informal village, but still survive on the urban spatial locality.

Keywords: uniqueness, complexity, space analysis, urban village, Bustaman

1. Introduction

The urban village have some classical problems related to the process of urban modernization in the era of globalization. Urbanization encourages changes in urban areas, especially urban centers. Urban halls have decreased the quantity and quality as places and densely populated urban residential neighborhoods. Clutter and slums are prominent characters in addition to the densely populated and poverty attached to the urban village. The urban village is understood as an urban burden because it is viewed negatively in the process of urban modernization. The apparent impact of urbanization on urban spaces in the city center resulted in slums, poverty, and chaos. The urban village shows the complexity of the problem as a place and neighborhood of the city center. The urban villages showed high occupancy density and this resulted in a complex pattern of activity and space constraints.

The urban village is the place where the population develops their economic activity. This small-scale activity is very important not only as the first provider of workplaces for migrants but giving urban villages a real role in tackling migration and unemployment. Urban village as a place of migrants trying to adapt themselves and learn to live in the city. In the process of becoming an urban resident the migrants in the village learn to live together and collaborate with other migrants with different ethnic, religious and cultural backgrounds. [1] (Jo Santosa, (2013) Freek Colombijn, et al, in his book Kampung Urban Indonesia: Anthropological-Historical Studies on Social and Urban Space Disparities, explains that urban kampongs are a historical construction and the most important social unit of the city. [2] The village was originally built as a space which then through a process of intensification of development to absorb more and more people moved to settle in the villages. [3] (Jellinek, Lea1991).

Bakti Setiawan (2000) explains that there is a misconception of the village in the urban area. [4] The urban village is considered an urban burden or judged negatively. Even academics give the name that urban kampongs are informal / illegal settlements, slums / squatter settlements, and spontaneous settlements. The urban village shows low-quality marginal settlements and is intended for low-income communities. Turner (1985) describes this as the complexity of a settlement environment. [5]

Improving the quality of urban villages through various approaches made by the government program has not been maximally successful. [6] (Silas Johan, 1996; Ministry of Public Works, 2010; in Sandyawan Sumardi, 2016) Various studies on urban kampongs show the complexity of urban spaces. Village towns are studied from various perspectives such as social issues, local wisdom, local economy, and urban village culture. Social aspects are the main focus of the study by urban researchers, such as residents' participation [7] (Sri Handayani, 2008), village creativity [8] (Iwan Kustiwan, 2015), social capital (cheap labor source) [9] (Jan Newberry, 2013). Physical environment issues are considered as the impact of socio-economic activities of the villagers. [10] (Triatno Yudo Harjoko, 2009).

Urban village as a place and settlement environment shows the complexity in the context of urban modernization. The urban village shows the culture of local residents and local wisdom of the town. [11] (Marcus Zahnd, 1999) The urban village shows spontaneous, organic, and irregular spaceship changes. That the urban spaces are shabby and spontaneous and irregular is an expression of form of socio-economic and cultural character of the village community. [12] (Marcel Hunziker, et al, 2007) It is necessary to control the spaces within the urban village. [13] (John Habraken, 2002)

Kampung Bustaman located in downtown Semarang has high occupancy density and high productivity level although it also shows negative label of slum and irregular city kampung and low occupancy quality standard. Bustaman as a city hometown also has a legendary historical value in addition to culinary processed goat meat. Bustaman villagers are aware of the limitations and weaknesses of their places and neighborhoods The use of the kampong to share functions and transfer activities can be seen clearly when we visit the village of Bustaman. This makes the movement of village activities complex and complex.

The complexity of urban village issues includes the social aspects of life, the economic aspects of the inhabitants, and the aspect of the living environment, in which aspects of the problem are interrelated. The character of urban kampongs with strong social aspects (social cohesion) is capable of reducing other problems, even overriding the quality and quantity of their dwellings. It is important to understand the value of importance of kampungkota to its inhabitants.

It is important to see and describe the urban spaces that show the uniqueness and complexity of the urban space architecture in Bustaman urban village. Critical analysis of the uniqueness and complexity is done by deepening the locality and culture of Bustaman villagers. The infrastructure constraints and the quality of urban resettlement environment are part of the novelty of this research. The results of the research can be used as data input development of local power-based villages.

2. Methods

The research method used qualitative research area with ethnography research strategy and technique used is critical analysis (critic ethnography). [14] (Spradley, 2007) Critic ethnography describes the culture of Bustaman villagers in carrying out daily activities using village artifacts in their daily activities. This technique is used by researchers because the researchers conducted a critical analysis and interpretation of the architectural phenomenon that occurred in the town. [15] (Cresswell, 2007)

The focus of the research is on the uniqueness and complexity of Bustaman village. The uniqueness of the village is based on the potential of artifacts, culture and culinary activities of goats while the complexity is a common character of urban kampong associated with the chaos and slums of Bustaman village hometown. Potential Bustaman village artifacts are ancient wells, RPH buildings, ten gedong, historical and cultural values of Bustaman residents, and goat meat culinary. The complexity of urban kampong is an architectural empirical phenomenon linked to chaos and village slums such as: overlapping function of kampung corridor room, changes of dwelling space, infrastructure limitations, lack of public space, residency of residential house, linkage of kampung room layout, and the pattern of Bustaman village movement.

Critical analysis is used to explore the related architectural phenomena associated with the uniqueness and complexity of Bustaman urban village. The analysis and interpretation are excavated through field measurements through a valid source of information (informants). The interpretation of the researcher and the informant information on the result of field measurement is processed into a concept of empirical base architecture knowledge through the process of information iteration until it gives an objective result.

3. Discussion

Kampung Bustaman is one of the old kampong located in the city center, the Purwodinatan sub-district of Central Semarang which has an area of 0.6 ha with a population density of 590 people per ha. Kampung Bustaman consists of 2 RTs, namely RT 04 and RT 05 with 65% of the total productive age population of approximately 400 people. This village has a phenomenal historical value associated with the painter legend Raden Saleh. The name of the hometown of Bustaman comes from Kyai Kertoboso Bustam, a famous figure of the Dutch government era as well as great-grandfather of Raden Saleh. Raden Saleh himself is a famous modern hybrid artist pride Indonesia who had studied to Europe.

Kampung Bustaman has a common character as a city in general in Indonesia. Urban village with negative attributes such as slum, chaotic, and inhabited by lower society. The urban village is a burden in the context of urban modernization. The uniqueness and complexity of Bustaman city hometown show empirical phenomenon as a surviving and independent urban village, although there are outsiders who encourage the development and preservation of Bustaman village. Behind all of the macro and micro villages, there must be a hidden motive for Bustaman village community with various efforts to maintain and share the life of the urban kampong.

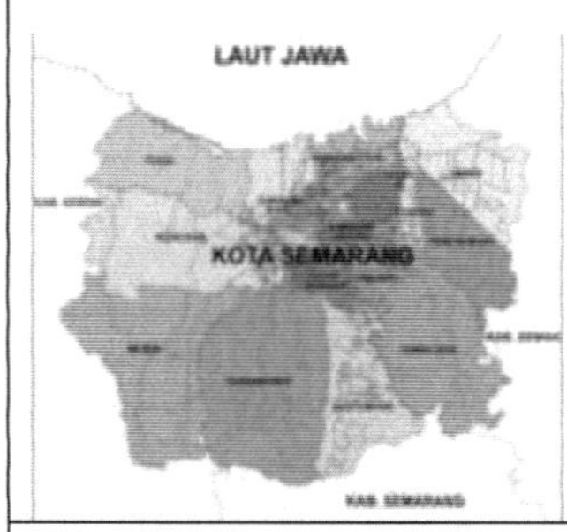

Figure 1 : Map of Semarang City

Source : petasmg.com

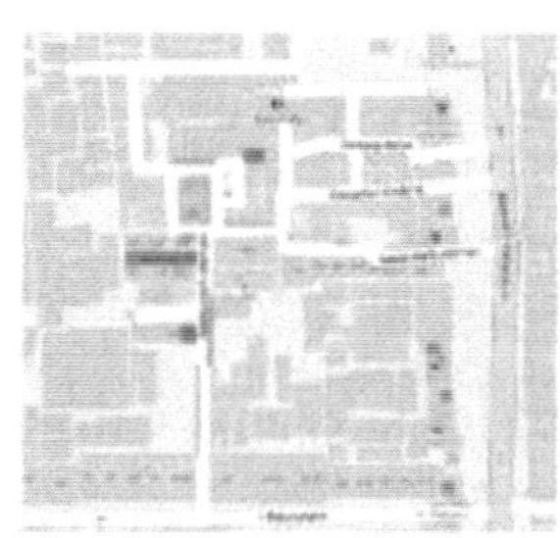

Figure 2 : Kampung Bustaman

Source : petasmg.com

3.1. The uniqueness of Bustaman village hometown

Some artifacts and kampong culture as well as the historical value of the village of Bustaman showing the uniqueness that can give positive value of Bustaman village. The results of data collection either directly through informants or measurements by researchers found explanation of the uniqueness as follows.

Kampung Bustaman has a source of clean water from the ancient Wells for water as sanitation and environmental sanitation. The joint MCK facilities take clean water from the ancient well, it called the community-based sanitation program or SANIMAS. Villagers from two RTs within Bustaman village use shared toilet facilities in turn. Morning and evening show very high usage of joint MCK. This MCK facility shows the limited function of water supply so that some villagers use clean water sources from outside the village. There are villagers of Bustaman who sell clean water with carts around to serve the needs of clean water.

Gedong Ten is one corner of a residential area that has a high occupancy rate. One house can be inhabited more than 2 families with the number of people living in it more than 100 people. Gedong ten has dimensions of small residential homes, coincident and facing the 1.5 meter wide front space shows the impression of a narrow space. The front porch is used as a kitchen and inner space is used as a resting place. Wet laundry activities hanging in front of the house terrace. Gedong ten does not have legitimate land tenure. Each occupant does not have a land certificate over the existing land.

RPH building is a goat cutting facility in Bustaman village. The present condition is not functioning and is used by the villagers as a warehouse for the village. RPH is used as a storage place for unused property. Bicycle residents are stored therein. RPH dispute status encourages this building not to be used as a cutting point. So that the cutting activity is done in front of the house or inside the private property, as done by yusuf and toni.

The name of the hometown of Bustaman comes from Kyai Kertoboso Bustam, a famous figure of the Dutch government era as well as great-grandfather of Raden Saleh. Raden Saleh is a famous modern hybrid painter Indonesian pride who has studied to Europe. Sri Hartatik is a direct descendant of the Raden Saleh lineage that is still alive today, and resides in bustaman. There is Harry Bustaman is also a community figure who still has a relation with the big family of Bustaman and also has a house in Bustaman village.

There are several Bustaman village cultures that have been preserved since a few years ago. The 'gebyuran' activity conducted before the month of Ramadhan, 'tengok bustaman' is an art and cultural event which is facilitated by Hysteria NGO. Look Bustaman is an activity that involves other communities outside Bustaman village. Several communities in Semarang were involved in the event for about 3 consecutive days. Tengok Bustaman become part to expand the existence of Bustaman village as a culinary tourism village.

Gule and sate cuisine from goat meat is typical of Bustaman village culinary. The whole production process is done by Bustaman villagers. Activities of cutting, processing goat meat, and cooking gule done by residents of the village. All these activities are in the village of Bustaman. There is a natural division of roles in local economic life. Each role function shows relationships in fulfilling the daily needs of life, and they also complement and support each other.

The specificity described above has given the character of the hometown of Bustaman city. Culinary goat meat is a peculiarity of kampung that is widely known by the people of Semarang city. Bustaman name became a marker of the existence of a village that has a linkage with the painter Raden Saleh lineage. The community-based sanitation program is awarded by the government for its successful management until now. Positive value of the urban village is awakened by the local wisdom of its citizens. NGOs and municipalities provide support for Bustaman's urban village to survive.

Figure 3 : RPH Building
Source : Field Survey 2017

Figure 4 : Gebyuran
Source : Field Survey 2017

Figure 5 : Gedong 10
Source : Field Survey 2017

IOP Conf. Series: Earth and Environmental Science **99** (2017) 012005 doi:10.1088/1755-1315/99/1/012005

Figure 6 : Sumur Tua
Source : Field Survey 2017

Figure 7 : H Yusuf
Source : Field Survey 2017

Figure 8 : Stage on Kampong
Source : Field Survey 2017

3.2. The complexity of Bustaman village hometown

The phenomenon of the urban hometown of Bustaman shows the chaos and the slums. Limitations of environmental infrastructure, high occupancy density, housing conditions, and welfare conditions of residents are latent constraints on the physical condition of urban spaces. There are several dynamics of urban space texture as follows: a) overlap of the function of the corridor room of the kampung, b) change of dwelling space, c) the absence of communal space, d) the feasibility of low-residential houses, e) spontaneous of spatial layout, f) Village landscape do not exist, and g) disruption of public areas, especially on road corridors.

Some of the width of the village road corridor is used for motorbike parking, although there is a special location for the parking lot, that is the way out of the village of Bustaman towards the street petudungan. The terrace of the house is used as a kitchen and consequently the socializing activities of the residents shifts to the road corridor of the village. Several nodes of the village road corridor are used for community gatherings, as happened at the mosque and in front of an MCK facility close to an old well, or in front of Harry Bustaman's house. Spacious communal space and able to accommodate the activities of citizens does not exist.

Housing terraces are used for various economic activities, such as food stalls, grocery stores, and areas for cooking and washing cooking utensils. Public spaces for the socialization of villagers do not exist, social interaction activities take place in several nodes of road corridors, village guard posts, or multipurpose kampong halls with limited capacity. The cooped house became the main cause of low occupancy quality, damp and dark. Plant and green layout not found, only pot plants that can be seen. The limitation of green space makes urban space unhealthy. The slums and chaos of the urban spaces are due to the socio-cultural limitations of the urban villagers. The character of the kampong does not show a formal pattern, but it is informal. Spontaneous and unplanned spatial layout makes the locality of Bustaman village very strong.

Figure 9 : Terrace
Source : Field Survey 2017

Figure 10 : Street corridors
Source : Field Survey 2017

Figure 11 : Parking Lot
Source : Field Survey 2017

3.3. The empirical findings of Bustaman village hometown

The uniqueness and complexity of Bustaman's urban village affect the quality of the existing kampong space. The uniqueness is able to give a positive meaning of the village space but the complexity is the opposite to show the negative value. This means that there is a balance phenomenon of Bustaman village space. The balance should be managed by the villagers and / or outside parties such as NGOs and municipalities. Citizen participation, the role of outsiders (NGOs and municipalities), community leaders must show a shared commitment.

Spontaneous, informal, compact, enduring, and local power are attributes of urban spaces attached to Bustaman village. Spontaneous is a symptom of space without careful planning, that every time the function of road corridors can change function in a short time. So it can be categorized as an informal space because the use of space on the basis of mutual agreement among the villagers. The result is efficient and optimal effectiveness of kampung space in the use of existing resources in Bustaman village. This can be cited as a compact system that can survive based on local strength and wisdom.

The uniqueness and complexity of the kampung space is the expression of space over the spontaneity of the activities of the citizens. Spontaneity of mutual agreement in the utilization of village resources. It aims to preserve and maintain Bustaman village. Existing symptoms still require outsiders such as Hysteria NGOs and municipalities. This is to accelerate and improve the quality of kampung space with a community-based conservation approach. This is an alternative to solve the uniqueness and complexity of Bustaman village.

4. Conclusion

The uniqueness and complexity of urban spaces Bustaman is influenced by internal factors and external factors. The internal character of the urban village space is built upon the socio-cultural construction of the residents of the Bustaman residential area and environment. External factors only as a catalyst and facilitator of the balance value of the village space, in order to remain preserved and exist. This has no effect on urban modernization and vice versa, that urban modernization is not able to influence Bustaman's urban space.

The balance of space occurs informally within the internal village of Bustaman. Positive value of kampong space can be a penetration of Bustaman village space shortage. The balance achieved is influenced by the nature of mutual tolerance and has the same sense of limitations among the citizens. The community's social values built among the citizens are due to the presence of Hysteria NGO facilitators as well as the city government. Including the local economic value of culinary goat meat also gives a positive impact.

5. Acknowledgments

Thanks to those who have corrected the writing of this study. Especially my thesis dissertation supervisor and dean of the diponegoro university's engineering faculty who provide research cost support. And also the villagers of Bustaman especially Harry Bustaman and Sugiono for all the information that has been informed to the researchers.

6. References

[1] Jo Santosa, (2013); "Understanding Urban Transformation in Asia: Case Study of Jakarta", Tata Loka Volume 15 No. 2, May 2013, 102-115.

[2] Freek Colombijn (2007); "Urban Village of Indonesia: Anthropological-Historical Study on Social and Urban Space Gaps ", International Conference on the Urban Kampong, Januari 19-20, Surabaya.

[3] Jellinek, Lea (1991). " The wheel of fortune: The history of a poor community in Jakarta , Sydney,: Allen and Unwin.

[4] Bakti Setiawan (2000) Bakti Setiawan (2010); "Kampung Kota And Kota Kampung: Challenges of Urban Planning In Indonesia"; Inaugural Speech of Professorship in Urban Planning Sciences Gadjah Mada University, Yogyakarta.

[5] Turner (1985); "Housing by People", MIT Press, Boston MA.

[6] Silas Johan, (1996); Ministry of Public Works, 2010; in Sandyawan Sumardi, 2016)

[7] Sri Handayani, (2008); "Urban Village Community Participation to Improve the Quality of the Neighborhood Environment (Case: Kampung Kota Settlement in Bandung)"; Bogor Agricultural Institute Dissertation.

[8] Iwan Kustiwan, et. al. (2015); “Identification of the Creative Capacity of Kampong’s Community towards Sustainable Kampong (Case Studies: Cicadas and Pasundan Kampong, Bandung): A Preliminary Study”, Procedia - Social and Behavioral Sciences 184 (2015) 144 – 151.
[9] Jan Newberry, (2013); ”Double Spaced: Abstract Labour in Urban Kampung”; Department of Anthropology University of Lethbridge 4401 University Drive Lethbridge, Alberta T1K 3M4
[10] Triatno Yudo Harjoko, (2009); ” Urban Acupuncture: An Alternative Purposive intervention to urban development to generate sustainable positive ripples for an ‘Aided Self-Help’ Kampung Improvement”, Informal Settlements and Affordable Housing 2009.
[11] Marcus Zahnd, (1999); "Integrated City Design"; Kanisius Series Strategy Architecture 2, Yogyakarta
[12] Marcel Hunziker, et al, (2007); “Space and Place – Two Aspects of the Human-landscape Relationship”, F. Kienast, O. Wildi & S. Ghosh (eds.) , A Changing World. Challenges for Landscape Research, 47– 62. © 2007 Springer.
[13] N. John Habraken, (2002); ” The Uses Of Levels”, Keynote Address Unesco Regional Seminar on Shelter for the Homeless Seoul 1988 As re-issued by Open House International Vol. 27 no. 2 2002
[14] Spradley, (2007) ; 'Ethnographic Method', Tiara Wacana, Yogyakarta.
[15] Cresswell, (2007); “Qualitative Inquiry & Research Design. Choosing Among Five Aproach”, Sage Publication, Thousand Oaks

Building resilience in heritage district: lesson learned from Kotagede Yogyakarta Indonesia

Dwita Hadi Rahmi
Gadjah Mada University, Yogyakarta, Indonesia

Corresponding e-mail: dwita_hr@yahoo.com

Abstract. Kotagede, a heritage district in Yogyakarta and famous as a silver town and a destination for heritage tourism, is one of the vulnerable heritage resources in Indonesia. Its history, dates back to the Old Mataram Kingdom in 16th Century, has inherited many heritage properties, including traditional settlement patterns and Javanese traditional architecture of houses. As a vulnerable heritage area, Kotagede experienced several disaster attackks; with the last one was the Java biggest earthquake in 2006 that destroyed more than 200,000 houses in Yogyakarta and dozens of traditional houses in Kotagede collapsed. Ten years after the big disaster, Kotagede has significantly developed with some former earthquake impact can still be found.

This paper aims to document and examine the way Kotagede built its resilience, particularly after the last earthquake, and how the reconstruction process relates to the broader concept of resilience. Descriptive and qualitative approaches are used based on historical data and field observation. This paper notes that although not as fast as other non-heritage areas, the reconstruction process in Kotagede has finally done. It is also argues that several factors contribute to the building resilience of Kotagede, and the most important factor is the collaborative actions among stakeholders in coping the disaster impact. Such collaboration can be effectively done when the local community have strong commitment and willingness to solve their problem and have a resilient Kotagede.

Keywords: resilience, heritage, earthquake, recovery, Kotagede

1. Introduction

Globally, concern on the resilience of cities to the impacts of natural and human-made crises has been increased. With its long history and cultural diversity, Indonesia is rich in heritage resources, both tangible and intangible. Meanwhile, Indonesia is also famous as a country in the ring of fire that makes it vulnerable. Under such condition, heritage resources face difficult situation, as they have to deal with disaster attack.

Yogyakarta is one of many regions in Indonesia that is considered as a disaster-prone area. This is apparent from its position toward Merapi volcano in the north and Opak vault in the south that causes Yogyakarta in a high risk of earthquake disaster. In the last decade, a number of disasters have occurred in Yogyakarta. The May 27th 2006, a 5,9 Richter scale of earthquake has ruined many Yogyakarta's and Central Java's buildings and district heritages. More than 200,000 houses in Yogyakarta collapsed and almost 3.000 people died. This disaster also collapsed local people's economics.

Kotagede was one of the districts in Yogyakarta that experienced major damage as the impact of the disaster. Hundreds of traditional houses were damaged, from minor to major damages, or even many of them were totally collapsed. In addition to collapsed houses, shops and markets also collapsed. Damage occurred because in traditional house construction there is no strong enough structural system of column-sloof-beam to withstand the earthquake jolting. Earthquake impacts not only on the physical damage, but also on the social and economic condition of the community. Earthquake has resulted in an economy paralyzed for a moment. The silver and brass craftsmen and food craftsmen had stopped their economic activity, because they had to think of their homes that collapsed. This situation has awakened many parties to help the post-earthquake recovery of the area. The question arises, whether Kotagede can withstand and cope with the impact of the disaster, and how Kotagede built its resilience, particularly after the last earthquake.

Resilience is popularly understood as the ability to cope with an unexpected danger when the danger of that happening, as well as the ability to learn to bounce back from or adapt to some kind of perturbation [1]. If resilient, a system has a degree of elasticity, allowing it to withstand a shock and reorganise itself when necessary and is thus forgiving of external shocks [2]. Godscalk also explains that a resilient city is an integration between physical and social systems. Physical system is a manmade and natural environment of the city, including infrastructure, geology, topography, and hydrology, whereas social system is a way of life of the community based on their tradition and believe. In the settlement, the Department of Planning and Community Development Victoria define resilient settlements as a concept that directs how the existing system in the neighborhood can accept the risks and is not too affected by the changes, but if necessary the existing system must adapt to those changes [3]. The ability to cope with various threats will be created if the community and institutions in the city can function effectively. The effective function is shown when the community and institutions are able to deal with unpredictable disturbances and adapt to the changes [4]

As an old traditional settlement in Yogyakarta, Kotagede has a unique physical condition and social life of the community that is apparent from their strong daily life tradition, especially *gotong royong* (working hand-in-hand among the people). Therefore, when their district was hit by the earthquake, the community of Kotagede had a big participation in the pot-disaster recovery. They have commitment and willingness to save their home and revive from the economic collapsed [5]. This paper aims to document and examine the way Kotagede built its resilience, particularly after the last big earthquake, and how the reconstruction process relates to the broader concept of resilience.

2. Methods

This paper is based on several writing about Kotagede post-disaster and my experience when involved in some activities during the recovery time of Kotagede. Two manual books that I am one of the authors are the basic references for this paper, which are: 1) Post-Disaster Conservation Manual for Kotagede Heritage District Yogyakarta Indonesia, and 2) Homeowner's Conservation Manual for Kotagede Heritage District. Descriptive qualitative approaches are used based on historical data and field observation.

3. Discussion

3.1 Kotagede as Heritage Ditrict

Kotagede lies 7 km Southeastern of Yogyakarta City, and administratively it belongs to both Bantul Regency and Yogyakarta City. Geographically, Kotagede lies in relatively higher area between two rivers, Gadjah Wong River in the West part and Manggisan River in the East. The Southern part of Kotagede tends to be green field areas, while the Northern part is high populated residential areas.

The existence of Kotagede is closely related with the history of Yogyakarta. Its history, dates back to the Old Mataram Kingdom in 16th Century, Kotagede used to be the capital city of Mataram Islamic Kingdom. The role of this region gradually faded along with the move of the capital to Pleret, then to Kerta, and finally to Surakarta. Nevertheless, the uniqueness of Kotagede reappeared along with its role as a trading and craft center during the Yogyakarta Sultanate and the Colonial Government of the Netherlands.

The historical value of Kotagede and its uniqueness landscape and culture can be seen from its traditional settlement patterns with the Javanese traditional architecture of houses and its social life. The narrow walled gangways and *jalan rukunan* (mutual gangways); the wooden construction of the Javanese traditional houses built hundred years ago; and the physical remains of the past such as old mosque, cemetery, ruin of fortress, are precious heritage of Kotagede. As the traditional sites,

Kotagede is also famous as a center for silver, gold and copper crafts. Along the main road which crosses the Kotagede, namely Jalan Mondorakan and Jalan Kemasan, silver jewellery shops are easily found. Tourists come to Kotagede see the heritage, buy the silver works, see the traditional houses, and feel the indigenous surroundings. 'Silver' at last is adopted to be the Kotagede's trademark, and Kotagede is called Silver Town.

The typical of Kotagede community has a long history. Formerly they were known to be a well-off or even rich community. Most of them were merchants, batik producers, goldsmiths and silversmiths. Their riches can be seen even nowadays from their big traditional houses and *Kalang* houses which mostly use high quality of tiles, stained glass, roof and even beautiful ornament. In this modern era, Kotagede community live as a traditional community harmoniously. It can be seen as the result of their residential place which is dense and closed with only few new comers from outside. Their pure social and cultural life are seen from their willingness to help each other and their close relationship can be witnessed from their *pintu butulan*, that is the door that connect each land border. This kind of door has a very high social function.

The activity of the community is varies, like activities that relate to religion, such as Islamic women's meeting for reciting Al'Quran and celebrating Islamic days. Activities that relate to Javanese believes, such as ceremonial things when building houses, having a new child, burying the death, and having marriage. In a daily life, a busy activity of Kotagede people can be seen in the Big Market which is mostly placed at the center of the residential places. The market is so full with not only people in that area but also those from the outside as well.

Today, with its rich heritage properties, Kotagede is one among several destinations for the heritage tourism in Yogyakarta, while also the home of about 33,300 people. The rich Kotagede in its landscape and culture proves that this area is a heritage district, a valuable asset of Yogyakarta. In 2011 Kotagede was set as a heritage district together with five other heritage areas in Yogyakarta under the Government Decree (SK Gubernur DIY No. 186/KEP/2011).

3.2 Post-Disaster Recovery of Kotagede

The post-disaster recovery of Kotagede was realized through Revitalization Program that was done simultaneously for both tangible and intangible heritages and by encouraging the community's economics to grow better. To recover the folk heritage and rebuild or encourage the betterment of people's economic; a comprehensive handling for multi discipline and multi sectors is needed. The recovery can only be done step by step gradually based on the priority and needs.

The post-disaster recovery of Kotagede heritage included rehabilitation and reconstruction activities of the area, which were the rebuilding of all the facilities and infrastructure, such as road, public buildings (mosques, market, schools, etc.), electricity, water, and drainage networks, and so on with retaining the value of Kotagede history, architecture and culture. This activity is also rebuilding the institutional community with the goal of continuing the development of the economic, social and cultural activities, and the rise of the role of the community in all aspects of community life.

Post-disaster recovery of Kotagede team was initiated by the Center for Heritage Conservation (CHC), Department of Architecture and Planning Gadjah Mada University, and Jogja Heritage Society (JHS). Behind this team were academicians who had many links with local, national, and international institutions for the idea and actions of the post-disaster recovery. The team then initiated to found the Society Concerned on Emergency (TSCE) or called *Pusaka Jogja Bangkit* (PJB/Jogjakarta Heritage Revival) for the recovery actions.

3.2.1 Founding the community organization

Many kinds of post-disaster recovery of heritage need the participation of the local community as well as community from outside the area actively. Two days after the earthquake striked Kotagede, PJB was founded and it grew from time to time. A volunteer of this PJB were organization of non-governmental, social, universities, private or individual which open heartedly help the earthquake victims to rebuild their area and encourage them to lead a new life. PJB then developed links with heritage conservation organizations all over the world to get support and funding.

On August 17th 2006, the local community organization called *Organisasi Pengelola Kawasan Pusaka (OPKP)* or Kotagede Heritage District Management has been founded. Its operational fund for the year 2007 was supported by the Netherlands Government. This organization was the bridge for many community organizations to communicate regarding the revitalization program of Kotagede. OPKP involved in selecting which houses needed to be reconstructed and it was as a community representative in all discussion and negotiation with the external institutions that were willing to help Kotagede. OPKP also encouraged local community activities such as *macapatan* training (traditional song that is full of philosophical learning), *keroncong* (classical-modern local music), and art performances for welcoming guests. OPKP was then expected to be able to involve in the heritage conservation, so that the member of OPKP were people who had capacity in conservation. There were some trainings to encourage OPKP on how to manage their heritage district intensively, and at that time were funded by JICA 2006/2007. Therefore, each village in Kotagede district was expected to have OPKP.

The quick formation of community organizations that care for disaster victims in the heritage area shows a concern for the conservation of Kotagede heritage district and of humanity, both the local community as well as other parties from outside Kotagede. PJB and OPKP worked together to solve the problem of recovery and build the resilience toward unpredictable disaster.

3.2.2 Reconstruction of heritage buildings

The condition of heritage assets in Kotagede was listed right after the earthquake by *Pusaka Jogja Bangkit* (PJB). It was done by doing damaged rapid assessment for the tangible as well as intangible heritage. The rapid assessment includes: 1) Identifying the heritage which is ruined; 2) Identifying the kind of the damages (total ruin, half or just minor damage); 3) Valuing the kinds of re-cure needed; 4) Cost estimation for each heritage. The rapid assessment done in June 2006 for Kotagede shows the following result:

Table 1 Rapid assessment for heritage building [5]

No.	Result of rapid assessment
1	The traditional houses with *joglo* roof type that have been evaluated are 88
2	The number of all *joglo* type houses in Kotagede is 150
3	Some houses are not yet evaluated
4	Among those which are already inspected: -8 houses are ruined (9%) -47 houses are severely damage and cannot be used (54%) -16 houses are partly damage, but can be occupied (18%) -17 houses crack (19%)

The next step, a detailed assessment of heritage buildings was done. This detailed assessment was needed to get more accurate information about the damage. The aims of the detailed assessment were: having documentation about kinds and level of heritage damage in detail; knowing the kinds of recovery and the most suitable handling activities; and preparation for treatment (experts, funds, picture, field labor, etc.).

After quick assessment towards the damage heritage was done, the dissemination about it to the international world was conducted through the PJB web-site to collect funding. Funds were the main factor in a series of activities to handle the emergency situation and recovery of Kotagede after the earthquake. A number of aids from individual, group of people, government and private agencies locally, nationally as well as internationally were collected for Kotagede, such as the Netherlands Government, JICA (Japan International Corporation Agency), Total Indonesie Perancis.

Several efforts to save Kotagede heritage post-earthquake, in the form of reconstruction of several heritage buildings, with funding support from neighboring countries are listed below [5][7]. They are:

1) The reconstruction of six traditional houses. In this case, the selection of which houses needed to be reconstruct was helped by OPKP as the representation of Kotagede community, and those six houses were:
   (a) Jatmiko's house: the *pendapa* reconstruction was funded by Netherlands Government funds; the survey and drawing was suppurted by JICA; and the *gandok* renovation were also funded by JICA and BAQ 2007 funds.
   (b) Edi Priyanto's house: the *pendapa* reconstruction was funded by Netherlands Government funds; the survey and drawing was suppurted by JICA funds.
   (c) Soediyo Prasetyo's house: the *pendapa* reconstruction was funded by Netherlands Government funds; the survey and drawing was suppurted by JICA funds.
   (d) Joko Nugroho's house: walls retrofitting was fully funded by Netherlands Government
   (e) Mukadi's house: the reconstruction is funded by Total Indonesie.
   (f) UGM's house: the *pendapa* reconstruction was suppurted by JICA funds; and the renovation of *ndalem* was supported by Total Indonesie.
2) The Kotagede community and UGM-JICA hold consultation, retrofitting and restoration of traditional houses.
3) The reconstruction of *Babon Anim* (Electric Center building), that was then functioned differently, was funded by Total Indonesie.
4) The renovation and reconstruction of monuments (gates, mosque's wall, mosques) were funded by the government (BP3).
5) Homeowner's Conservation Manual by Jogja heritage Society was funded by UNESCO Jakarta.

Figure 1. *Joglo* reconstruction of UGM's house

Figure 2. The ruin of Edi Priyanto's house before reconstruction

Figure 3. *Gotong royong* to reconstruct the *joglo*

3.2.3 Rehabilitation activities in Kotagede post-earthquake
Basically, the post-disaster recovery of heritage has a long agenda after emergency situation. The recovery needs a long time and is done continuously. The more damages happen and the less preparation for handling it, the more complex the problem will be. This condition will influence the resilience of the heritage from disaster impact.

In Kotagede, two years after the disaster, the economic revival of the community began. This condition was influenced by the environmental conditions that were already better, and support of many foreign agencies for the revival of the local economic activities including silver craft production [6]. After the emergency situation, some rehabilitation activities for Kotagede were conducted [6], such as:

1) Livelihood recovery
   This activity was called Kotagede Craft Revitalization Post-Earthquake 2007, supported by Exxon Mobil, from March 2007 until February 2009. This activity aim to reactivate the small craft production by empowering 13 silver crafters with low income because of the low access to market their products. This included to help designing the more fashionable design and to market the new products in various places; and strenghtening the crafters managerial and marketing capacity.
2) Rehabilitation and reconstruction of settlements-based disaster mitigation

This activity begun in 2008 under World Bank support and was known as Java Reconstruction Fund (JRF) that includes some countries (Uropean Commission, the Netherlands, British, Finland, Canada, and Denmark Governments) and run by Ministry of Public Work. This program was called REKOMPAK and the output was the planning document for the medium term community-based development that was prioritized on the disaster mitigation. The aim of this program was to improve the environment quality of the villages in Kotagede, and at the same time preserving the heritage assets.Three areas in three villages that already had OPKP were chosen to be rehabilitated and reconstructed, which were: Soka Tumenggungan area in Jagalan Village; Sopingen area in Prenggan Village, and Between Two Gates area in Purbayan Village. Elemens to be rehabilitated and reconstructed were traditional houses, roads, street furnitures, and public open spaces.

Various rehabilitation activities in Kotagede were held continuously until several years after the disaster. Support and fundings for these activities came from various sources, both the government and private agencies, nationally as well as internationally. The local community always fully participated in all the rehabilitation activities, and step by step they could be able to manage their environment and social life.

3.3 Resilience of Kotagede

Many lessons can be learned from the experience after the 27th of May 2006 earthquake left terrible damages for tangible and intangible heritages in Kotagede. The recovery of the heritage was continuing until a long time after the disaster. In line with the recovery, the disaster mitigation, which is a serial action to decrease the risk of the disaster can be implemented physically as well as by encouraging the community's awareness and skill to overcome the disaster. The disaster mitigation is becoming one of the approaches in Disaster Risk Management (DRM), and this management should be owned by all areas in Indonesia. The DRM should consist not solely to keep the community aware how to face disaster and its mitigation, but also increase their awareness to do emergency things in the long run.

Although not as fast as other non-heritage areas, the recovery process in Kotagede has finally done successfully. These activities of heritage recovery post-earthquake in Kotagede were based on the principles and approaches of DRM, which are: 1) Well coordinated collaboration between public and private networking, 2) The comprehensive choice of the priority of heritage assets to be recovered, and 3) Broad networking, especially with those involved in expertise, funds and aids. Those three points contribute to the building resilience of Kotagede.

Finally, the resilience of Kotagede in facing the disaster is built by three aspects, which are:

1) The resilient local community to face the disaster is supported by their good community system. The community already have a strong social capital, such as *gotong royong* and a sense of togetherness, so when a disaster comes, they immediately help each other hand by hand and work together to cope with the disaster. Managing Kotagede heritge district will not be successfully done if the local people do not actively involve and do their work voluntarily and responsibly. Their participation can be accommodated through Community's Organization or Non-Governmental Organization (NGO). The existence of OPKP shows that the community have good motivation, solidarity and social linkage as an action capital.
2) The various parties from the outside (community and institution) who voluntarily helped Kotagede, have a very big role in supporting the resilient of Kotagede. They have supported the physical rehabilitation as well as strengthened the social system of the local community.
3) The resilience of Kotagede can be judged from the effective collaborative action between the local community and outside parties in facing and coping with the unpredictable disturbances and capable to adapt to change. Such collaboration can be effectively done when the local community have strong committment and willingnes to solve their problem and have a resilient Kotagede. Although unconsciously understood by the local community, the whole process of recovery can be seen also as an important resilience building.

4. Conclusion

This paper concluded that Kotagede is one example of the success story in managing post-earthquake recovery, both in improving the physical heritage quality and reviving the local economic activities of the community. Post-disaster recovery program of a heritage district is not a short time program, but it needs yearly to bring it back to the normal condition or even better. Building the resilience toward disaster can be achieved if there are voluntarily commitment and willingness among stakeholders (local community, government, private agencies, university, and other parties) to work together in coping the disaster. When a heritage district is resilient, it will be able to cope with an unexpected disaster and easy to adapt to the changes.

5. References

[1] Gopalakrishnan, Kasthurirangan & Peeta, Srinivas, 2010, *Sustainable and Resilient Critical Infrastructure Systems*, Dordrecht, The Netherlands, Springer.

[2] Godschalk, David R., 2003, Urban Hazard Mitigation: Creating Recilient Cities, *Natural Hazard Review*, Vol. 4, No. 3, August 1, 2003.

[3] Department of Planning and Community Development Victoria, 2008, *Settlement Vulnerability Assessment*, Victoria. www.dpcd.vic.gov.au

[4] Butsch, Carsten; Benyamin Etzold; dan Patrick Sakdapolrak, 2009, *The Megacity Resilience Framework,* United Nations, University Institute for Environment and Human Security (UNU-EHS). http:www.ehs.unu.edu/article/read/policy-brief

[5] Rahmi, D.H. and Handayani, T., 2009, *Post-Disaster Conservation Manual for Kotagede Heritage District Yogyakarta Indonesia*, Jogja Heritage Society, Yogyakarta

[6] Wijayanto, P., 2012, *Model Pengelolaan Kawasan Kotagede Pasca Gempa 2006*, unpublished thesis, Program Studi Magister Perencanaan Kota dan Daerah, Universitas Gadjah Mada, Yogyakarta

[7] Jogja Heritage Society, 2007, *Homeowner's Conservation Manual for Kotagede Heritage District*, Jogja Heritage Society, Yogyakarta

Study on the alternative mitigation of cement dust spread by capturing the dust with fogging method

Jaka Purwanta[1*], Tjukup Marnoto[2], Prabang Setyono[3], Ari Handono Ramelan[4]
[1,2] Universitas Pembangunan Nasional "Veteran"Yogyakarta, Yogyakarta, Indonesia
[3,4] Universitas Sebelas Maret Surakarta, Solo, Indonesia

*Corresponding e-Mail: jakapurwanta_tlupn@yahoo.co.id

Abstract. The existence of a cement plant impact the lives of people around the factory site. For example the air quality, which is polluted by dust. Cement plant has made various efforts to mitigate the generated dust, but there are still alot of dust fly inground either from the cement factory chimneys or transportation. The purpose of this study was to conduct a review of alternative mitigation of the spread of dust around the cement plant. This study uses research methods such as collecting secondary data which includes data of rain density, the average rains duration, wind speed and direction as well as data of dust intensity quality around PT. Semen Gresik (Persero) Tbk.Tuban plant. A soft Wind rose file is used To determine the wind direction propensity models. The impact on the spread of dust into the environment is determined using secondary data monitoring air quality. Results of the study is that the mitigation of dust around the cement plant is influenced by natural factors, such as the tendency of wind direction, rain fall and rainy days, and the rate of dust emission from the chimney. The alternative means proposed is an environmental friendly fogging dust catcher.

Keywords: dust cement, mitigation, fogging method

1. Introduction

According Fityatur (2015) where the cement plant impact on the live soft people around the factory site as public revenue and public health. Similarly, the existence of the activities of PT. Semen Gresik (Persero) Tbk. Tuban plant also caused some impact, both positive and negative. Some positive effects that create jobs; increase opportunities in the community such as the establishment of food stalls, grocery stores, boarding houses/rented, and others, to support the fulfillment of the daily needs of employees; as well as rising incomes as a result of subsequent/follow-up of work opportunities and increasing the business opportunity. On the other hand, the activities of PT. Semen Gresik (Persero) Tbk. The Tuban plant had a negative impact in the form of air pollution, especially dust. The emergence of dust from the production process can be dispersed into the environment around the plant site. In effect, then the environment will be exposed to dust. Exposure to dust is inhaled dust particles public both out doors and indoors. Exposure to this dust can interfere with the respiratory community outside the home.(Thaib et al, 2014). The purpose of this study was to determine the effect of the wind against the spread of dust and find alternative means of environmentally friendly dust catcher. One is a cement manufacturing technology with Dry Process. Lower operating cost sand greater production capacity required a major consideration using Dry Process Cement factory, but the negative impact is the emergence of dust which if not managed the nitwould be harmful to the environment. (Duda,1983).

According Dimitriou and Christidou (2011), air pollution is one of the important environmental issues that contribute to the effects of high temperatures that affect public health, animal life, natural ecosystems and ma-made environment. Air pollution is also responsible for climate change, green house effect, acid rain, and others.

According Nurbiyantara (2010), exposure via inhalation of dust particles in the respiratory tract will cause a variety of lung function impairment. The dust particles will accumulate in the respiratory tract. According to the nature of chemistry, physics, and biology of the dust. According Naqpure (2014) termed TSP dust are defined as PM (Particulate Material) with aerodynamic diameter not more than 30 µm. Dust is one type of solid aerosols formed by a material separation process mechanically like the crushing, grinding and blasting. This process occurs because of the friction material with strong winds or shift to other materials. For example, cement dust (cement dust) and dust from the metal elements (metallurgical). Dust particles of solid material is regarded as finely divided with sizes ranging from 0.1 µm to 100 µm.

Dust particulate characteristics including size, size distribution, shape density, adhesiveness, corrosive properties, reactivity and toxicity. One of the most important charakteristic of dust particle suspensions is particle-sizedistribution of aerosols. Generally, particles with a diameter of less than 2.5 µm is considered fine and larger particles 2.5 µm is considered rude (Suhariyono etal, 2003).

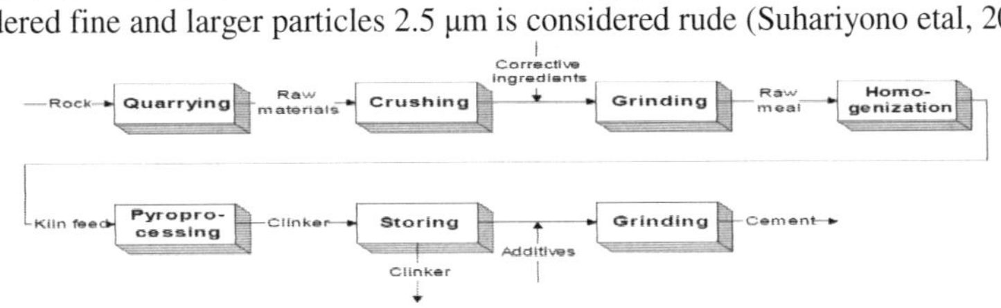

Figure1. The process of making cement
(Source: Ibrahim et al, 2012)

Control of dust emissions in the cement industry generally use mechanical collectors/collectors mechanical / dust collector, electrostatic precipitators, fabric filters, or a combination of such equipment (Kinsey, 1987). According Zimwara et al (2012), air pollution control technology that is Flexible Pulse Jet Filters, Electrostatic Pricipitators, and Wet scrubbers. In wet scrubbers, the gas stream is polluted taken and contact with the liquid that is sprayed with a liquid,and the liquid will catch dust.

2. Methods

This research are : determining the trends in wind direction by collecting secondary data rainfall intensity, long average rainfall, and wind speed and direction in Tuban last few years. The data was then entered in to the file soft Wind Rose to know the trend of the wind direction and designing a dust catcher equipment fogging method.

3. Discussion

3.1. Determining the trends in wind direction

Table 1. Monthly rainfall during the year 2011-2015

Month	Rainfall (mm/month) at Year........					Average
	2011	2012	2013	2014	2015	
January	185.13	264	344	152	17	192.4260
February	107.79	219	161	187	45	143.9580
March	214.50	267	147	312	15	191.1000
April	205.25	0	232	170	36	128.6500
May	213.13	123	148	45	83	122.4260
June	20.54	61	52	86	-	54.8850
July	21.88	0	93	117	-	57.9700
August	0.67	0	-	26	-	8.8900
September	9.29	0	-	-	-	4.6450
October	47.04	77	16	-	-	46.6800
November	273.17	27	150	57	49	111.2340

Month	Rainfall (mm/month) at Year........					Average
December	274.54	435	257	527	6	299.908
Amount	1,572.93	1,473.00	1,600.00	1,679.00	251.00	1,315.1860

Source : DPU Kabupaten Tuban (Bidang Pengairan) dalam Kabupaten Tuban dalam angka 2012-2016

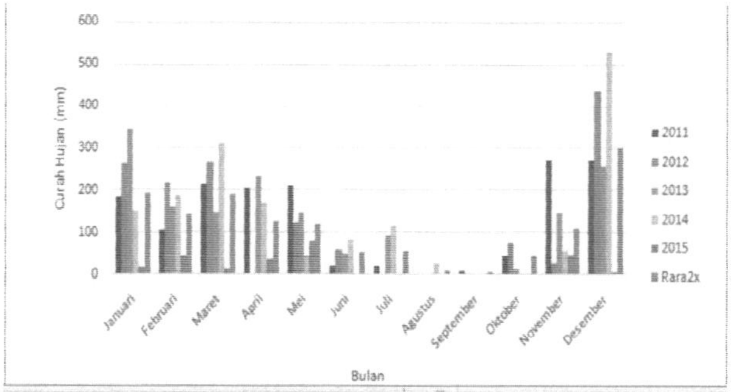

Figure 2. Graph rainfall monthly average in Tuban for 5 years (Years 2011-2015)
(Source: data processing constituent, 2017)

Based on the chart above can be observed that for the highest rainfall in December 2014, because it is the peak of the rainy season, while the lowest rainfall in September in the year 2012-2015, when drought.

Table 2. The number of rainy days during the Year 2011-2015

Month	Number of rainy days /month in Year...					Average
	2011	2012	2013	2014	2015	
January	15.79	15	11	8	5	10.9580
February	8.54	12	7	6	12	9.1080
March	12.83	14	5	8	4	8.7660
April	11.58	0	11	5	9	7.3160
May	9.88	4	12	3	2	6.1760
June	1.50	2	9	4	0	3.3000
July	1.75	0	7	3	0	2.3500
August	0.08	0	0	1	0	0.2160
September	0.08	0	0	0	0	0.0160
October	0.92	3	1	0	0	0.9840
November	13.42	2	11	2	2	6.0840
December	14.50	9	18	13	7	12.3000
Amount	90.87	61	92	53	41	67.5740

Source : DPU Kabupaten Tuban (Bidang Pengairan) dalam Kabupaten Tuban dalam angka 2012-2016

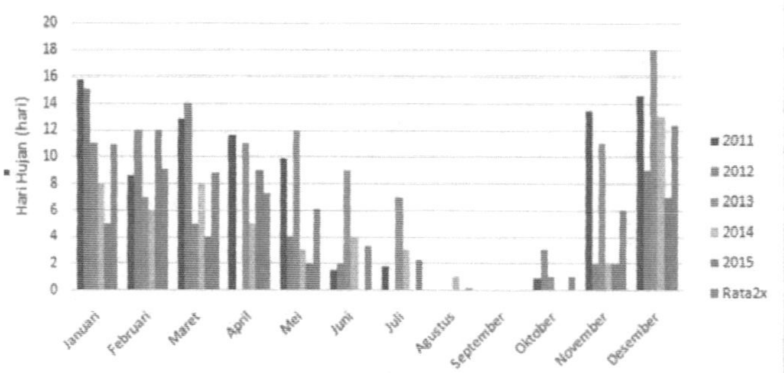

Figure 3. Graphic number of rainy days during the years 2011-2015
(Source: data processing constituent, 2017)

Based on the picture above it can be seen that the highest number of rainy days in the month of December 2013 which is 18 days, so in about the month, the dust can not be spread with maximum because rain water is affected. While the number of rainy days was lowest in the few months in 2012-2015 with the number of rainy days 0 days of rain. In the months with the minimal number of rainy days it will contribute to increasing the deployment area of cement dust.

Table 3. Data monitoring wind speed and direction were done by. Semen Gresik (Persero) Tbk. Tuban at Some village in the district of Kerek, Tuban, East Java Province (2016)

No	Measurement Locations (Village)	Measurement Time (km/jam) (Village)	Wind Speed (km/jam)	Wind Direction	No	Measurement Locations	Measurement Time	Wind Speed	Wind Direction
1	**Temandang**	07-08March	0.90	East	13	Temandang	01-02 August	0.60	North
2	**Tlogowaru**	08-09March	0.70	East	14	Tlogowaru	01-02 August	1.20	West
3	**Kasiman**	11-12March	0.70	West	15	Kasiman	02-03 August	1.00	West
4	**Margomulyo**	14-15March	1.40	East	16	Margomulyo	02-03 August	1.40	North
5	**Sumberarum**	15-16March	1.30	East	17	Sumberarum	03-04 August	1.10	West
6	**Karanglo**	16-17March	1.30	East	18	Karanglo	03-04 August	1.20	West
7	**Temandang**	09-10May	0.90	West	19	Kasiman	20-21 Nov	1.30	East
8	**Tlogowaru**	09-10May	0.50	South	20	Sumberarum	24-25 October	1.30	West
9	**Kasiman**	11-12May	1.10	North	21	Karanglo	25-26 October	1.40	North
10	**Margomulyo**	11-12May	0.90	North	22	Margomulyo	26-27 October	1.30	East
11	**Sumberarum**	12-13May	2.00	South	23	Temandang	17-18 October	1.50	East
12	**Karanglo**	12-13May	1.00	South	24	Tlogowaru	18-19 October	1.30	East

Source:Laporan Pelaksanaan RKL-RPL PT. Semen Indonesia Semester I&II 2016, 2016

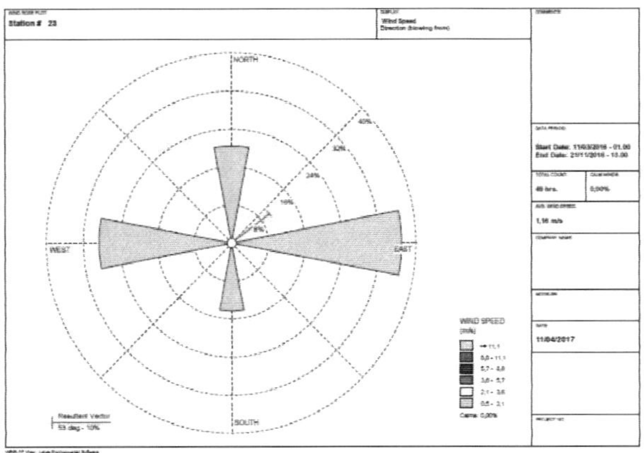

Figure4. Wind rose on March-November 2016
(Source:data processing constituent, 2017)

In the picture above it can be seen that the dominant wind speed in the area of 0.5 to 2.1 m/s and the average wind speed is 1.16 m/s. The tendency of the wind direction is east. This means that the spread of dust tends to the east so that the area on the east side impact sources will be potentially affected by the spread of dust more than the territory in the other direction.

3.2 Designing tools dust catcher with fogging method

The spread of dust can be overcome by using a dust catcher fogging method is environmentally friendly appliance because it uses the mistatomized water demand will be less than with other methods, can catch cement particles whose size is very small and Clots/floc capture results of this appliance can be recycled. To determine the concentration of dust particles, base do the data results of sampling and substituted into the following equation: $C= [(M_1\text{-}M_0) /(t.v)](\mu g/m^3)$ .... (1)

Information:
C : The concentration of dust particles measured
($\mu g/m^3$) M_1: Weight filter after dust measurements (pg)

IOP Conf. Series: Earth and Environmental Science **99** (2017) 012007 doi:10.1088/1755-1315/99/1/012007

M_0: Heavy dust filter before measurement (g)
t : Duration of measurement (hours)
v : velocity of volumetric air (m^3/ h), in this study v= 1,698 m^3/ h

Cement dust concentrations obtained in equation (1) to them odel equations Canter conversion to obtain cement dust concentration at the time of sampling/sample for 24 hours so in accordance with Government Regulation No. 41 of1999. $C_1= C_2.[(t_2/t_1)\ p]$.. .... (2)

Information:
C_1:The concentration of cement dust with long average sampling instance t_1(μg/m^3)
C_2: an average dust concentration at the sampling time instance t_2 (μg/m^3)
t_1 : The duration of sampling sampel 1 (in this study=24 hours)
t_2 : The duration of sampling sample 2 (h)
p : conversion factor value between 0.17 to 0.2.

Thep-value in equation (2) obtained from the Government Regulation No.41 of 1999 with C_1=150 μg/m^3, t_1= 1 day, C_2= 50g/m^3 and t_2= 365 days in order to obtain the value of p = 0.186. Factors temperature, humidity, and pressure greatly affects the air concentration including the Concentration of dust. Therefore the weather conditions recorded and accounted for in this study either temperature, humidity, wind direction, wind speed, and the season. (Suhariyono, 2003). Equation (2) corrected for the effects of T, P, and RH normal to the following equation.
a. Normal wet air condition (≥50% RH): Cn,f= CB.[T /TN][PN/P] .... (3)
b. Normal dry air conditions (RH<50%): CN, tr= CN,f. [100 /(100-F)].......... .... (4)

Information:
C N, f: The concentration of dust in normal conditions (μg/m^3)
CB: The concentration of dust in the air in the normal condition at the measurement time= C_1μg/ m^3. C N, tr =concentration of dust in the dry air conditions (μg/m^3).
T = temperature at the time of measurement (°K).
TN= air temperature in normal conditions (°K) = 25°C = 298,15°K
P N= The air pressure in normal conditions (Pa)= 760mmHg = 1Atm.
P = absolute air pressure at the time of measurement (Pa).
F = humidity at the time of measurement/TN][PN/P]

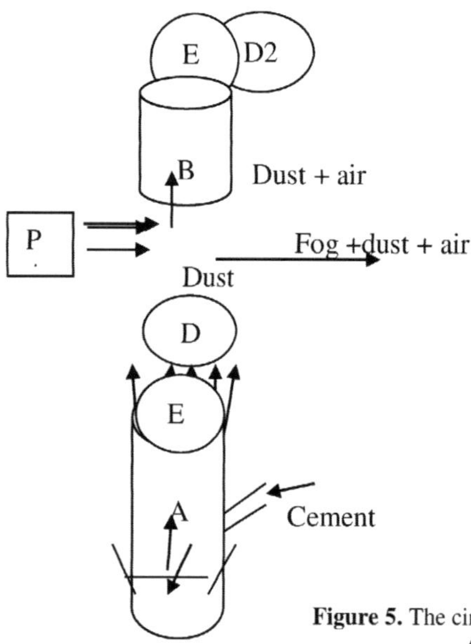

Note: A : Pipe A
B : Pipe B
D : Dust collector
E : Exhauster
F : Fogging tool

Figure 5. The circuit schematic tool in fogging method
(Source: the author, 2016)

This research will use the variabel of nozzle diameter (Dn) and distance betwen betwen sprayer toward column (L).

4. Conclusion

The trend of wind direction identical with the trend towards the spread of dust which is to the east with average wind speed is1,16 m/s. Found alternative means dust catcher with foggging method is more environmentally friendly.

5. Acknowledgments

I look forward to another opportunity to develop this research, namely by examining other variables to obtain optimal operating conditions for the dust catcher with the foggging method.

6. References

[1] Anonim 2017 DPU Kabupaten Tuban (Bidang Pengairan) dalam Kabupaten Dalam Angka 2012sampai 2016 *BPS Kabupaten Tuban*

[2] Anonim 2016 Pelaksanaan RKL&RPL Penambangan Batu Kapur dan Tanah Liat Semester 2016 *PT. Semen Indonesia (Persero) Tbk Tuban*

[3] Anonim 2017 Laporan Pelaksanaan RKL&RPL Izin Lingkungan Kegiatan Industri SemenPortland Di Desa Sumberarum, Kecamatan Kerek, Kabupaten Tuban Semester II 2016 *PT. Semen Indonesia (Persero) Tbk Tuban*

[4] Dimitriou A. and Christidou V 2011 Causes and Consequences of Mix Pollution andEnvironmental Injustice As Critical Issues For Sciences and Environmental Education *TheImpact of Air Pollution on Health, Economy, Environment, and Agricultural Sources, In Tech* 218

[5] Duda WH 1983 Cement Data Book, International Process Engineering in The Cement Industries*Bauverlag GMBH Weisbaden and Berlin* **2nd edition**

[6] Fityatur R 2015 Dampak Sosial-Ekonomi Pabrik Semen Puger Di Kecamatan Puger Kabupaten Jember

[7] Ibrahim HG, Okasha AY, Elatrash MS, and Al-Meshragi MA 2012 Emissions of SO_2, NOx, and PMs from cement plant in Vicinity of Khoms City in North western Libya *Journal of Environmental Science and Engineering A 1 (2012) ISSN1934-8932*620-628.

[8] Kinsey JS 1987 Lime and cement industry particulate emissions: source category report *Cement Industry, Air and Energy Engineering Research Laboratory Research Triangle Park*. **Volume II**

[9] Kumar TJ and Armani A 2012 Environmental impact analysis: a case study of acccementsPlant *Journal of EnvironmentalResearch and Development* **Vol.7**

[10] Naqpure AS 2014 Human Health Risks In National Caiptal Territory Of Delhidueto AirPollution *Atmospheric Pollution Research India* 373

[11] Nurbiyantoro S 2010 Pengaruh Polusi Udara Terhadap Fungsi Paru-Paru Polisi Lalu Lintas DiSurakarta *UNS, Surakarta* 44-45

[12] Suhariyono, Sri Saeni M S and Bay A 2003 Analisis Tingkat Bahaya Partikel Debu PM10 Dan PM2,5 Terhadap Kesehatan Penduduk Di Sekitar Pabrik Semen, Citeureup, Bogor *P3TM Batan Yogyakarta*

[13] Thaib Y P 2014 Hubungan Antara Paparan Debu Dengan Kejadian Gangguan SaluranPernafasan Pada Masyarakat Kelurahan Kairagi Satu Lingkungan 3 Kota Manado *Universitas Sam Ratulangi Manado*

[14] Zimwara D, Mugwagwa L, and Chikowore TR 2012 Air pollution control techniques for th cement manufacturing industry : a case study for Zimbabwe Town South Africa *CIE 42 Proceedings 16-18 July 2012* 37-1 until 37-13

Resilient design in the conservation of Johar market heritage building

Totok Roesmanto*

Department of Architecture, Faculty of Engineering, Diponegoro University, Semarang, Indonesia

*Corresponding e-mail: kanyaka_prajnaparamita@yahoo.com

Abstract. Johar Market building based on Law of Repubic of Indonesia No.11 of 2010 is a heritage building. It was built in 1936 located near traditional square of Pasebaan and Aloon-Aloon was built by Dutch merchants union in 1678. Semarang based on decentralized policy by the Dutch Government in 1906 became a trading city. Rapid development of trading activities has caused the city planning policy by Semarang Government since 1970 made Aloon-Aloon become market buildings. Johar Market and market buildings in surrounding were on fire in 2015. Basing on Law No.11 of 2010 Semarang Government plans to conserve Johar Market heritage building and reconstruct Aloon-Aloon based on proposal of Roesmanto in 2016. The architect bureau assigned by the Semarang Government designed a new building in the middle of South-Johar Market to accommodate Johar merchants. This study aims to evaluate the design of new building by the architect bureau considering that since 2012 the city of Semarang including earthquake prone areas. The revitalization of Johar Market should use resilient design in order to prevent future damages to heritage buildings located nearby and new building must be spaced sufficiently against surrounding heritage buildings. This research uses descriptive qualitative method base on the field data after Johar Market burned and design planned bureau architect. The results of this study found that the distance between new building and heritage building is less wide.

Keywords: resilient design, heritage building, conservation, Johar Market, Semarang

1. Introduction

Based of ancient map of Paan van het Fort in omlegende Cituatie van Samarangh [1] can be seen that in Samarang or Semarang in the 17th century there are places called Pasebaan, Dalem, Javaan te tempel, Negery and Fort. Based on another an ancient map of Plan Platte Grond van Samarang met dies Environs op een Afstand in [1] can be known that Vereenigde Oost-indische Compagnie or V.O.C. as the union of Dutch merchants who had been given authority by the Mataram kingdom to reign in Semarang since 1678 built a De Vijfhoek or Fort of Vijfhoek and then a new square called Aloon-Aloon. Aloon-Aloon was located near the traditional square of Pasebaan and to the north of the Dalem as the residence of Semarang Regent. Semarang Regency was a part of Mataram Kingdom situated in the Pesisir Wetan or Eastern Coastal area. Aloon-Aloon and Pasebaan were located on the west side of the Semarang River and between them was the Masigit Semarang or Semarang Mosque. The Semarang Mosque was located in the northeast part of traditional square of Pasebaan [1], but then the mosque burned during the Geger Pacina or Chinese War in 1740-1743. Until the mid 18th The V.O.C. had reigned in the area along the North Coast of Java [2], hence the displacement of the Semarang Mosque from its original presence in the south-eastern part to the south-western part of

Aloon-Aloon was intended to distance it from the Semarang River as the centre of major trade and transportation activities. Local architects designed New Semarang Mosque or Masjid Besar Kauman with the Nusantara-Java Architecture based on Demak Mosque architecture model, which might be taught by Sunan Kalijaga [3]. The New Semarang Mosque used wood construction, its size was enlarged and its location was adjusted to the size and location of Demak Mosque, but the roof of the mosque was struck by lightning in 1885. Renovations were made by Gambier on the orders of Nicolaas Hartingh who served as the director of the V.O.C. on the East Coast or Eastern Java while still utilizing wooden pillars in the south gallery and north gallery that were still intact.

After the Chinese War is over Chinese community and merchants were moved by the V.O.C. to the east of the Semarang River. The building of the salt storage shed after the Chinese War was moved closer to the Fort of Vijfhoek in order to make trading activities easily controlled by the V.O.C. for its commercial interests. Trading activities along the Semarang River grew rapidly, and on the west side appeared Pasar Sebandaran or Sebandaran Market, Pasar Pedamaran Lor or North Pedamaran Market and Pasar Pedamaran Kidul or South Pedamaran Market. Fort of Vijfhoek was extended eastward become the Little Netherland or Semarang Oud as an embryo of Semarang City, and now called the Kota Lama. In 1799 the V.O.C. collapsed, the wall of Little Netherland fortress was dismantled. The main road in Little Netherland was connected to Daendels Street limited the northern part of Aloon-Aloon.

During the brief reign of Thomas Raflles little change occurred in Semarang city. Significant developments affecting Semarang city occurred in the late quarters of the 19th century when the first railway line in the government of Hindia Belanda or Dutch East Indies Government was built. The construction of the railway was intended to facilitate the transportation of crops from the Vorstenlanden region where the crops were used as export commodities to Europe for the Dutch East Indies through the port of Tanjung Mas. Prison buildings were built in some areas of the Pasebaan and Aloon-Aloon became the Main Square in Semarang city.

Since 1906 Dutch Government implemented decentralization policy. As a result, Semarang city was developed into a trading city. At that time in Semarang there were three railway companies namely N.I.S., S.J.S. and S.C.S. The Semarang Express train line of S.J.S. crossed Aloon-Aloon. In 1914 the largest international exhibition of Koloniale Tentoonstelling was held, its impact was the construction of Nieuw Tjandi in 1916 as an expansion of Semarang city based on the design of Thomas Karsten. In 1930 the northern prison building was changed into traditional market of Johar. In 1933 until 1936, Dutch Government built Central Market of Johar with design of Thomas Karsten to united Sebandaran Market, Pedamaran Market and the traditional market of Johar . In 1957 Central Market of Johar expanded to the south by the addition of the South Johar Market built on Pasebaan. Central Market of Johar became the largest modern market in Southeast Asia in the mid of 20th century. In 1970 the Semarang City Government built Pasar Yaik Baru or New Yaik Market in the southern of Aloon-Aloon to accommodate the growing number of merchants, also built hotel, bank and mall in the northern of Aloon-Aloon, and the area of Semarang Regent also changed into parking building and Pasar Kanjengan. Aloon-Aloon as Semarang Square became lost. Central Market of Johar then known as Johar Market.

On 9 May 2015 the Johar Market and surrounding markets were on fire. Johar Market only leaves the building elements made of concrete, but New Yaik Market burned down. According to Law of Republic of Indonesia No.11 of 2010 Johar Market is classified as an important heritage building in Semarang because it is more than 50 years olds and has special meaning for the history of Semarang city development. Johar Market is a masterpiece and first central market in Indonesia and applies tropical architecture in accordance with Indonesian geographical conditions. Johar Market is a building that implements modern architecture, and has the best mushroom construction in Indonesia.

Semarang City Government plans to preserve the building of Johar Market. The Architect Bureau assigned by Semarang City Government plans the Conservation of Johar Area with a parking lot by demolishing the South Johar Market. Totok Roesmanto on Focus Group Discussion-1 to Conservation of Johar Area in early 2017 proposed reconstruction of Aloon-Aloon and on Focus Group Discussion-2 proposed conservation of the heritage building of Johar Market including South Johar Market. The Architect Bureau assigned by Semarang City Government cancelled the plan to demolish the South Johar Market Heritage Building, and planned to conduct a revitalization by adding a new building in the middle for a trading place for merchants.

Based on Standar Nasional Indonesia/SNI or National Standard of Indonesia 1726-2012 about Eartquake Resistant Planning Procedures city of Semarang including earthquake prone areas since 2012, the revitalization of heritage building of South Johar Market should use resilient design.

This study aims to evaluate plans made by architect consultant who will revitalize Johar area by adding new building in the middle of the heritage building of South Johar Market whether it has considered the use of resilient design. There has been no previous research on the need resilient design in the revitalization program for heritage building conservation. This study is based on Law No. 11 of 2010 that regulates the conservation of heritage buildings. Employing descriptive method, data were collected on the damage of Johar Market Heritage Building after burning on 9 May 2015. The results of this study will be useful for the Semarang City Government which will rehabilitate and conserve the heritage building of Johar Market and revitalize the surrounding in 2018. This study is based on Law No. 11 of 2010 which regulates the conservation of heritage buildings and considers the rule of SNI 1726-2012 for new building not to fall on heritage building during an earthquake.

The research method used is descriptive qualitative by doing observation object in South Johar Market, analyzing plan made by architect bureau based on Law No.11 2010, SNI 1726-2012, and basic principle of conservation and revitalization.

2. Methods

The method used in this research is descriptive qualitative, which is done with observations in the field and other sources, especially the new building design proposed by the architect bureau. The observation to Johar Market especially South Johar Market is intended to find the existing data about the building after a fire on 9 May 2015. The architect's bureau plans to add new building in the middle of the South Johar Market should be known. The design proposal for a new building made by architect bureau is based on the needs of the floor area that will be used as a place to accommodate Johar Market merchants especially those who originally trade in South Johar Market. By combining exiting data from South Johar Market and the plan of new building designed by architect bureau it will be able to analyze the distance between the two buildings.

Based on Law No.11 of 2010 can be analyzed whether the building of South Johar Market including heritage building should be conserved or an additional building on the Johar Market that can be demolished. Based on SNI 1726-2012 Semarang including earthquake prone areas, since 2012 all buildings both heritage buildings and new buildings will be affected.

Resilient design is a design that has the ability to make buildings have resilience. Johar Market is a heritage building, because it has been burned and the elements of the building made of concrete material is still intact it must be conserved. Consider that Johar Market including South Johar Market can not function as a place of trade the building must be revitalized.

3. Discussion

3.1. Conservation of the heritage building of Johar Market

The Based on Law No.11 of 2010 Article 5 the building can be proposed as heritage building if it has criteria : 1.aged 50 years or older; 2.represents a period of style at least 50 years; 3.has special meaning for history, science, education, religion, or culture; 4.has a cultural value for strengthening the nation's personality [4].

Johar Market was originally called Centrale Pasar [5] is merger of Sebandaran Market, North Pedamaran Market, Central Pedamaran Market, South Pedamaran Market, Jurnatan Market, Beteng Market and Johar Traditional Market conducted Gemeente Semarang or Semarang Government in 1936. Baldinger argues that the merger of several markets has been done to add to the beauty of Semarang city and improving hygiene in the marketplace [6]. Semarang Government then makes a requirement for modern market based on Johar Market design ie.1.the room inside the market must be sufficiently light and air; 2.all floors and walls should be easy to clean; 3.all buyers and sellers must be protected from direct sunlight; 4.the aisles for circulation of buyers should be wide enough; 5.the distribution of the sale plots should be transparent and clearly visible [6].

Johar Market is the most important modern architectural building in Semarang. It was designed by Thomas Karsten who is famous architect and planner as the pioneer of the urban renewal and urban planning design in Indonesia. Johar Market is designed by applying tropical architecture in accordance with its existence in tropical climate, air circulation and natural lighting into the building very well. The building equipped with an elevated floors for bargaining activities while sitting between buyers and merchants. The bargaining activity between buyers and merchants is a typical culture in traditional markets in Indonesia. Because of the potential of Johar Market, the building is a very important heritage building in Semarang and must be conserved.

According to Orbasli [7] conservation is the process of understanding, protecting, when necessary, maintaining, repairing, restoring and adapting historic buildings to preserve the cultural value. According to Law No. 11 of 2010 Johar Market Building is a heritage building, hence after the fire damage on May 2015 the conservation of the building is necessary based on Article 58 of Law No. 11 of 2010. Based on Article 53.3 of Law No. 11 of 2010, the building should be returned to the original condition.

3.2. Reconstruction of Semarang Square

Law North Johar Market and Central Johar Market are built on Aloon-Aloon, and South Johar Market is built on Pasebaan. Previously at Aloon-Aloon there are also some office buildings and bus terminal. The remaining field is called Aloon-Aloon Semarang or Semarang Square. Since 1971 South Semarang Square was changed to market buildings for the sale fruits and clothes, and North Semarang Square for local transport terminal. Then the local transport terminal replaced the building for the bank in 1977 and the hotel of Metro in 1981. At the Semarang Regent area built Kanjengan Market and parking building. Since 1981 Aloon-Aloon and Pasebaan have been lost.

At the time of Johar Market on fire 9 May 2015 also burned the building market around it. The number of merchants who lost their trading places was approximately 4000 people. They have been relocated and will be partially returned to Johar Market later on. The bureau architect assigned by Semarang Government plans to revitalize Johar area covering Johar Market, South Semarang Square and Semarang Regent area.

Roesmanto submitted a proposal to reconstruct North Semarang Square as a part of Aloon-Aloon at the Focus Group Discussion-1 held in October 2016 [8]. In Focus Group Discussion-1 architect explained his proposal for revitalization of Johar Area. The architect planned New Semarang Square as a substitute South Semarang Square based on the reconstruction proposal by Roesmanto. New Semarang Square will be reconstructed in 2018, and some merchants will return to the South Johar Market Heritage Building. New Semarang Square has a semi-basement that will be used for parking area. The semi-basement of New Semarang Square will facilitate merchants and buyers in Johar Market go to pray to the Semarang Mosque comfortably. In the southern part of the New Square of Semarang several units of three-stories buildings will be built to accommodate merchants.

3.3. Revitalization of South Johar Market

South Johar Market was built in 1957 using a mushroom construction as designed by Thomas Karsten. The building plan of South Johar Market is different from North Johar Market and Central Johar Market. The heritage building of North Johar Market and Central Johar Market are two-stories building and shaded by concrete roof from mushroom construction. South Johar Market Heritage Building that is also a two-stories building has an O Trapezium-Shaped plan. The South Johar Market was built on Pasebaan. Based on Law No.11 of 2010 South Johar Market over the age more than 50 years and has relationship with North Johar Market and Central Johar Market is a heritage building.

On Focus Group Discussion-2 in December 2016 architect planned to demolish the South Johar Market Heritage Building to be replaced by a multi-stories building for vehicle parking. Based on Law No.11 of 2010 Article 51 because South Johar Market Heritage Building is not destroyed and does not change shape and style then the building can't be demolished.

Roesmanto opposed the destruction of South Johar Market Heritage Building and proposed to revitalize it. Because South Johar Market Heritage Building is a two-stories building has a O Trapezium-Shaped plan, so that in the void a new building can be built to accommodate merchants.

Other than that the architect bureau must plan a parking building in the former Semarang Regent area [8]. According to Murtagh historic buildings should not be demolished and changed [9].

3.4. Semarang prone to eartquakes

Based on SNI 1726-2012 since 2012 Semarang has been added to the list of areas prone to earthquakes. The decision is based on the results of the evaluation of many natural disasters that occurred in Indonesia before 2012. The Johar Market consists of the North Johar Market and Central Johar Market that were built in 1936, and South Johar Market that was built in 1957. Johar Market Building is a concrete constructed building using mushroom konstruktie or mushroom construction, that has been proven to be sturdy and not collapse until now. The construction strength of the Johar Market Building after the fire damage can certainly decrease. Considering that since 2012 Semarang city has been included in the list of earthquake prone areas, the reconstruction, rehabilitation and revitalization of Johar Market Heritage Building must be planned with resilient design.

South Johar Market Heritage Building is two-stories building with mushroom construction. South Johar Market Heritage Building is known by its popular nickname Pasar Maling [Thief Market], because traded goods are sold at very cheap prices. The height of the building on the second floor that is the distance between the floor surfaces to the concrete roof surface is 2.80 m. The height of the first floor building that is the distance between the first floor surfaces to the second floor surface is 1.90 m. The condition of the first floor surface is in the form of not only a pile of cumulative burning remains, but also stockpile of new materials. Merchants collected the stockpiling for an adjustment to changes in the types of traded goods. According to the rules of the Law No. 11 of 2010, the first floor needs to be returned to its original shape. All materials that have been stockpiled on the surface of the first floor should be discarded. The Johar Market Heritage Building will return to its original form like when it was built in 1957.

3.5. Re-placement of old merchants

Before Johar Market was burned on May 9, 2015 there was a crowded and large number of merchants. Alleys between the circulation ways designed by Karsten were quite spacious and fine. However, about half width of the alleys was used as new trading places or old merchants expanded the area towards the front.

According to Law No. 11 of 2010 and Karsten's design the alley section must be returned as a circulation path for buyers. Considering that after the burning the strength of concrete construction of Johar Market Heritage Building also decreased, re-functionalization needs to be carried out to transform the market into exhibition and souvenir center with a selective and limited number of merchants. Likewise, the number of merchants resettled in South Johar Market Heritage Building should be limited. Due to the limited Johar area owned by Semarang Government, the architect bureau proposes the addition of a new building in the middle of the void in South Johar Market Heritage Building.

The architect bureau designed a new building in the middle of the void of the South Johar Market Heritage Building. Architect plan the distance between the new building in the middle of the void and the surrounding heritage building is only 4 meters so that the capacity to accommodate old merchants can be maximized. Based on measurement data in South Johar Market Heritage Building on void can only be made new building of size [30-4-4] meters x [30-4-4] meters or 22 meters x 22 meters. That number of merchants to be accommodated should be based on the height of new building that can be built.

3.6. Resilient design for heritage building

Resilience is the ability to adapt and remain able to firm in difficult situations. In the opinion of Eskew resilience in built environment is concerned communities and not about building. Eskew-Dumez-Ripple define resilience is the preservation of communities through ongoing planning for the capacity to learn, adapt, and change in the face of present-day and future, both predictable and unknown [10].

The area of plot for Johar merchant communities replaced to the South Johar Market Heritage Building and the new building in the middle of the void must be in accordance with the Karsten design of 1936. Johar merchants who are not accommodated in South Johar Market Heritage Building and the new building should be placed in new building in the former Semarang Regent area.

American Institute of Architects define resilient design is the design that can adapt to changing conditions to rediscover its function and vitality to deal with impending troubles. According to the Rockefeller Foundation resilient design relates flexibility to the ability change, adapt to alternative strategies for facing disasters, accept new solutions, and also the ability to re-establish and re-organize function [10]. The upcoming problem that must be faced by all buildings including heritage building is earthquake.

Even though the Johar Market Heritage Building has survived until now, has been burned in 2015, since 2012 until now has experienced several earthquakes, but in the future the building structure and construction will be reduced and vulnerable to the earthquakes. Merchants who are planned to be replaced to new building in the middle of South Johar Market Heritage Building will adapt to their new place.

The height of South Johar Market Heritage Building after the reconstruction of its first floor to the original condition is 2 x 2.80 meters = 5.60 meters. The South Johar Market Heritage Building was built in 1957 before Semarang City was considered as an area prone to earthquakes. The new upcoming building on the void of the South Johar Market Heritage Building is planned based on SNI 1726- 2012. From the difference in the construction strength of the building, in case of major earthquake buildings form South Johar Market will be more affected first than the new upcoming building in the void.

With a height of 5.60 meters, if the South Johar Market Heritage Building collapses due to the earthquake, the debris of the ruins will be slammed down and sideways as far as 5.60 meters. If the ruins of the South Johar Market Heritage Building are reflected again irregularly to the side about 1 meter, then the open space area that should be provided between South Johar Market Heritage Building and the new building is 5.60 meters + 1 meter = 6.60 meters. This means that new buildings that can be built in the void part of the South Johar Market Heritage Building are sized [30-6.60-6.60] meters x [30-6.60-6.60] meters or 7.30 meters x 7.30 meters. Building with the measurement of 7.30 meters x 7.30 meters cannot accommodate many merchants.

The resilient design of South Johar Market Heritage Building in the conservation will be restored to its original form in 1957. The void of South Johar Market Heritage Building in 1957 was functioned as a place for fruit and vegetable stalls attaching shaded umbrellas that could be closed and opened. The quality of the stalls on the void was then improved with the construction of wooded roof with tajug [a pyramid-shaped roof]. The shape of the building imitated Karsten's work in Pasar Gede Hardjonagoro or Gede Market Heritage Building in Surakarta it was built in 1930. From these considerations, the void of the South Johar Market Heritage Building will become a new building that is not storied and is constructed from wood with a tajug shaped roof. In order to prevent future earthquake damages in the new building in the void of South Johar Market Heritage Building, the position of the first floor surface should be raised.

The higher of the first floor surface position of the new building, the less affected it will be by the ruins of the debris of the South Johar Market Heritage Building. The maximum position of the first floor surface of the new building is equal to the half of the height of the first floor surface position of South Johar Market Heritage Building (2:8:2) = 1.4 meters. The distance between the open space between South Johar Market Heritage Building and the new building in its void part is [5.60 – 1.40] meters = 4.20 meters.

4. Conclusion and Recomendation

Based on the results of the analysis of the revitalization plan conducted by the architect bureau, the resilient design should be used in order to prevent future impacts of natural disasters. The results are as follows:

a. The distance between the South Johar Market Heritage Building and the new building in its void should be widened, it is at least 4.50 meters and should be an open space; with the following requirement the position of the first floor surface of the new building should be 1.40 meters higher than the surrounding surface;

b. The new building in the middle of the void of South Johar Market Heritage Building must be designed with structure and construction based on SNI 1726 of 2012;
c. The reconstruction of the New Semarang Square is necessary and it must remain a green open space; in case of an earthquake buyers and Johar merchants can save themselves to the New Semarang Square;
d. The conservation of Johar Market Heritage Building should be carried out by increasing of the strength of the mushroom pillar construction; the mushroom pillar should be reinforced to last longer;
e. The height of the new building in the middle of the void of South Johar Market should not interfere with the view to the South Johar Market Heritage Building, maximum 3-stories building.

5. Acknowledgments
I would like to thank the Semarang Government for approving my proposal on the reconstruction Semarang Square to remains green open space and preserve the South Johar Market Heritage Building, to Timmy Setiawan as an architect who has cancelled the plan to remodeled the South Johar Market Heritage Building into parking building; and also thanks to Rif'an Achmad as a coordinator compiling team of existing data of Johar Market Heritage Building after Fire 2015.

References
[1] Roesmanto T. [Bangunan] PasarJohar + [Upaya] Konservasi / Johar Market [Building] + Conservation [Effort]. paper. In: Arsitektur Johar dan cagar budaya / Johar architecture and cultural heritage, Architectural discussion forum, Department of Architecture, Faculty of Engineering, Diponegoro University, Nuri Cafe, Semarang: 18 May 2015..
[2] Yuliati, Dewi. Industrialisasi dan segregasi sosial dalam kajian sejarah Semarang pada era kolonial 1850-1930 / Industrialization and social segregation in the historical study of Semarang in the colonial era 1850 - 1930, Inaugural speech of professorship of Faculty of Cultural Sciences, Diponegoro University, Undip Publisher Agency. Semarang: 4 August 2015.
[3] Roesmanto, Totok. Persepsi arsitektural-planologis dan arkeologikal rekonstruksi tata ruang Kraton Bintoro Demak / Perception of reconstruction of spatial space pattern of Bintoro-Demak Palace based on architectural, planological and archaeological reconstruction. international seminar. Urban Department of Archeology, Faculty of Cultural Sciences, Gadjah Mada University, Yogyakarta: 30 March 2011Karsten, Thomas. Iets over de Centrale Pasar. Locale Techniek 7e Jaargang Nummer 2. Indisch Bouwkundig Tijdschrift. Technisch Orgaan v/d Vereeniging Voor Locale Belangen. Maart-April 1938. Batavia: 1938:63-66
[4] Undang-Undang Republik Indonesia Nomor 11 Tahun 2010 tentang Cagar Budaya / Law No.11 of 2010 about Cultural Heritage. Jakarta: 2010.
[5] Karsten, Thomas. Iets over de Centrale Pasar. Locale Techniek 7e Jaargang Nummer 2. Indisch Bouwkundig Tijdschrift. Technisch Orgaan v/d Vereeniging Voor Locale Belangen. Maart-April 1938. Batavia: 1938:63-66
[6] Baldinger, H.Th. Het pasarwezen in de Stads-Gemeente Semarang, Locale Techniek 7e Jaargang Nummer 2. Indisch Bouwkundig Tijdschrift. Technische Orgaan v/d. Vereeniging voor Locale Belangen. Maart-April 1938, Batavia: 1938:48-53.
[7] Orbasli, Aylin. Architectural conservation, U.K.: Blackwell Science, 2008.
[8] Roesmanto, Totok. Penolakan revitalisasi Pasar Johar Selatan untuk bangunan parkir bertingkat / Rejection of the revitalization of South Johar market building for low-rise parking building, internal meeting of Board of Building Expert and Cultural Heritage, Semarang City Government, Semarang: 2016
[9] Murtagh, William J. Keeping Time: the history and theory of preservation in America. United States: John Wiley & Sons, 2005.
[10] Eskew, Dumez, Ripple. A framework for resilient design. 2014 [Internet].[cited 2017 August 8]. Available from: www.eskewdumezripple.com

Socio spatial adaptation as a resilience form of native unplanned settlement in confrontation with new planned settlement development pressure (case study: enclave native settlement in Serpong, Tangerang)

Mohammad Ischak[1*], **Bambang Setioko**[2], **Dedes Nurgandarum**[3]

[1]Department of Architecture, University of Trisakti, Jakarta, Indonesia
[2]Department of Architecture, Faculty of Engineering, University of Diponegoro, Semarang, Indonesia
[3]Department of Architecture, University of Trisakti, Jakarta, Indonesia

*Corresponding e-mail: mohischak@yahoo.com

Abstract. Urban growth refers to expansion of a metropolitan into sub urban areas as the surrounding environment, with no exception of Jakarta city due to limited availability and high price of land within the city. The city of Jakarta, as a metropolitan, carries of expansion in its surrounding environment including Tangerang. Privat developers may an important role in this urban growth through their large scale of new settlement development project. The formation of establishment of enclave native unplanned sub urban settlement scattered within planned new settlement in Tangerang is to be an consequence of Jakarta urban growth. This fenomena could be comprehended as a form of resilience native settlement in confrontation with the new planned settlement pressure. The aim of this research, presented in this paper is to understand the socio-spatial concept of those enclave native settlement as an adaptation form to the new planned settlement pressure. Through descriptive qualitative research method, with indepth interview as a main research instrument, this research could depict or uncover the facts that there are various form of socio-spatial adaptation as the main theme of resilience native suburban settlement formation.

Keywords: socio-spatial, enclave, native unplanned settlement, planned settlement

1. Introduction

1.1. The existence of native enclave settlements

The growth of urban population in Jakarta due to urbanization and growth naturally supported by economic growth resulted in the development of Jakarta City toward the suburbs, especially for the provision of shelter. The agglomeration of Jakarta and its surrounding cities constitutes the megacity of Jabodetabek. Data from the UN states that Jakarta city (Jabodetabek) ranks second with a population above 29 million (figure 1).

IOP Conf. Series: Earth and Environmental Science **99** (2017) 012009 doi:10.1088/1755-1315/99/1/012009

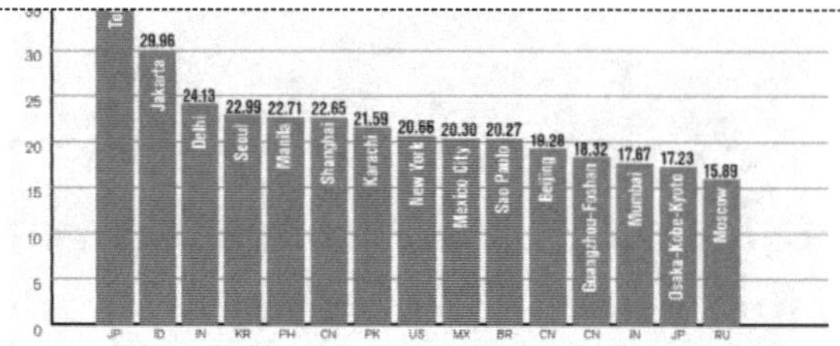

Figure 1 : The order of cities that are included in the classification of megacity based on population

Source : *United Nation* [26], 2014

The process of urban growth in developing countries is more preceded by the high growth rate of urban population which is the beginning of the emergence of problems in terms of decent housing. So the process of growth of new cities around big cities starts more from the construction of housing, then is followed by the movement of population from downtown to the suburbs, and developed with the completeness of facilities in the form of social facilities and public facilities. For other cases, it starts from the industry, and then came the workers who need housing [20]

Thus the growth of cities around Jakarta is basically the impact of the growth of Jakarta city as the center. According to Sujarto [22], there are three strategies that can be done to overcome the occupancy needs for the urban population, such as urban intensification, urban extensification, and new town development. Out of these three strategies, new town development is considered the best strategy to meet current needs. Thus, a large area of land is needed, which is available in the suburbs of Jakarta, one of which is Tangerang area.

Provision of formal housing by developers is very diverse, both from the type of house offered and in the context of providing a residential area complete with facilities and infrastructure supporting the settlement. At the beginning of the development of the property industry, developers only provide housing supported by the provision of public facilities and limited social facilities. In the present development, many developers provide housing while offering an urban concept, in which almost all supporting facilities are provided in a single area, such as commercial functions from shop level to shopping center, education function from study group to university, big hospitals, entertainment function in the form that varies from building scale to amusement park (park), and office functions from home office scale up to high rise building with function as rental office and apartement.

To meet the concept of development of a region that is shaped like a miniature of the city requires a very wide land. As an illustration, the development area of Gading Serpong which is controlled by two developers (Paramount Land and Summarecon) has a development land of 1500 Ha. With the background of business calculations, developers get this vast land by freeing land in the suburbs, one of which is Tangerang region consisting of Tangerang Regency, Tangerang City, and South Tangerang City. Tangerang has an area of 129,468 hectares, consisting of 18,378 hectares of the city and 111,090 hectares of regencies.

In an effort to acquire a vast land area, major developers undertake land acquisition, both vacant and uninhabited land. In the process it turns out that not all communities of land and house owneres accept the offer of the purchase of land and buildings that have been inhabited for generations. This resulted in the present phenomenon of the existence of native enclave settlements located in the midst of new settlement development areas (figures 2A and 2B). The enclave settlements on a daily basis show a landscape that is in contrast to the new development areas, both in terms of appearance of buildings and daily life of the local people.

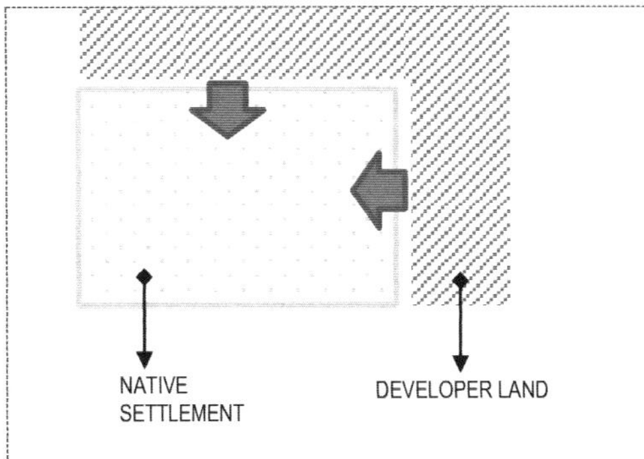

Figure 2A : Pressure from Developer to Buy the land and housing

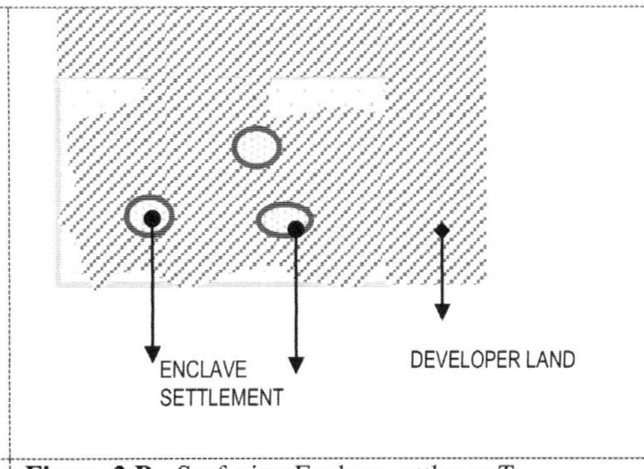

Figure 2 B : Surfacing Enclave settlemenT

Against the background of business considerations to raise the value of new development areas, the developers fence the enclaves with high walls and provide only in and out access road (Figures 3 and 4).

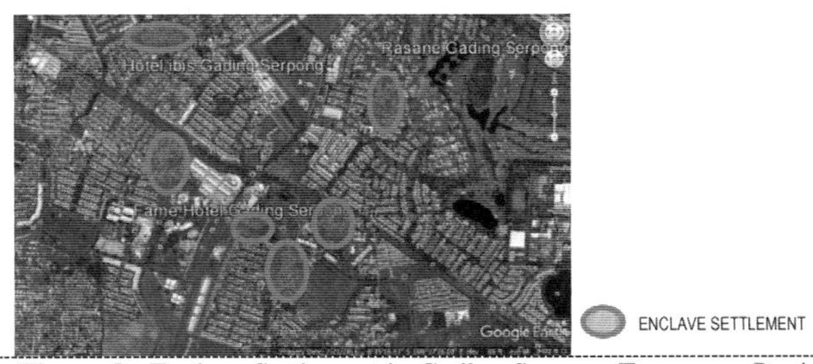

Figure 3 : Enclave Settlement in Gading Serpong Tangerang Regional
Source : Modification from Google earth, 2017

Figure 4 : The massive wall that segregate among the new settlement and enclave settlement
Source : research field, 2017

Such condition increasingly creates the impression of bordered settlement, either physically spatial, or socially in the form of relationship with the environment outside the enclave of the settlement. Even though the change of the landscape and daily lilfe occurs differently than before, the enclave settlement still exists until now. This becomes the basic of this research which is to find out the answer of how is the adjustment pattern of space utilization in the settlement as the adjustment towards space development which changes a lot?

The research reviewing the positive and negative influence towards the presence planned settlement to the native unplanned settlement has been conducted in plenty of times, some of which are the positive influence in the form of economic growth and the availability of job opportunity triggered by road infrastructure development which connects the center of the city and the other supporting cities, and the completeness of supporting facilities (Widyo [28]; Siahaan, et.al. [17]), the emerge of space integration which connects the planned settlement dwellers and the local people (Setioko [19]; Prihanto[14]).

The negative influence found is related to the social and spatial segregation between the planned settlement and the native unplanned settlement (Firman[2]; Wulangsari[30]; Diningrat[1]). On the other side, the research analyzing the adaptation form done by the native settlement towards the planned one has not been conducted yet, especially the research inside the settlement trapped inside new (city) settlement.

1.2. Theoretical perspectives on settlement resiliences and socio spatial

1.2.1. Settlement resilience.

The definition about urban resilience is mostly related to the city resistance towards the threat of natural disasters like the climate change, earthquake, flood, tsunami, etc, while the threat on the the resistance of a city or a part of city is not only to the disaster caused by natural factors, but what is also important to be analyzed is the resistance of a city towards the environmental change caused by humans, one which is in the form of intervention of new settlement construction which frees the native settlement and region, done by the big developers by buying the whole or the part of the existing settlement especially in the suburbs. In the context of the intervention of urban area to suburb area, Hejiman[6] introduced the concept of rural resilience that refers to the capacity of a rural region to adapt to changing external circumstances in such a way that a satisfactory standard of living is maintained, while coping with its inherent ecological, economic and social vulnerability. On the other side, there is a perception that the concept of resilience is part of the model proposed by Harrison[4]. The model is mathematically derived based on the potential action of a stressor and ultimately aimed at providing specific measurable indicators of stability. The four indicators are (1) resistance (range of fluctuations determined by a stressor), (2) resilience (speed of returning to the equilibrium state), (3) persistence (duration of maintaining the regular values of state variables during the action of a stressor), and (4) variability (frequency of changes during the action of stressors).

A settlement often responds in the form of adaption in defending themselves from outside threat[18]. In the context of ecological change in their surrounding, the effort of human self defense through adaptation is known as non-equilibrium paradigm, in which it is the system ability to adapt and adjust themselves to the internal or external process change by analyzing the ecological linking and social heterogeneity and function[13].

Through the analysis focusing more on the effort of the local people in responding against the external threat, a resilience community term is known, which is the the existence, development, and engagement of community resources by community members to thrive in an environment characterized by change, uncertainty, unpredictability, and surprise [8]. This means that well developed economic, institutional, social, cultural and natural domains are crucial for resilient communities. Relation between resilience and adaptation process described by Thulstrup[23] states that Resilience is understood as a measure of the level of access to endowments of capital – financial, natural, physical, social, and human – that can be mobilized in order to respond and adapt to environmental change. Habraken[3] mentioned it as the search of stability by a built environmentwhich is being analogized as a built environmental game, where a process is started from a simple situation and unbalanced condition then it processes through gradual transformation to reach more complex balance with lots of players. Therefore, built environment is explained as a life configuration operating on different level. In spatial context, spatial resilience capacity is a competence of a region to respond, recover, overcome and adapt and even develop when it faces an unexpected condition, sudden disturbance or stressor[11].

From those definitions, it can be concluded that the resilient effort of a settlement can be observed through community social behavior which is the response and adaptation process due to landscape change on its surrounding as the impact of the construction of new planned settlement. The social behavior and activity in community life resulting in the shape and model of space is known as sosio-spatial analysis[27].

2. Socio Spatial

Social groupings exist independently of space, and do not require spatial reorganisation to make them better. This view originates more in sociological studies, and tends to be held more by planners and others concerned with social and economic processes, rather than with detailed physical arrangements of space through architecture[5].

To describe the relationship between constructed shape and social organization, two types different spatial order formulation is needed: first: space arrangement by the community differentiating the function and the border which is known as architectural identity. It is usually in the form of building: house, special building, village, city, etc. Second, space arrangement which is called as pattern; a community arrangement in a space where every member of a community sets himself in a room[5].

Correspondence model needs a physical arrangement in a space by a community (what activity is done is certain location) like house, neighbor, city and exactly to reflect the grouping physically, like family, interest group, certain community, where those things are resulted out of social categorization forms. However, according to Strauss[21], the form of relationship always exists albeit unclear and the group configuration can spatially reflect on social structure. This shows that there is connection between space and social life.

Socio-spatial concept is a concept regarding social relation activity among humans in order to fulfill their needs implying on the effort of space arrangement (arrangement on the length, width and height of room) as the place to held the activity [27]. In settlement context, a room is basically a product of an activity or social structure [7];[25]. Social acitivity and life have an important role in space, place, and city structure forming [12]; [20].

The discussion about human activity in the settlement either individual or social, cannot be separated from the related discussion with the territory. At present, territory theory has become almost the means by which we think about space. To deny it appears to be a denial of space itself[5]. Rapoport [15] states that all of us consider that territory instinct which is the need to identify and occupy is the basic and ever-lasting need. Lynch[9] mentions firmly that: human is territory animal which uses their personal space to arrange their individual activity. Every person will control more than one territory.

However, it cannot be ignored that in the effort in analyzing social life in the context of settlement, architectural analysis cannot be separated from space analysis due to the fact that basically architectural ontology is a space[5]. Even Castels clearly states that space is not a photocopy of society, it is society. Spatial forms and process are formed by the dynamics of overall sosial structure[25].

3. Method and Scheme of Research Conduction

The research was conducted with the purpose to see the resilient process of the enclave settlement in responding to the landscape change of physical and non-physical environment which happens around the settlement due to the new planned settlement construction. The first step is to determine the physical and non-physical object which would be observed as the first material of the analysis. Because the observed object involves non-physical materials, the method used in the research was descriptive explorative. Therefore, the search of social data is done by interview. The variable determination, the materials needed in the meaning and analysis, and the observed object determination to measure the resilient level of enclave settlement in this research can be seen in scheme (Figure 5):

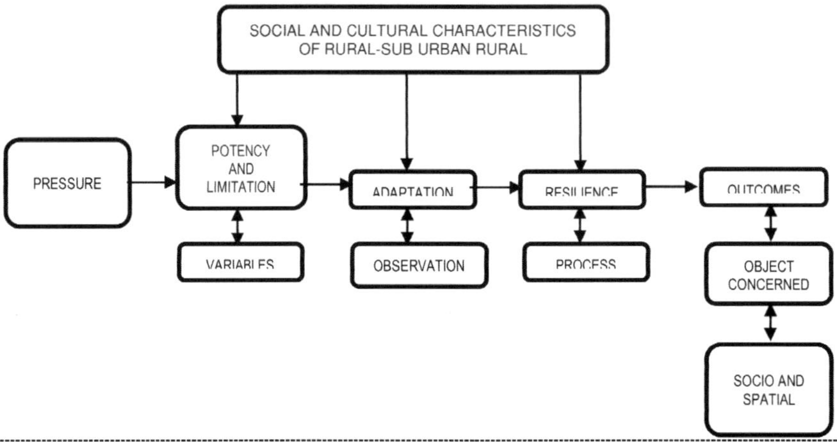

Figure 5 : Scheme of the research process

4. Discussion

4.1. Research Location

The research was conducted in Kampung Curug Sangereng which are RW 01 and RW 02 in Curug Sangereng Village, Kelapa Dua District, Tangerang regency. Looking from its name, this is the main kampung in the village so it has the same name that is Curug Sangereng village. Regionally, Curug Sangereng village is one of the village whose most of the regions is developed by two big developers, they are PT. Summarecon and PT. Paramount Land (Figure 6).

Figure 6. Locus of research at RW 01 and RW 02 Kampung Curug Sangereng Kelapa Dua District, Tangerang Region
Source : Google Earth 2017

4.2. Characteristic of Rural-Urban Community

In suburb community context, Kusumastuti[12] states that suburb community has social capital potential known as social capital bonding which is the social capital among individuals in a group by seeing into the inward looking. The social capital keeps being looked after through individual interaction everytime formally or informally. In some cases, this social capital is concluded as the first step in responding to every change outside the community which directly influences community daily life. The adjustment mechanism is known as adaptation mechanism.

Related to community daily social activity in research location, it can be categorized as Sunda ethnic society who lives in semi-urban region with farmer dominated. Before land discharge is done by the developers, the land is dominated by rice farming and rubber plantation[24]. With such background, the research location has characters which can be seen as potential like shown in following table 1:

Table 1 : Rural-Urban Community Characteristics

NO	COMPONENTS	CHARACTERS
1	Public Environment and orientation to nature	Strongly related to nature; guided by faith and natural laws
2	Livelihood	farming traditionally and not efficient or not common or what is called as subsistence farming
3	Community size	Rural community is smaller than urban area
4	Population density	Population density is smaller than urban area
5	Homogenity	Similarity of social and psychological characters, language, faith and tradition
6	Social Interaction	Solidary is raised by the habit, purpose and experience similarity
7	Social Levels	Not as steep as urban community
8	Social Mobility	Rural community is lower than urban ones
9	Social Monitoring	Stronger because individual contact is informally intensed
10	Leadership Pattern	Decided by personal quality and based on generation criteria
11	Life Standard	Simple orientation and life standard
12	Social Solidarity	Stronger and unrewarded cooperation
13	Value System	Strong religion value, simple economic value

Source: Wijayanti[29]

4.3. Planned Settelement Pressure on Unplanned Settlement

Due to physical enclave settlement bordered by walls and spreat out in regions mastered by developers, the dwellers access in enclave settlement towards social and public facilities like education, health, population administration, and market is separated physically and non-physically. Such condition is felt by dwellers who conform that they occupy separated settlement (Figure 7).

Figure 7. Position of research locus from the Village Head Office
Source : Google Earth 2017

From social perspective, massive bordered by developers causes relative and social bonding in a village which previously feels closed gets separated due to distance. Because a half of population moves us since their land and houses are discharged by developers. Bigger pressure form felt by dwellers of settlement enclave is the offer coming from developers to prepared in buying land and houses to be occupied, especially the land and building which direcly borders on lands owned by developers with the price above usual market prices. Those pressures faced by dwellers of enclave settlement everytime until now. The effort to hold out from such condition is traced by analyzing social adapatation forms done by the society occupying the enclave settlement which then ends in space creation and utilization

4.4. Spatial Adaptation Mechanism

To analyze the adaptation form spatially, referring to Rapoport [16] who elaborates a space which is basically formed out of 3 elements: 1) Fixed-feature elements (Buildings, floor, wall, and others) are elements that rarely experience spatial change which can be organized into sizes, location, order and structure; 2) Semi-fixed elements (diverse interior and exterior facilities) are elements which can feel the change in certain element types like street, stores, shop window, and other elements; 3) non-fixed elements (human, their activites and behavior) are elements that directly connect to behavior and attitude shown by humans.

The observation based on RW administrator data and the interview of important roles which direct the analysis on adaptation form done by dwellers of enclave settlement can be seen from non-fixed elements; elements in the form of behavior and economic activity and social dwellers caused by landscape change around the settlement due to formal settlement construction as shown in Table 2:

Table 2 : Matrix of socio-economic change of the dwellers ending at spatial adaptation

NO	CHANGE / NO CHANGE			SPATIAL CONSEQUENCE	CHANGE AND SPATIAL ADAPTATION
	INITIAL CONDITION	CAUSE	NOW		
1	Economy				
	Occupation:				
A	Farmer – Plantation	No land	Sales / shop keeper	Meetings are commonly done in informal space	Change: only on working time component
					Adaptation :
			Lanscaper	gathering point for picking up service	Change: spatial needs gatereing in front of settlement access
					Adaptation: utilizing space at the end of the access entering the settlement and formal housing to become the part of territory
		Houses to be boarding house	Entrepreneur	Amount of time at house, more intensive interaction among dwellers	Physical change of houses and settlement, and economic adaptation
					Adaptation: House layour to take place as boarding house
			Food Selling	Houses at the edge of street which mostly take function as stalls	Change: for building utillization
					Adaptation of stall terrace which is at night used as gathering space
B	housewife / farming/ doing plantation	• No land • Egalitarian culture with the other dwellers • Dense distance between houses	Housewife	often meet the neighbors so they form places to gather around the house	Change: more time at home, more often gathering with others so informal communal spaces are formed.
					Adaptation: more intensed territory in neighbor environment
			Landscaper	Gathering point for picking up service	Change: space needs to gather in front of the settlement access
					Adaptation: utilizing existing space at the end of the access entering the settlement and formal houses being the part of territory
			Housemaid/ formal clothes washer in formal housing		Change: access to formal houses
					Adaptatin: formal houses to be the part of territory
2	Social culture				
A	Interaction among children				
		Decreasing land	Gathering in	common space in	Change: more time to play with

	Interaction on village territory		neighbor territory	neighbor scale	children Adaptation: utilizing house terrace as public space
B	Interaction among teenagers				
	Interaction on village territory	Graduating from SMP/SMA, working from informal sector	Gathering informally in enclave territory	Formal: house of important figure, musholla	Change: frequent meeting intensity, territory around RW Adaptation: utilizing formal and informal space as communal room
			Gathering informally	Informal : Field, stall	Perubahan : interaksi di luar waktu kerja Adaptation: creating communal space by utilizing the existing one in neighbor territory
C	Interaction among women				
	Interaction on village territory incidentally if there is any social event	•Guidance from subdistrict •Kinship and religious one	Gathering formally (reading Quran, PKK, Posyandu)	RT/RW houses, musholla	Change: more interaction frequency so building as interaction space will appear Adaptation: squared terriroy in RT, RW and village level causing function specification of social building
		Not working	Informal gatering (taking care of children, shopping and religious events)	House yard, street, stall	Change: more time to take care of children, frequent interaction, creating existing space as interaction place Adaptation: utilizing house terrace, house yard, stall and street as public space
D	Interaction among men				
	Interaction on village territory	Neighborhood value (stabile)	Working together, patrol	Informal communal space	Change: smaller interaction on RT/RW level due to smaller settlement area Adaptation: more intense interaction on neighborhood territory creating informal communal space
		Religious value (stabile)	Reading Quran	Musholla	Change: Stabile value system, change on interaction level gets narrower Adaptation: house takes function as communal space to accommodate reading Quran event in turn
E	Interaction among all local people				
	Interaction on village territory	Neighborhood value system (stabile)	Big day celebration and events	Opened space (street, house yard)	Change: stabile value system, change on narrower interaction environment. Interaction can happen with relatives or neighbors who have moved to other settlement location Adaptation: because there is no wided opened land, utilizing street as incidental communal area
		Religious value system (stabile)	Celebration on Muslim Big Day	Musholla, mosque	Change: stabile value system, change on narrower interaction environment. Interaction can happen with relatives or neighbors who have moved to other settlement Adaptation: because there is no wide opened land, utilizing street as incidental communal space

Source: analysis based on interview and observation, 2017

From the matrix, we can explain them as follow: pressure in the form of change involving physical change of settlement spaces, surrounding landscape, the lack of village communities, and community social structure on farmer-plantation-based is responded by the settlement dwellers with adaptation on each social life level which eventually ends on spatial change and adaptation. In the adaptation process recorded, social life is applicable and not changed even though it is explicitly declared by the dwelllers. The value system is the heritage of applicable social community life when the change has not happened yet, and it becomes base or norms or social life even if it is not written or spoken but it is consciuously or unconsciously inherited from one generation to next one. The resistant social value includes kinship value either because of solidarity or because of the common experience from the enclave settlement dwellers, helping each other, cooperate each other, and religious system.

The landscape and the utilization change reforms the occupation type of the dweller directly. The presence of planned settlement on one side offers advantage by providing many job opportunities, either formal or informal, and it is done inside and around the house because there are many oursiders coming as workers in planned settlement. The form of business as substitution of previous job like opening boarding house, rented house, and stalls (figure 8).

Figure 8. economic activity by the resident
Source : researcher, field survey 2017

On the other side, job in planned settlement area is absorbed as office employee, shop officers and park caretaker. The change of job characters and physical change of settlement cause the change in characters and interaction pattern among dwellers. The factors include time availablility, narrow opened land, and dense distance between houses influencing the form and intencity of interaction. Those things cause physical change of space either individual spaces (houses) or public space (figure 9).

Figure 9. Patterns of interaction in the communal space
Source : field survey 2017

Social, cultural and economic activity of dwellers of enclave settlement done daily which is basically an adaptation process towards pressure in the form of physical and non-physical change inside and outside their settlement which ends in space creation and utilization. Each of the adaptation steps can be elaborated through variables as seen in the scheme (Figure 10).

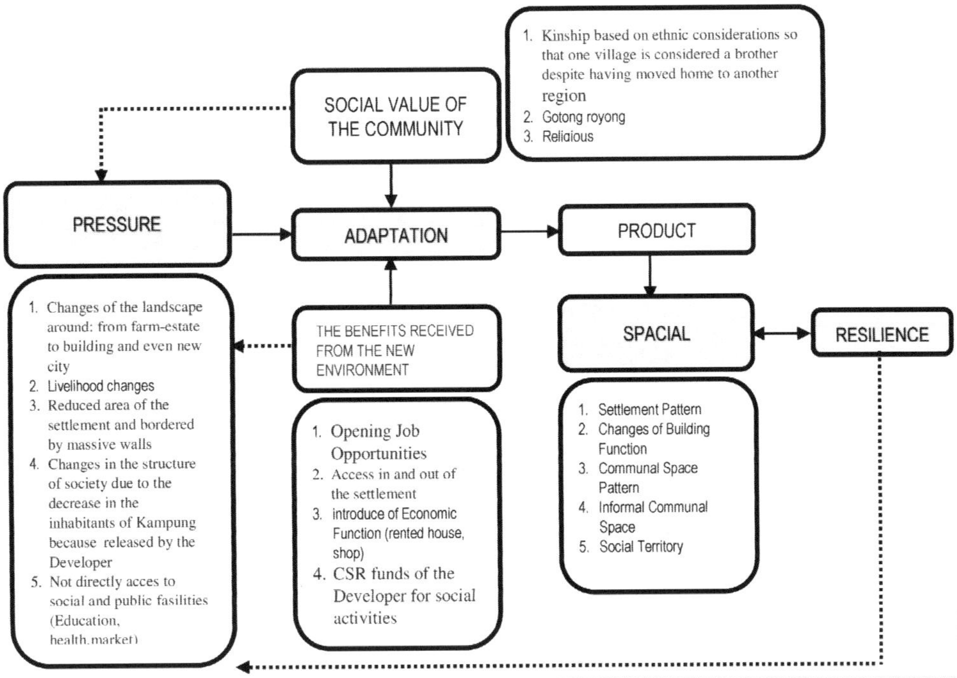

Figure 10. Scheme of adaptation process that leads to spatial pattern
Source : researcher, 2017

The change and other adaptation form also can be the supporting factor of resilience community in enclave settlement which is related to territory. Before the discharge and construction of land by the developers, social territory of community members is generally divided into three forms: 1) family territory, 2) village territory as manifestation of village community members, and 3) working territory which involves house, village and farming or plantation land as working asset (Figure 11A). Because of their settlement change, as elaborated before, it makes each of individual territories change with the emerge of smaller territory on neighborhood level, and big territory without physical border connecting solidary between relatives or neigbors who have moved out to the other settlement due to land or house discharge by the developers (figure 11B).

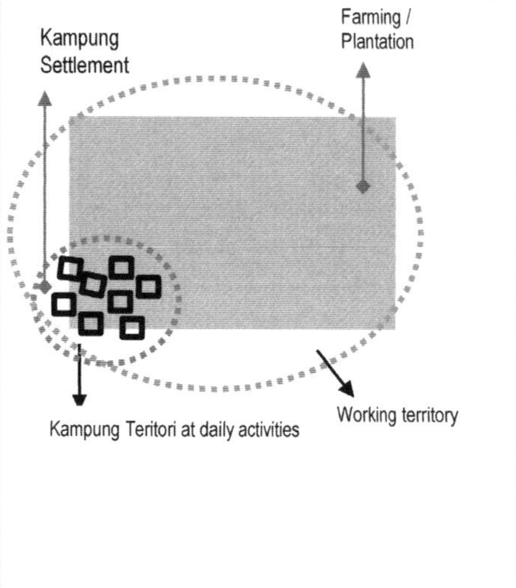

Figure 11.A An overview of the individual territories of the inhabitants of the settlement prior to the envieonment change
Source : Analysis from field, 2017

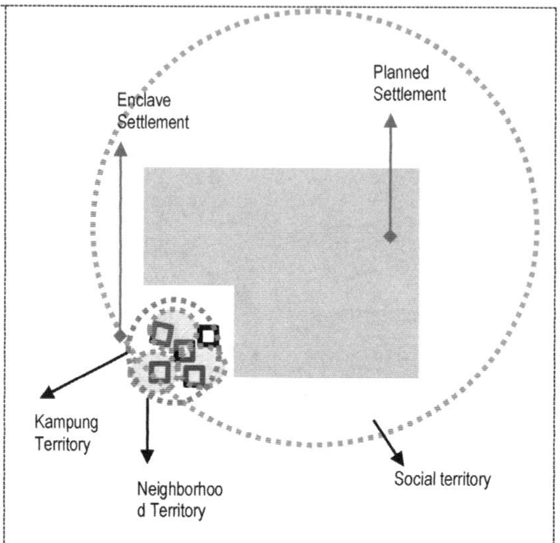

Figure 11.B An overview of the individual territories of the inhabitants of the settlement in this time
Source : Analysis from field, 2017

5. Conclusion

1) Landscape change inside and outside the settlement has been considered as threat because all social activities as neighborhood bonding still exists even though the adjustment process continues.
2) The adjustment form includes individual activity and social interaction behavior. This is supported by the applicable social value system and neighborhood and the creation of solidarity bonding with the relatives who have moved out ouf village due to land discharge by still inviting them on certain social events.
3) On the other side, the presence of formal settlement becomes the opportunity and takes advantage economically because they provide job field and business opportunity for every age-range dweller, men, women, or teenagers.
4) Eveny social, economic and cultural activity can still go on even though every adjustment happens because there is still inside and outside acess of settlement by developers.
5) Therefore, settlement enclave can still take hold for future with any intervention from related parties alike developers and regional government

References

[1] Diningrat R A 2014 Ketergantungan Kota Baru Kota Harapan Indah terhadap Kota Jakarta dan Wilayah Sekitarnya Jurnal Perencanaan Wilayah dan Kota ITB, vol. 25 no. 3 192-212 Desember 2014
[2] Firman T 2004 New town development in Jakarta Metropolitan Region: a perspective of spatial segregation Habitat International 28 349–368
[3] Habraken NJ 1998 The Structure of the Ordinary, Form and Control in the Built Environment MIT Press
[4] Harrison G W 1979 Stability under Environmental Stress: Resistance, Resilience, Persistence, and Variability The American Naturalist 113 (5) 659-669
[5] Hanson J and Hillier B 1987 The Architecture of Community: Some New, Proposals on the Social consequences of Architectural and Planning Decisions Arch. 8 Confort /Arch.Behav., Vol. 3 (3) 251 -273
[6] Heijman W, Hagelaar G and Heide M 2007, Rural resilience as a new development concep

EAAE seminar Serbian Association of Agricultural Economists, Novi Sad, Serbia.
[7] Hillier B and Hanson J 1984 The social logic of space Cambridge University Press Cambridge
[8] Kelly C, Ferrara A, Wilson GA, Ripullone F, Nolè A, Harmer N, and Salvati L 2015 Community resilience and land degradation in forest and shrubland socio-ecological systems: Evidence from Gorgoglione, Basilicata, Italy", Land Use Policy 46 11–20
[9] Lynch K 1981 A Theory of Good City Fom MIT Press Cambridge Massachusetts
[10] Kusumastuti A 2015 Modal Sosial dan Mekanisme Adaptasi Masyarakat Pedesaan dalam Pengelolaan dan Pembangunan Insfrastruktur MASYARAKAT: Jurnal Sosiologi, 20 (1) 81-97
[11] Muntele I and Bănică A 2013 Romanian Functional Urban Areas: Between Polarization and Spatial Relisience proceeding, 4th Global Forum on Urban Resilience & Adaptation, Bonn, Germany 31 May – 2 June
[12] Nas PJM 1984 Kota di Negara Ketiga, Pengantar Sosiologi Kota-Bagian Kedua Penerbit Bhratara Karya Aksara, Jakarta.
[13] Pickett STA, Cadanasso ML, and Groveb JM 2004 Resilient cities: meaning, models, and metaphor for integrating the ecological, socio-economic, and planning realms Landscape and Urban Planning 69 369–384
[14] Prihanto T 2010 Perubahan Spasial dan Sosial Budaya Sebagai Dampak Mega Urban di Daerah Pinggiran Kota Semarang Jurnal Teknik Sipil dan Perencanaan Universitas Negeri Semarang Vol. 12 (1) 131 – 140
[15] Rapoport A 1969 House Form and Culture Prentice Hall New Jersey
[16] Rapoport A 1977 Human Aspect of Urban Form Pergamon Press Ltd
[17] Siahaan, Alexsandro, Nasution, Zulkifli, Purwoko, dan Agus, 2014 Analisis Pengaruh Pembangunan Perumahan Terhadap Pengembangan Wilayah Kecamatan Siantar Marimbun, Kota Pematang Siantar, Jurnal Ekonom, Vol 17 (3)
[18] Schouten M, Van der Heide M, and Heijman W 2009 Resilience of Social-Ecological System in European Rural Areas : Theory and Prospects 113 th EAAE Seminar Belgrade Republic of Serbia December 9-11
[19] Setioko B 2010 Integrasi Ruang Perkotaan di Kelurahan Meteseh Kawasan Pinggiran Kota Semarang Disertasi Program Doktor Teknik Sipil dan Perkotaan Universitas Diponegoro, Semarang
[20] Sieverts T 2003 Cities Without Cities Spon Press London
[21] Strauss L C 1972 Structural Anthropology One Penguin Harmondsworth
[22] Sujarto D 1995 Perencanaan Kota Baru Penerbit ITB Bandung
[23] Thulstrup WA 2015 Livelihood Resilience and Adaptive Capacity: Tracing Changes in Houshold Acces to Capital in Central Vietnam', World Development, Vol 74, 352-362
[24] Tim Peneliti Kota Tangerang, 2015 Sistem Sosial Profil Sosial Budaya Masyarakat di Kota Tangerang Laporan Penelitian Program Studi Perencanaan Wilayah dan Kota, Institut Teknologi Indonesia
[25] Tjahjati B ed 2005 Bunga Rampai Pembangunan Kota Indonesia dalam Abad 21, Konsep dan Pendekatan Pembangunan Perkotaan di Indonesia Lembaga Penerbit Fakultas Ekonomi Universitas Indonesia Jakarta.
[26] United Nations, Department of Economic and Sosial Affairs, Population Division, 2014, World Urbanization Prospects: The 2014 Revision, Highlights, United Nation, New York
[27] Wibawa DNG 2014 Perancangan Permukiman Kota dan Kontrol Sosio Spasial Penerbit Universitas Trisakti Jakarta
[28] Widyo W 2013 Perwujudan Permukiman Terpadu dalam Pengembangan Wilayah, Studi kasus : Kawasan Permukiman Rungkut Surabaya, Laporan Penelitian Jurusan Teknik Arsitektur Institut Teknologi Adhi Tama Surabaya (ITATS)
[29] Wijayanti TA 2014 Masyarakat Desa Kota Universitas Negeri Yogyakarta. www.themegallery.com
[30] Wulangsari A 2014 Tipologi Segregasi Permukiman berdasarkan Faktor dan Pola Permukiman di Solo Baru, Sukoharjo Jurnal Pembangunan Kota Universitas Diponegoro, Volume10(4):387-399

Waste processing building with incineration technology

Wasilah Wasilah*, Muh. Zaldi Suradin

Alauddin Islamic State University, Makassar, Indonesia,

*Corresponding e-mail: wasilah@uin-alauddin.ac.id

Abstract. In Indonesia, waste problem is one of major problem of the society in the city as part of their life dynamics. Based on Regional Medium Term Development Plan of South Sulawesi Province in 2013-2018, total volume and waste production from Makassar City, Maros, Gowa, and Takalar Regency estimates the garbage dump level 9,076.949 m3/person/day. Additionally, aim of this design is to present a recommendation on waste processing facility design that would accommodate waste processing process activity by incineration technology and supported by supporting activity such as place of education and research on waste, and the administration activity on waste processing facility. Implementation of incineration technology would reduce waste volume up to 90% followed by relative negative impact possibility. The result planning is in form of landscape layout that inspired from the observation analysis of satellite image line pattern of planning site and then created as a building site pattern. Consideration of building orientation conducted by wind analysis process and sun path by auto desk project Vasari software. The footprint designed by separate circulation system between waste management facility interest and the social visiting activity in order to minimize the croos and thus bring convenient to the building user. Building mass designed by inseparable connection series system, from the main building that located in the Northward, then connected to a centre visitor area lengthways, and walked to the waste processing area into the residue area in the Southward area.

Keywords: energy, waste processing facility, landscape layout, incineration technology

1. Introduction

Waste problem is the part of the dynamics of human life. The waste indeed results from every movement and human activity. Undoubtedly, the garbage will never apart from pollution that currently always disturb the health and society comfort. This definitely will issue about a discussion for the solution of waste phenomenon. Solid waste management is a challenge for the cities authorities in developing countries mainly due to the increasing generation of waste, the burden posed on the municipal budget as a result of the high costs associated to its management, the lack of understanding over a diversity of factors that affect the different stages of waste management and linkages necessary to enable the entire handling system functioning [1].

According to technical plan, support in Mamminasata Regional landfills stated that, Makassar city as capital city of South Sulawesi Province is one of metropolitan city in Indonesia that has rapidly developed and functioned as a gateway of Eastern Indonesia. Makassar City develops into a major city with back region or interrelated satellite city based on the city development. Based on the result of Metropolitan regional study of Makassar, Mamminasata is regional metropolitan area of Maros, Makassar, Sungguminasa, and Takalar. One of impact in city development is an increasing in waste volume number, required improvement on facility and infrastructure against limited land. As a result, there is a requirement on good and efficient on waste processing management although the cities in the area of Mamminasata previously have managed their own waste.

Based on Regional Medium Term Development Plan data of South Sulawesi Province 2013-2018, total volume of waste production from Makassar City, The Regency of Maros, Gowa, and Takalar is 4,231.23 m^3/day and transported garbage amount is 3,735.78 m^3/day. The standard of metropolitan waste pile is 0.0035 m^3/person/day which calculated by total population of 2,593,414 people. Afterwards, the estimation of waste pile level is 9,076.949m^3/person/day. This number definitely required further management in integration and comprehensive waste processing for the scope of Mamminasata cities.

Based on this reality, innovation in city waste management process is required. Waste incineration, introduced originally for volume reduction and hygienic reasons, went through a long and intense development. Together with prevention and recycling measures, waste to energy (WTE) facilities contribute significantly to reaching the goals of waste management. Sophisticated air pollution control (APC) devices ensure that emissions are environmentally safe [2].

One of the solutions is converting the waste into electrical energy or waste to energy. Implementation of incineration technology is by the fuel from the main problem all this time, the waste and turn into strategic option in reducing the waste. The incineration technology also proves to be an efficient solution with a relatively higher efficiency (25%) and lower operational cost ($1.5–2.5/ton) [3]. Potency in reducing the waste by incineration technology is highly effective and principally produced an output for electrical energy. This incarnation technology substantial help a lot the Indonesian Stated-Owned Company in providing the electric for the society.

Therefore, waste management facility planning with incineration technology (waste to energy) expected would provide an innovation and indicator for the problem of energy. Additionally, waste processing facility plan will offer the best solution in waste management in the area of Mamminasata cities.

2. Methods

In order to achieve the result of the design approaches to energy saving, the methods applied in designing process are:

- Literature study: In form of data collecting that related on city waste processing, implementation of incineration technology in Mamminasata, and requirement and space standard for waste processing facility
- Observing of the location: Gathering the information on social-economy potency, and physical environment of location that supports on designing achievement.
- Analyzing the design approach: analyzing and simulation of the potency of designed location, application strategy analyzing on form, material, structure, and utility.
- Defining the concept: defining waste processing building concept by implementing the incineration technology that integrated with the environment of design site.
- Design result: Produced the planning of waste processing facility by implementation of incineration technology that will minimize the waste problems and maximize the potency of waste to produce the energy.

3. Discussion

3.1. Landscape Layout

Landscape layout inspired from observation analysis of satellite image line pattern in the planning area and then formulated as building site pattern. Considering the building orientation, wind analysis and sun path analyzed by autodesk project Vasari software. Given the challenges to design Net Zero Energy Buildings (NZEBs) the use of Building Performance Simulation (BPS) tools during early design phases has been indispensable [5]. Therefore, based on the consequences for the site form follows the site pattern and then oriented lengthways to the Southern in order to maximize the wind blows direction and sun pathway.

Figure 1. Landscape layout of planning site

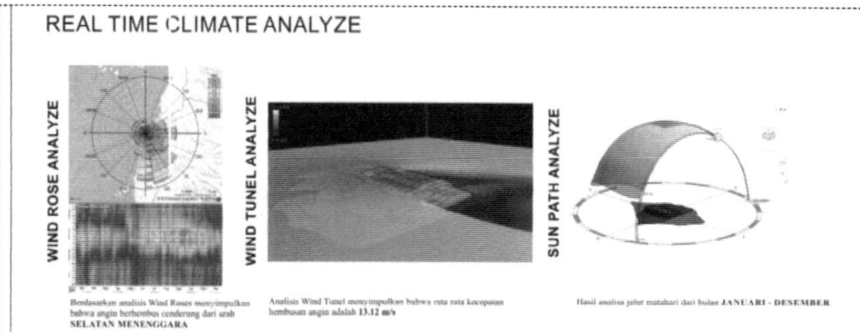

Figure 2. Real time analyses of planning location

3.2. Building Mass

Waste processing facility planning based on mass. The design creates integration areas that interconnected based on the function of the building. This facility not only focuses on waste processing system, but also integrates to the education process for the society. Therefore, the facility offers the explanation on waste processing process that would convert into the energy.

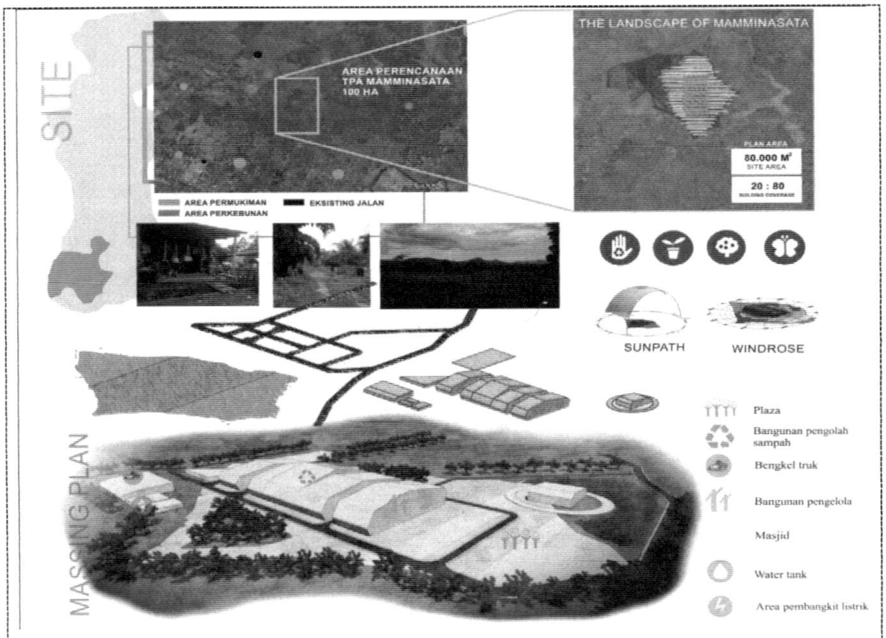

Figure 3. Building mass layout of planning site

3.3. Footprint design

Footprint designed by separate circulation system between waste processing facility interest and visitor activity for the society. The design is minimizing the croos that would present convenient to the building user. Circulation system has divided into two parts in the main gate area, between public vehicle and dump truck. The public vehicle will pass the main parking post and the dump truck will through the weight bridge post. The vehicle paths separated with boulevard as circulation marker. The marker for public vehicle will direct to the parking area and the dump truck will direct to the main building of waste processing. In addition, the pedestrian will directed to the pedestrian precinct, offering secure and convenient impression for the building users.

Building mass designed by a series of integrated system, from the main building located in the northward that connected by visitor center area lengthways that follow the waste processing area until the residue area in the southward. Therefore, the visitor will have visual access on was processing process and become an interesting education facility. Green layout design spread into all the area of to create an ecological area. The design will raise a convenient green open space for the building user.

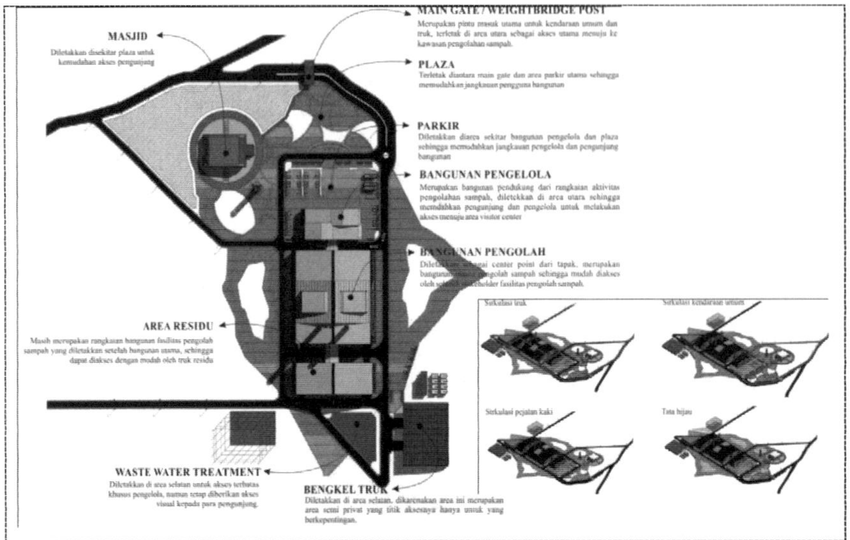

Figure 4. The design of footprint planning

3.4. Form

Aim of the building form plan is to create expressive building that appropriate for the function of waste processing building. As a result, the planning would remove a negative image from the society on the waste processing itself. As waste processing facility, the building design creation focus on process of environmental integration. An interesting effort in application process is delivering an interesting education facility, avoiding the filthy and dirty image. Therefore, design application involved to improve the image of waste processing facility.

During the finding of the inspiration, form followed up of landscape pattern application the continued by the combination of geometry transformation into the hill-like form. The hill-like form is an integration of surrounding environment and continued by reforming process. For building display in northern facade area, organic pattern added to create natural expressive impression.

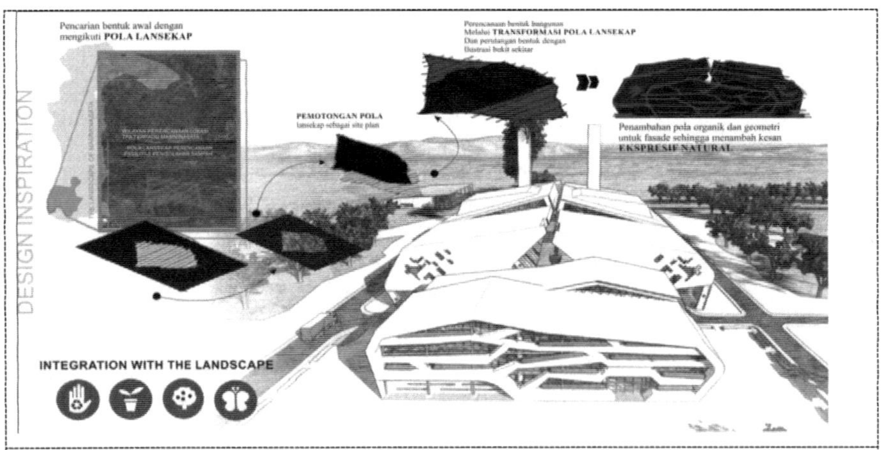

Figure 5. The concept of form transformation

Figure 6. The perspective of planning site

3.5. *Interior layout*

Interior layout consists of the layout of characteristic relation. During the planning process, zoning performed to present restricted area. In the process of planning, four sites zoning applied in this study are public, semi-private, private, and exclusive zone.

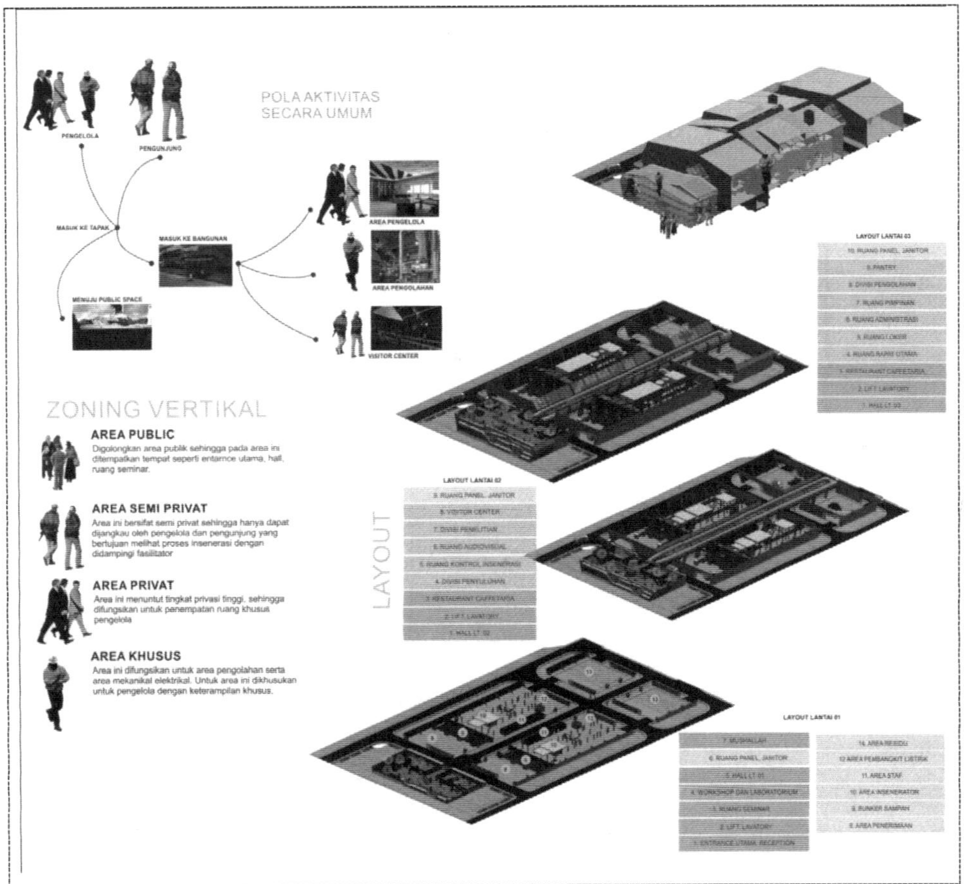

Figure 7. Zoning concept and interior layout

3.6. *Waste processing concept*

Waste production of Mamminasata cities that would be processing is 576 ton and the moisture content is about 60 %. Additionally, reducing moisture content about 50% performed by placing for draining method inside the bunker for 3-5 days. Minimum combustion temperature with incinerator is 850 C. Dioxin will create when the temperature is too low. [6]

Building design planning specification will be able burns the waste is up to 700 ton per day by incineration machine with two units electrical output 10,000 KWh. Incinerator operational operation is 50% of electrical total output.

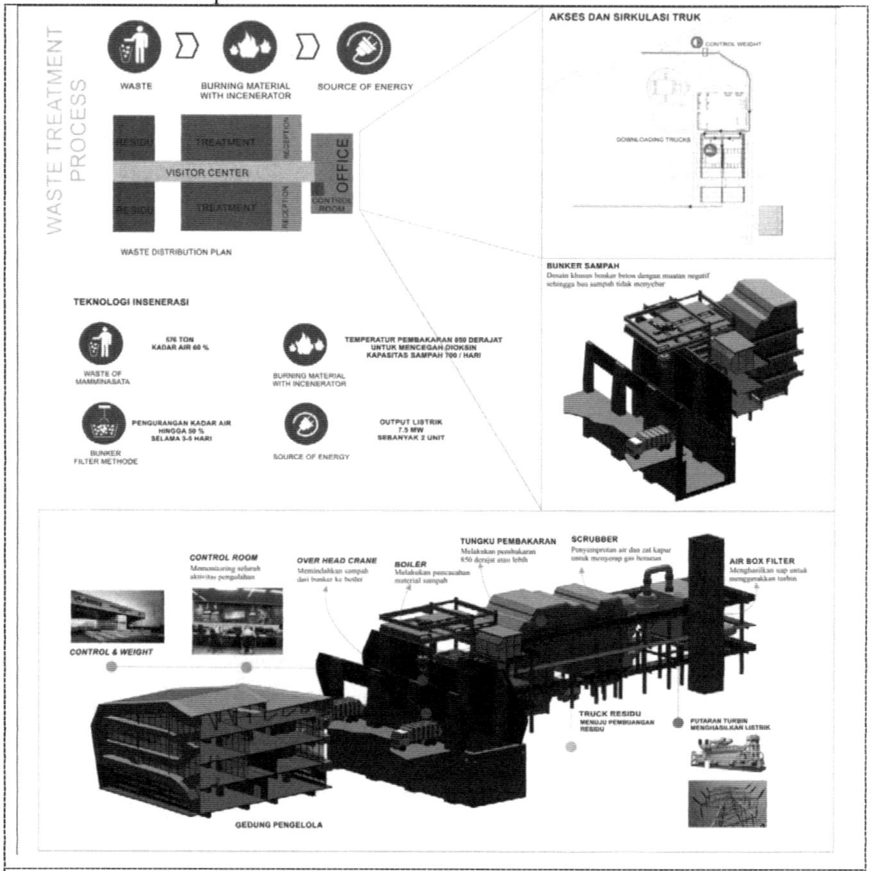

Figure 8. Waste Processing Concept

4. Conclusion

In fact, city waste processing into electrical energy has implemented by European and Asian developed countries. Implementation of incineration technology is by the fuel where the waste is the main problem all this time and turn into strategic option in reducing the waste. Potency in reducing the waste with incineration technology is highly effective. The main point is presenting electrical energy output.

More in depth study is required in the technology to produce alternative energy. For architecture study, target environment design, then in the concept design planning, the application of design concept directed to the energy saving approach and the approach in waste conversion into the energy. As a result, this orientation would combine by the approach of design method that integrated with the environment. Additionally, this combination will support the effort on generating the energy and for the environment itself.

References

[1] Guerrero, Lilliana Abarca, Ger Maas, and William Hogland 2013 Solid waste management challenges for cities in developing countries *Waste management* **33**.1 Elsevier: 220-232.

[2] Brunner, Paul H, and Helmut Rechberger. 2015 Waste to Energy–key Element for Sustainable Waste Management *Waste Management* **37**. Elsevier: 3–12.

[3] Ouda, O K M, S A Raza, A S Nizami, M Rehan, R Al-Waked, and N E Korres. 2016 Waste to Energy Potential: A Case Study of Saudi Arabia *Renewable and Sustainable Energy Reviews* **61**. Elsevier: 328–40.

[4] de Souza, C. B. 2013 Studies into the use of building thermal physics to inform design decision making *Automation in construction* **30** Elsevier: 81-93.

[5] Vysotskiy, A., Makarov, S., Zolotova, J., & Tuchkevich, E 2015 Features of BIM Implementation Using Autodesk Software. *Procedia Engineering* **117**, 1143-1152.

[6] Allegrini, E., Vadenbo, C., Boldrin, A., & Astrup, T. F 2015 Life cycle assessment of resource recovery from municipal solid waste incineration bottom ash. *Journal of Environmental management* **151.** Elsevier: 132-143.

Typology of social space in Kauman Kampong Semarang

Edward Endrianto Pandelaki*, **Atiek Suprapti, Satriya Wahyu Firmandhani**

Department of Architecture, Faculty of Engineering, Universitas Diponegoro, Semarang, Indonesia

*Corresponding e-mail: epandelaki@yahoo.com

Abstract. Social space is various forms of spaces used by community in conducting social interaction. These kinds of spaces should be given more attention since they serve as catalyst in the implementation of good social cohesion in community. The effort includes giving concern toward their various implemented form. This concern will give benefits in the creation of resilient built environment. Kauman Kampong in Semarang, is an old urban kampong and still exist up until now. During its development, the inhabitant live and conduct their activities in good shape. Therefore this kampong is an appropriate place to learn and explore social spaces which is formed and utilized by the community who conduct their activities in this kampong. The aim of this research is to find out forms and typology of social space in Kauman Kampong in Semarang. Qualitative method is used in this research since the nature of this research is explorative. There are various social activities in Kauman Kampong in Semarang, such as religious, trading, and other social interaction, which have formed various social spaces. These social spaces have their own physical characteristics and with various intensity of activities. Based on collected data in field survey, the typology of social spaces that could be inferred are: permanent, temporary, and dynamic social space.

Keywords: Social space, Typology, Kauman Kampong Semarang

1. Introduction

Indonesia is facing the fast urbanization condition and the vast need of housing provision. Areas are changing fast become urban and movement of people to urban area is unavoidable. Data of Statistic Bureau shows that in 2015 percentage of population living in urban area is already exceeding population living in rural area about 53,3% [1]. Meanwhile in the housing development sector, data of Indonesian Government shows that backlog of housing needs is increasing and in 2007 about 5,8 million units is needed [2].

Urban kampong has become a place of first destination for people from rural area when they first start their living in urban area to stay. In kampong, people still has their former living style from where they belong, where togetherness and kinship are reflected in their life. Therefore social relation between inhabitants is created and cared in many urban kampong. A good social cohesion is one characteristic that could be met in urban kampong. This condition will be different compare to the higher income where individualism will be thicker felt in their neighborhood.

The need to be always in contact with the other member of inhabitant is a basic need for human as a social creature. Concern need to be given toward creating a good social cohesion condition of the inhabitant as an effort to develop a built environment. Social cohesion could be a push factor in the implementation of sustainable living environment for the inhabitant. In the concept of sustainable development, social aspect is an aspect that need to receive a proper and balance concern, besides the economic aspect and ecological aspect. Social condition of the inhabitant, such as level of education, health, participation, and good social cohesion have become indicators of successfulness of development.

Kauman Kampong is a kampong in the urban area of Semarang City. It is an old kampong which has been existed and last up until now. During its development, various ethnic live and conduct their activities in good condition. This kind of condition makes Kauman Kampong Semarang as an appropriate place to learn social cohesion condition and its related physical aspect.

Activities of the inhabitant and their interactions are conducted in certain time and place. The place for the inhabitant to conduct their activities will have various forms. Likewise when the inhabitant conducting social interaction. There will be various kind of place created and utilized by the inhabitant. The various form of place which is utilized by the inhabitant to conduct social interaction is interpreted as social space in this research. Concern need to be given to these social spaces, since social space could have a function as catalyst in the creation of good social cohesion among the inhabitant [3]. From the various kind of social space which is existed, created, and utilized by the inhabitant in kampong, concern toward typology of the various social spaces is necessary. Therefore it could be used in the development and maintenance of kampong in order to create and enhance good relationship in the society.

Kauman Kampong Semarang has been developed up until now and still become one of economic activity center of the city with a good and maintained social cohesion condition. The forms of social space which is created and utilized by the inhabitant are various. Therefore this research aim to explore the various forms of social space in Kauman Kampong Semarang and find typology from those various forms of social space.

The method used in this research is qualitative method, since the nature of this research is explorative as an effort to probe the diversity in local context about social space created in Kauman Kampong Semarang. The data needed in this research is secondary data of basic map of Kauman Kampong Semarang, and primary data of social spaces in Kauman Kampong Semarang altogether with information from source person about the utilization of the social space. The collected data are then analyze and interpreted. Typology analysis is conducted through data categorization based on certain characteristic occurred.

2. Kauman Kampong Semarang

Kauman Kampong Semarang is a living area occupied by Kauman community since a long time ago. This kampong has different characteristic with other kampong or settlement which implement thicker Islamic value. This condition could be seen on various Islamic religious activities which is held in Kauman Kampong, such as activity of pesantren and the Great Mosque of Semarang [4]. Furthermore, Suprapti (2012) explains that Kauman Kampong Semarang has the nature of kekaumanan, such as consanguinity of Kauman people, Islamic obediently, mostly Hajj, and live in the surrounding settlement of the Great Mosque [5].

Kauman Kampong Semarang is located in downtown area, about one kilometer from the center of Semarang City. It is in a strategic location beside the Johar Market which is one of the center of economic activity in Semarang. As research locus, Kauman Kampong administratively located in Bangunharjo Sub District (Kelurahan) and Kauman Sub District. Basically the surrounding area of this kampong is business or economy activity area. In the North side, Kauman Kampong adjoin with shops in Pemuda street, in the East side adjoin with shops in Kanjengan street and West Alun-alun street, in the West side adjoin with the PHI block, and in the South side adjoin with shops in Wachid Hasyim street. In 1985, Kauman Kampong Semarang administratively located in Kauman Sub District, which is then area consolidation was conducted by the government and it consist of two Sub Districts of Bangunharjo Sub District and Kauman Sub District. Even though it is divided in two sub districts, the characteristics of Kauman is still reflected physically and non-physically. The non-physically characteristics could be seen from the inhabitant activities which is inspired by Islamic value, such as pesantren or Islamic education system, recitation in each neighborhood on every Friday night, and the Great Mosque has become a magnet of various religious activities. Whereas physically could be seen from characteristic of entrance gate of each street of Kauman Kampong which is resembled the gate of Great Mosque. The utilization of these gates is planned by the inhabitant of Kauman Kampong. It is meant as an effort to preserve the characteristic of Kauman Kampong to be continuously integrated with the Great Mosque, according to Mr. Arief the head of RW (neighborhood organization) 1 of Bangunharjo Sub District.

Kauman Kampong Semarang has an area of about 13 Ha which consist of smaller kampongs. In Bangunharjo Sub District there are kampongs of Suromenggalan, Guntingan, Suroyudan, Grajen, Patian, Kepatihan, Kepatihan Barat, Glondong, Kauman Barat, Bok, Getekan, and Jonegaran. In Kauman Sub District, there are kampongs of Krendo, Mustaram, Butulan, Pompa, Kauman Timur, and Pungkuran.

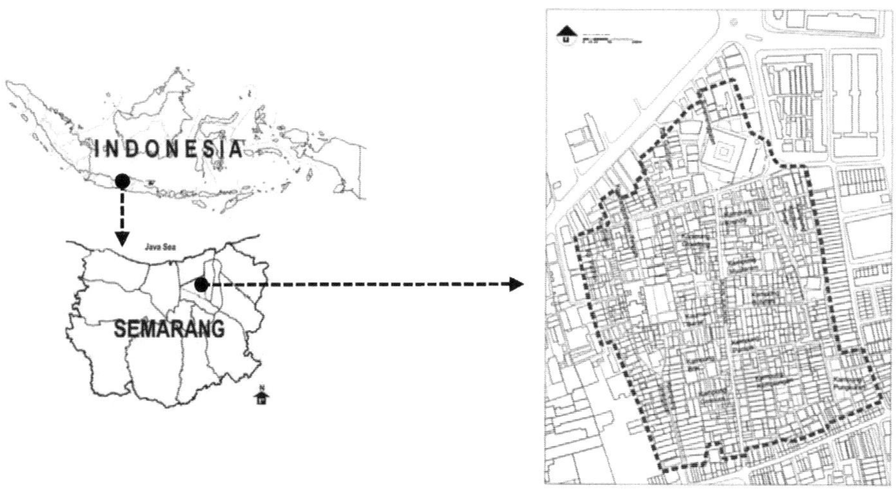

Map of Kauman [4]

Figure 1. The location of Kauman Kampong Semarang

3. Findings and discussion

Social space is sociological facts that create space as well as a space in physical meaning [6]. Therefore the observation on social space in the field survey is conducted through observing objects of physical space as social space as well as social activities of the people which create a boundary. The social activity which create boundary basically has a catalyst or trigger for social activity [7]. The trigger of social space can be a permanent or temporary property, as well as a form of activity that draw attention of people to gather and have social interaction.

Based on field observation, there are various activities conducted in Kauman Kampong Semarang related with social spaces as follows: 1. Social community activities, where people communicate and interact directly, such as children playing, chatting, sports, and conducting community services; 2. Religious activities, which is the characteristic of Kauman Kampong. This activity is dominated with pesantren or Islamic boarding school activity. The other activities are recitation for children or their parents, reading Quran, and Islamic study; 3. Economic activities, which can be felt strongly in Kauman Kampong, not only in the main street but also in many alleys of this kampong, such as inhabitants who open their stall in front of their houses, various shops in the main street, and peddlers. Locations of these activities are Mosque, small prayer room, Islamic boarding school, alley, main street, front yard of houses, and area in front of schools. There are properties that support these activities, such as sitting group, chair and desk, vehicle, floor mat, and charts.

Social spaces which are formed in Kauman Kampong Semarang, can be categorized based on its characteristic of physical nature and the intensity of activities that occur. Based on its physical nature, there are three types of space that could be inferred, that is concrete space, semi concrete, and non-concrete space. Characteristic of physical nature has the meaning of appropriation of condition of a social space with the contained interaction activities. From empirical condition in Kauman Kampong, formed social spaces are triggered with activities contained. These activities has space or location right where the social spaces is formed. In other words, social space is formed through utilizing space which is used by an activity. Therefore observation toward existing environmental condition is necessary. There are properties or equipment which support activities in those spaces. As an example is a food stall in front of a house equipped with sitting group which making it possible for people to have a chat while enjoying the food. This kind of social space is formed and supported by its surrounding environment, therefore it is known that there is an adjustment effort of the physical environment to accommodate social activities which will be held there. This social space has the nature of concrete, or concrete social space, because the space is implemented to accommodate a certain social activities thus forming a social space. On the other side there are activities held in a space which is not specified for certain activities or a multipurpose space, such as neighborhood hall, house terrace, and neighborhood open space. The social space is created through temporary adjustment made to accommodate a social activity of the inhabitant. As an example is a neighborhood open space in Glondong Kampong. This neighborhood open space could be used as sports activities, commemoration of Independence Day activities, as well as community religious study. In the case of house terrace, sometimes it is used as social gathering for local community and selling children toys. These kind of spaces has the nature of semi-concrete, because the space is not only designed and intended for a certain social space, even though whenever another social activity occurs in this space it will not disturb the initial function and its surrounding space. Furthermore there is non-concrete social space. This kind of social space is formed in a container which is basically not allocated as social space and there is no social space in any particular physical form. Street corridor and alley are the place where this kind of social space often happened, therefore it has possibility to make a burden to other activities and causing the actual space dysfunction. As an example is social space which is formed during activity of competition on the Independence Day anniversary held by the inhabitant in the street or alley in August. It cause the street or alley impassable, therefore the street user should search for alternatives.

Results from field observation found that based on the characteristic of intensity of activities that occurred, the social space can be categorized into: continuous social space, periodical social space, and uncertain social space. Continuous social space is a social space created for a long time to accommodate activities that is long and continuous in everyday life and conduct at least by one or several individuals alike. Trading activities that always invite interactions between individuals such as buying and selling in stores from morning to evening and even night, rice seller packs in the afternoon until the evening, is an example of continuous activity. The social activities that are formed here take place continuously every day. Periodical social space occurs when social interaction activities with specific and scheduled periods are conducted by the community. This social activity is common in the home terrace and social facilities. This is in accordance with the semi-concrete nature of the space that is only at a certain time used for social space which does not interfere with the function of the surrounding environment. There are two kind of periodical activity, that is : the planned or scheduled at certain time (monthly, weekly, or yearly), such as monthly community gathering, religious study, and commemoration of Independence Day, and the unscheduled, such as receiving guest in the terrace, and having chat in public wells, toilet, and bathroom. The uncertain social space occurs upon social activities that is naturally and incidentally without formerly planned. It is commonly occurred in un-concrete space such as street corridor and alley. Example of activities that make up the uncertain social space is children playing in the street as well as itinerant street food vendors.

Based on field observation and analysis on collected data, the social space is implemented through process of interaction activities between people and container which hold those activities. This is in accordance with Carmona et al (2003) which expressed that in understanding the formation of social space requires container and activities which is contained, and space as container should provide conducive environment in order to fulfil the terms of interaction [8]. In its formation process, social space will be shaped because of there is people along with activities or interactions done together and there is space that accommodate the activities, as already mentioned above in intensity of activities and physical nature of space. In this framework of implementation process characteristics of social space, it could be inferred three types of social space, that is: permanent social space, temporary social space, and dynamic social space. Permanent social space is formed and not easily disappear because the activity occurs continuously and periodically with the support of concrete space. This social space results from the correspondence of concrete spaces with continuously and periodical social space. Examples of permanent social spaces that take place in Kauman Kampong Semarang are routine recitation activities at inhabitants' houses and Mosque's porch, and buying and selling interactions in permanent shops. The second type of social space is temporary social space. Temporary social space is formed but is easily disappear due to the nature of activities tend to be periodic and indeterminate, and not supported by a concrete space. Temporary social space occurs in semi-concrete space that has multifunctional properties, therefore the existence of social space can change or disappear at certain times and every time. The third type of social space is dynamic social space. Dynamic social space is formed from indefinite activity in non-concrete spaces and has a tendency to move to adjust the existence of the trigger of its activity. Dynamic social space is the result of non-concrete social space correspondence with uncertain social space. Examples for this kind of social space are social space which occurred from the mobility of itinerant street food vendors and mobility of children playing.

Figure 2. Permanent social space

Figure 3. Temporary social space

Figure 4. Dynamic social space

4. Conclusion

Space is formed by the existence of physical elements of space and people with activities that contained in it. Based on the characteristics of the physical nature of space and the nature of the activities intensity that occur, the process of formation of social space in Kauman Kampong Semarang can be inferred into three types, namely: 1. Permanent Social Space. This social space takes place in concrete spaces and occurs continuously or planned periodically. Therefore this space is always there and its existence is not easily change or disappear; 2. Temporary Social Space. Temporary social space is formed but could easily disappear due to the nature of activities tend to be periodic and indeterminate, and not supported by a concrete space; 3. Dynamic Social Space. Dynamic social space is formed from indefinite activity in non-concrete spaces and has a tendency to move to adjust the existence of the trigger of its activity.

References

[1] Central Bureau of Statistics, 2015. *Persentase Penduduk Daerah Perkotaan menurut Provinsi, 2010-2035*, Central Bureau of Statistics, Republic of Indonesia.

[2] Mungkasa, Oswar, 2008. *Pembangunan Perumahan dan Permukiman di Indonesia,* Directorate of Settlement and Housing, Ministry of National Planning and Development of Republic of Indonesia.

[3] Pandelaki, Edward Endrianto, 2013. *Menuju Pembangunan Perumahan yang Berkelanjutan di Kota: Rumah Susun dan Kohesi Sosial*, Proceeding of National Conference on Sustainable Urbanism 2nd, Culture, Social and Technological Approach". February 28th, 2013. Doctoral Program of Architecture and Urbanism, University of Diponegoro, Semarang.

[4] Suprapti, Atiek, et al, 2016. *Dialog Spasial antara Seochon Village dan Kampung Kauman Semarang dalam hal Pengembangan Konservasi. Research report.* Faculty of Engineering, University of Diponegoro, Semarang.

[5] Suprapti, Atiek, 2012. *Kontrol Protektif Pada Ruang Bermukim Komunitas Kauman Semarang*, Dissertation, Doctoral Program of Architecture and Urbanism, University of Diponegoro, Semarang.

[6] Fearon, David. Janelle, Donald ed., 2004. Georg Simmel, The Sociology of Space. CSISS Classics . UC Santa Barbara: Center for Spatially Integrated Social Science. Accessed from: http://escholarship.org/uc/item/7s73860q, accessed 07/03/2016.

[7] Lang, J. 1987. *Creating Architecture Theory.* Van Nostrand Reinhold Company : New York

[8] Carmona, et al. 2003. *Public places–urban spaces, the dimension of urban design. Architectural press.*

Algae façade as green building method: application of algae as a method to meet the green building regulation

Heru W. Poerbo*, Widjaja Martokusumo, M. Donny Koerniawan, Nissa Aulia Ardiani, Susan Krisanti

School of Architecture, Planning and Policy Development,
Institut Teknologi Bandung, Bandung, Indonesia

*Corresponding e-mail: heru@ar.itb.ac.id

Abstract. The Local Government of Bandung city has stipulated a Green Building regulation through the Peraturan Walikota Number 1023/2016. Signed by the mayor in October 2016, Bandung became the first city in Indonesia that put green building as mandatory requirement in the building permit (IMB) process. Green Building regulation is intended to have more efficient consumption of energy and water, improved indoor air quality, management of liquid and solid waste etc. This objective is attained through various design method in building envelope, ventilation and air conditioning system, lighting, indoor transportation system, and electrical system. To minimize energy consumption of buildings that have large openings, sun shading device is often utilized together with low-E glass panes. For buildings in hot humid tropical climate, this method reduces indoor air temperature and thus requires less energy for air conditioning. Indoor air quality is often done by monitoring the carbon dioxide levels. Application of algae as part of building system façade has recently been introduced as replacement of large glass surface in the building façade. Algae are not yet included in the green building regulation because it is relatively new. The research will investigate, with the help of the modelling process and extensive literature, how effective is the implementation of algae in building façade to reduce energy consumption and improve its indoor air quality. This paper is written based on the design of ITB Innovation Park as an ongoing architectural design-based research how the algae-integrated building façade affects the energy consumption.

Keywords: algae, building system façade, green building regulation, ITB Innovation Park

1. Introduction

The Local Government of Bandung city has stipulated a Green Building regulation through the Peraturan Walikota Number 1023/2016. Signed by the mayor in October 2016, Bandung became the first city in Indonesia that put green building as mandatory requirement in the building permit (IMB) process. At provincial level, the government of DKI Jakarta has their green building regulation earlier in their Peraturan Gubernur Number 38/2012. More cities will follow the example of Jakarta and Bandung, because green building regulation has become a national policy in the ministry of public works and housing Peraturan Menteri Pekerjaan Umum dan Perumahan Rakyat No. 02/PRT/M/2015. The expansion of regulatory framework indicates the growing importance of green building application in Indonesian large urban areas. This is in line with the findings that building sector is a major energy consumer. Research indicated that carbon-dioxide emissions from the building sector (when embodied energy is considered) exceed those of the industry and transport sectors [1].

Introducing and implementing green building regulation in Indonesian cities is not a simple task. There are green building standards and regulations in other countries that can be adopted or used as reference for Indonesia. However, those standards are set according to the local climate that might be far different from the hot humid tropical climate in Indonesia. The economic level of the countries must be taken into consideration as well, because some measures of green building method may be too costly for Indonesian cities. Hence the green building regulation in Jakarta or Bandung might have been limited to control certain aspects of the building only for its ease of implementation by local government and building owners. There is still a chance that new or more advanced method of green building with appropriate technology to be implemented in this situation, provided that the method has been studied beforehand.

Green Building regulation is intended to have more efficient consumption of energy and water, improved indoor air quality, better management of liquid and solid waste etc. These objectives are attained through various design methods in building envelope, ventilation and air conditioning system, lighting, indoor transportation system, and electrical system. To minimize energy consumption of buildings that have large openings, sun shading device is often utilized so that indoor temperature rise due to reduction of solar heat gain through radiation. Another popular measure to reduce heat gain in buildings with large opening is by using low-E glass panes. For buildings in hot humid tropical climate, this method reduces indoor air temperature and thus requires less energy for air conditioning.

Indoor air quality is often done by monitoring the oxygen and carbon dioxide levels, among others in the building. Innovation in improving the air quality by installing a photo bioreactor or algae bioreactor has been experimented. Such bioreactor absorbs CO2 from the indoor air and releases more oxygen during photosynthesis process because algae are photoautotroph organisms. Algae photo bioreactor has been installed at the BIQ house in Hamburg, Germany and the GSA federal building in Los Angeles among others.

Figure 1. The BIQ House: first algae-powered building in the world. [2]

Figure 2. GSA federal building in Los Angeles. [3]

The application of algae bioreactor as sun screening device in building façade has not been studied extensively in terms of its solar heat transmittance or luminance. An architectural design-based research on the use and role of algae as part of the building façade system is currently being undertaken at the School of Architecture, Planning and Policy Development in Insitut Teknologi Bandung. Based on that research, this paper is intended to provide a description of possible alternative technology in the form of an algae bioreactor and the argumentation for its implementation to meet the green building regulation in Bandung city.

2. Methods

This paper is written based on a study conducted as part of an ongoing research project on building façade systems. The research project investigates the possibility of using an algae bioreactor as part of a building façade. The case study of the research project is a new Innovation Park building of ITB that is currently in the preliminary design phase. The new building is located in Jl. Ganesha across the ITB main campus. It sits on a site with its narrow side on the North and relatively long side on East/West. Large areas of the building's façade are facing East and West, thus exposing the building to more sun radiation due to its large windows on these sides. The research investigates the possibility and the advantage and disadvantage of using algae bioreactor in the building façade.

Figure 3. The design of ITB Innovation Park on Jalan Ganesha: the four storey building in the center.

The study uses a three stages method. It begins with a critical review of the existing green building regulation in Bandung city. The review is done to recognize the objectives of the regulation, as well as some background information regarding pertinent research findings that are used as the base for the green building regulation in Bandung. The background information will help to understand why these standards need to be applied in Bandung. The next step is to identify the requirements to meet those objectives, particularly to determine which parts of the building regulation are relevant to the algae bioreactor façade experiment.

The third or final stage is performing an experiment as an effort to meet the regulation. The experiment is performed in two ways. One experiment is done through computational modelling, and the other experiment is performed with physical measurement in the laboratory. Analysis is done by comparing the performance of the designed façade using conventional design features and the façade with algae bioreactor. This is performed by calculating the energy requirement for cooling of the building in three alternatives: glass windows, window with sun shading device or horizontal screen, and window with algae bioreactor as screen. Energy requirement for building cooling is used as the parameter in this experiment because the objective of green building is clear enough, i.e., to reduce the consumption of resources such as electicity, water and air.

3. Background Study on Green Building Regulation in Bandung

The Bandung city green building regulation has main principles that the buildings establishment shall not burden the environment beyond its carrying capacity; and shall promote efficient use of natural resources for the purpose of building construction and operation (As stated in the Article 4 of the Bandung Green Building Regulation). The regulation applies to two categories of buildings: the first category is for all new buildings or building-addition with a total floor area over 5,000 sqm, and the second category is for new building under 5,000 sqm.

IOP Conf. Series: Earth and Environmental Science **99** (2017) 012012 doi:10.1088/1755-1315/99/1/012012

Most of the energy in buildings in Indonesia is consumed by the HVAC system, regardless of the type of building. Office building, retail center, hotel, hospital, apartment building and educational facilities (as presented in Fig 4) show indoor cooling as the main energy consumption. The diagram indicates that HVAC contributes about 37% to 54% of building total energy consumption. Artificial lighting contributes 20% to 39% of total energy consumption. Therefore, reducing energy consumption for HVAC and artificial lighting through passive and active design can effectively reduce the overall building energy consumption. [4] The study of Innovation Park building focuses on the reduction of energy consumption for HVAC and artificial lighting.

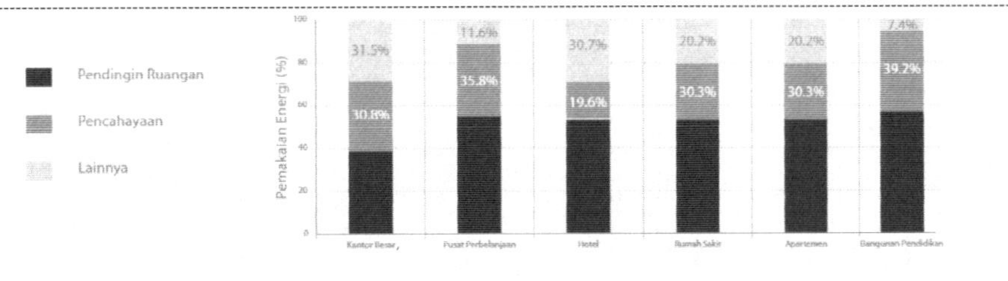

Figure 4. Details of energy consumption for various types of buildings. [5]

The Innovation Park building has a total floor area of approximately 2,000 sqm including its semi basement, so it falls into the category of new building under 5,000 sqm. The Mandatory Requirements for New Building construction with floor area less than 5,000 m^2 including basement is as following:

Article 39 [6]

The compulsory requirement of new green buildings for constructions with floor space of less than 5,000 m^2 (five thousand square meters) including the basement as referred to in Article 6 paragraph (5) letter b, including:

a. Energy efficiency;
b. Water efficiency;
c. Indoor air quality; and
d. Land management.

Article 40

Energy efficiency in building buildings as referred to in Article 39 letter a, covering:

a. System of building envelope;
b. Air conditioning system; and
c. Lighting system.

Article 41

1) The requirements of the green building building envelope system as referred to in Article 40 letter a, following the provisions referred to in Article 9.
2) Requirements of the green building building envelope system as referred to in paragraph (1) shall be required in buildings with functions other than residential houses.

In order to comply with Articles 39, 40 and 41 the Innovation Park building must have energy efficiency and indoor air quality by utilizing building envelope system, air conditioning and lighting because the building is used for non-residential purpose.

As shown in Figure 5, the external heat gain from the windows and walls of a typical office building in Bandung is about 63%, while the internal heat gain from equipment, lighting and occupancy is about 37%. This demonstrates enormous energy savings opportunities through carefully designed and precise building envelopes to reduce air cooling loads. [7] The Innovation Park building is mainly used as office, thus it is expected to have those values of external heat gain from window and walls.

Article 9

1) The requirements of green building building envelope system as referred to in Article 8 letter a, is intended to make the cooling load more efficient.
2) The system of building envelope referred to in paragraph (1), carried out by air conditioning system shall plan the building envelope by calculating the PMD / OTTV and PMA / RTTV values.
3) The PMD / OTTV value and PMA / RTTV value as referred to in paragraph (2), in buildings using air conditioning systems, shall be taken into account not exceeding 45 Watt / m2 (forty five watts per square meter).
4) The calculation of PMD / OTTV and PMA / RTTV as referred to in paragraph (3) shall use spreadsheet calculator or graph with solar radiation factor PMD / OTTV and PMA / RTTV areas provided by SKPD.
5) Provisions concerning the local solar radiation are listed in Attachment I which is an inseparable part of this Mayor Regulation.

4. Background Study on Green Building Regulation in Bandung

The design alternatives are tested and evaluated using computational modelling with Open Studio-Energy Plus/OS-EP software. Three design alternatives of building envelope system are evaluated using this method, whereas the alternative with algae bioreactor is physically tested in the laboratory. The three building envelope design alternatives are façade with conventional glass window panes, façade with horizontal shading device, and façade with brise-soleil. Weather data for Bandung is not available in the Open Studio-Energy Plus software, hence we used Singapore weather data as the basis for the simulation because it is the closest location.

For the simulation, several thermal zones are applied in the building, which has been designated according to its function. The use of air conditioning in this experiment is required to measure the amount of total energy (in kBtu) in the building with OS-EP software. For this experiment, thermal zone 3 was tested. Thermal zone 3 accommodates the main facilities such as co-working spaces, rental offices, and multi purposes/exhibition hall.

As the benchmark for the other façade design alternatives, the original design with glass panes windows is tested. Figure 6 displays the simplified model that was used for the computational modelling. The OpenStudio-Energy Plus result is shown in Table 1.

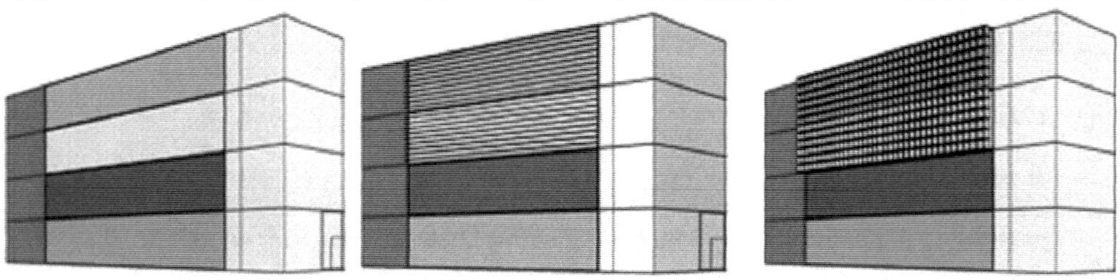

Figure 5. Simplified model of façade configurations (left to right): original design with glass panes windows, alternative with horizontal shading device, alternative with brise-soleil.

Table 1. Building Summary Original configuration window with glass pane.

Information	Value	Units
Net Site Energy	1,300,945	kBtu
Total Building Area	17,443	ft^2
EUI (Based on Net Site Energy and Total Building Area)	74.58	$kBtu/ft^2$

The second façade design alternative uses shading device in the form of fixed horizontal shutters in front of the glass window. Simplification of the building design for computational modelling was performed in order to enable quicker calculation of the main openings in the façade, while retaining the overall building shape and dimensions. Figure 7 below shows the model of façade configuration with horizontal shading device.

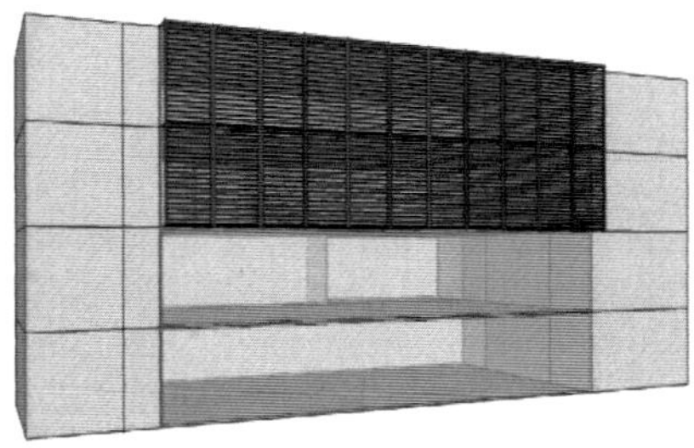

Figure 6. Model of façade configuration with fixed horizontal shutter shading device.

The OpenStudio result for the second façade design alternative is shown in Table 2 below

Table 2. Building Summary window configuration with horizontal shading device.

Information	Value	Units
Net Site Energy	1,281,051	kBtu
Total Building Area	17,443	ft^2
EUI (Based on Net Site Energy and Total Building Area)	73.44	$kBtu/ft^2$

The third façade design alternative uses brise-soleil, comprised of lightweight concrete with waffle pattern (square shaped holes) that is mounted in front of the glass window openings

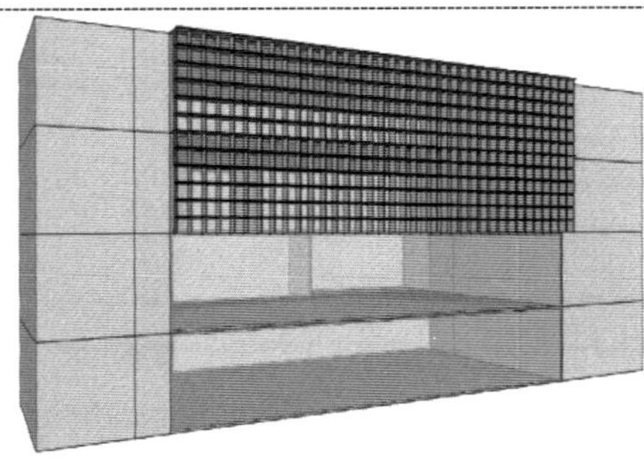

Figure 7. Model of façade with brise-soleil.

The OpenStudio-Energy Plus result for the third façade design alternative is shown in Table 3

Table 3. Building Summary window configuration with brise-soleil.

Information	Value	Units
Net Site Energy	1,278,700	kBtu
Total Building Area	17,443	ft^2
EUI (Based on Net Site Energy and Total Building Area)	73.31	$kBtu/ft^2$

The result of the experiment with computational modelling shows that energy consumption for indoor cooling of the Innovation Park building will be reduced 1.53% from 1,300,945 kBtu to 1,281,051 kBtu if we install horizontal shading device as compared to the glass windows in the original design. This confirms the suggestion in the green building regulation to apply shading device as a measure to reduce building energy consumption. However, changing the façade screening from the horizontal shading device to brise-soleil only reduces further the energy consumption by 0.18% from 1,281,051 kBtu to 1,278,700 kBtu.

5. Algae Bioreactor Experiment

The second experiment of measuring the heat properties of the algae bioreactor has not been performed in the research. Physical model construction is still undergoing and the bioreactor will be ready by end of July 2017. Measurement of light and heat transmittance through the reactor – that uses green algae – can be performed later in the ITB laboratory.

The model of façade as a sample in the experiment of algae bioreactor is made of thin glass block as a prototype of Algae building facade. A wooden box is attached behind the glass block. Measurement of the air temperature and the intensity of the incoming sunlight will be performed inside the wooden box that can be regarded as a representation of the indoor space of the building.

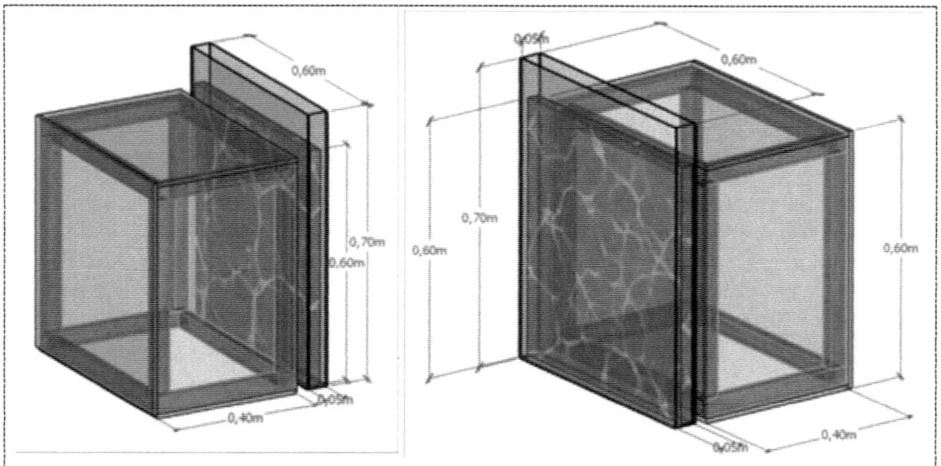

Figure 8. Model of façade with algae bioreactor.

Preparation of the bioreactor testing equipment was conducted through several trials. The first uses a 5mm thin glass (Figure 9 left) but it could not hold the weight of the water, so it had to be rebuild from a thicker glass that has 8mm thickness (Figure 9 center). The black box has no leaking or structural problem in this stage (Figure 9 right)

Figure 9. Construction process of algae bioreactor.

The Bioreactor is moved to the roof top of the Inter-University Center to determine the best placement position for the bioreactor test apparatus (Figure 10)

Figure 10. Placement of algae bioreactor on roof top.

Apparently after reconsideration, the Experiment tool can still be optimized before testing. The optimization is planned to be done with more air bubble pump modification and algae preparation, so the tool is temporarily taken to the Aquatic Ecosystem Analysis Lab., School of Life Sciences and Technology ITB.

After 10 days, the algae have also been grown and ready to be put into the photo bioreactor at the time as we mount it on the rooftop of the Inter-University Center. However, there are still some tools to acquire for the photo bioreactor, which are:

1) A larger aerator because the one installed is not sufficiently powerful
2) Rock aerator (3 pieces) with a diameter of 8cm
3) Clear silicone glue
4) Black paint for wooden stand (figure 11).

Figure 11. Temporary placement of algae bioreactor in the Aquatic Ecosystem Analysis Laboratory.

6. Concluding Remarks

The experiment has confirmed the Green Building Regulation's suggestion to apply sun shading or screening device in building façade, particularly the façade in East and West orientation. The ITB Innovation Park building has larger East/West façade than its North/South façade, hence implementation of shading device is important for this office building.

The experiment of algae bioreactor as alternative for sun shading device is not yet completed at the time this paper was written, due to some setback during the preparation of the experiment device. The result of this algae bioreactor experiment can be compared to result of the computational modelling. According to the expert in Microbial Biotechnology Research Group ITB, another advantage of the bioreactor is the conversion of CO2 into O2 during its process that can be utilized to improve the air quality in the building. This is done by blowing the air from inside of the building into the bottom of the bioreactor. The O2 rich will come out of the top of the bioreactor air. Indoor air quality is one of the requirements stated in the Bandung's Green Building Regulation Article 39.

Acknowledgments

The ongoing research at the Architectural Design Research Group is being funded by the special research grant of ITB (P3MI). The research team expressed their gratitude to Dr. Gede Suantika of the Microbial Biotechnology Research Group, School of Life Sciences and Technology, Institut Teknologi Bandung for his support and thoughtful remarks on algae bioreactor.

References

[1] Laurenzi M P (ed.) 2007 Building Energy Efficiency. Why Green Buildings are Key to Asia's Future Hong Kong: The Asia Business Council US Energy Information Administration
[2] Source: http://www.buildup.eu/en/practices/cases/biq-house-first-algae-powered-building-world
[3] Source: http://buildipedia.com/aec-pros/featured-architecture/hok-vanderweil-process-zero-concept-building-as-green-asalgae
[4] Panduan Pengguna Gedung Hijau Bandung Vol. 1 Selubung Bangunan p. 2
[5] International Finance Corporation (IFC). 2015. Bandung Building Energy Efficiency Baseline and Saving Potential: Sensitivity Analysis
[6] Peraturan Walikota Bandung Nomor 1023 tahun 2016 tentang Bangunan Gedung Hijau
[7] Panduan Pengguna Gedung Hijau Bandung Vol. 1 Selubung Bangunan p.3
[8] International Finance Corporation (IFC). 2015. Bandung Building Energy Efficiency Baseline and Saving Potential: Sensitivity Analysis

Resiliency and affordability of housing design, Kampong Cieunteung-Bale Endah in Bandung Regency as a case study

Allis Nurdini*, Wanda Yovita , Patriot Negri

Housing and Settlement Research Group, School of Architecture, Planning and Policy Development, Bandung Institute of Technology

*Corresponding e-mail: allis@ar.itb.ac.id

Abstract. Recently the discussion about resiliency and housing design has been fast developed, including design for housing on riverfront as marginal area that usually occupied by low income people. The low income people generally will be the worst affected community in flood disaster situation, because of their un-affordability to achieve quality housing design. In other side, the funding support capacity especially from the local government is limited, so that the resilient environment also need to be supported by the community itself. In this context, the study about low income people's affordability to achieve resilient housing design is essential. This study was conducted to identify two important points: the design choice and the affordability level of resilient housing from the community viewpoints. Kampong Cieunteung, in Bale Endah Area, Bandung Regency is chosen as a case study, because this area annually experience severe flood from overflow of the Citarum River branch. In preliminary research phase, approximately 60% of the Kampong Cieunteung's resident need to stay and become indication that the community need resilient housing design to accommodate their live hood. The next phase, the contingent valuation method was implemented to gain resilient design choice and affordability perspective from the community. It is concluded that the community have ability to choose the resilient housing design based on their aspiration and based on their ability to pay. The result indicates that resilient housing design should have character of optional, module co-operational, and incremental to be afforded by the low income people.

Keywords: resiliency, affordability, housing design, kampong

1. Introduction

Resilience is the ability of a social or ecological system to respond a problem while maintaining the basic structure and ability to adapt the distractions and changes. One of the ideas emphasized in the principle of resilience is the need of urban or settlement systems, especially for low-income people who are often exposed to natural disasters, to increase their capacity to deal with the impact of the disaster (ARUP [1], Folke [2], Moench [3]). Resilience settlements means housing settlements that have the capacity to maintain their basic functions and structures so that communities who live and work in site, in particular low-income communities affected by disasters, can survive and adapt in the situation of distractions and changes, including climate change. From the resiliency agenda perspective, the low income people will become the important actor.

The plan and design concept of resilient settlement has grown widely today including the idea of adaptive settlements in locations that are often hit by floods like riverfront kampong. The low income people generally will be the worst affected community in flood disaster situation, because of their un-affordability to achieve quality housing design. In other side, the funding support capacity especially from the local government is limited, so that the resilient environment also need to be supported by the community itself. In this context, the study about low income people's affordability to achieve resilient housing design is essential.

In general, the idea of affordability related to the ability to pay of requested product by the user. It is still arguable that paying 30 percent of income an appropriate threshold for defining affordability, especially for the low income people (Vale, et. als, [4]). The rational way to understand affordability of specific product, especially built environment product, is directly measure the user preferences and their willingness to pay.

As the case study, Kampong Cieunteung in District of Bale Endah, Bandung Regency was selected. This location is in the riverfront area of the Citarum River. Several settlements or kampongs have been growing for decades along the Citarum River and are dependent on this river for agricultural activities including Kampong Cieunteung. Along with the decreasing of rice field in Bale Endah and changing with the dense building around Bale Endah, this kampong experience annual flood and increasing intensity since 2000. This study was conducted to identify two important points: the design choice and the affordability level of resilient housing from the community viewpoints. Those points identify through contingent valuation method or willingness to pay (WTP) method.

2. Methods

2.1. The Case Study

One of the early generation kampong that annually experience the worst flood is Kampong Cieunteung, located in District Bale Endah. This kampong is selected as the location of case study. Rain that fell within 10 to 30 minutes was able to cause flooding in this kampong. Flood height up to 30 cm is a common thing for the residents. The worst condition during the last decade is the flood that reaches 2 meters and could not recede for a week in 2014. The worst location in this village is in Block 20 (called as Rukun Warga/RW 20) (Figure 1).

Kampong Cieunteung RW 20 consists of 4 sub block (called as Rukun Tetangga/RT) or 350 families and is always affected by the worst floods since 2000. Based on preliminary research data (the research was conducted from 2014 until 2016), the majority of Kampong Cieunteung RW 20 residents refused to be relocated because they have been living in the area for decades (60%). In general, people are aware that their current settlement is a potential flood area. The residents generally argue that the floods that have been increasingly frequent in recent years caused by the Citarum River silting. In Figure 1, the condition of major houses in Kampong Cieunteung especially RW 20 has been in poor condition due to flood and some are not occupied by the owner. From the interview question, the majority of residents in RW 20 are land owners or get inheritance land from their parents. The majority of residents have worked at services or informal sector like as employees of textile factories located around Bale Endah, and there is no one become farmer as a main profession.

For spatial improvements, the Bandung Regency Government implement building control such as demolish a number of buildings attached to the Citarum River area and in collaboration with the Board of River Territory (Balai Besar Wilayah Sungai/BBWS) to handle the siltation. In 2014, land acquisition from BBWS has begun for the construction of polder or river water retention ponds in Kampong Cieunteung which is at the lowest contour land and directly adjacent to the Citarum River. The majority of people whose land and houses are not affected by this polder acquisition area are the respondents for further study. In reality, they are still affected by floods until nowadays, choose to stay in the kampong and have to survive on site with self-improvement quality. The residents' housing option to stay at this location is further studied with the Contingent Valuation Method approach.

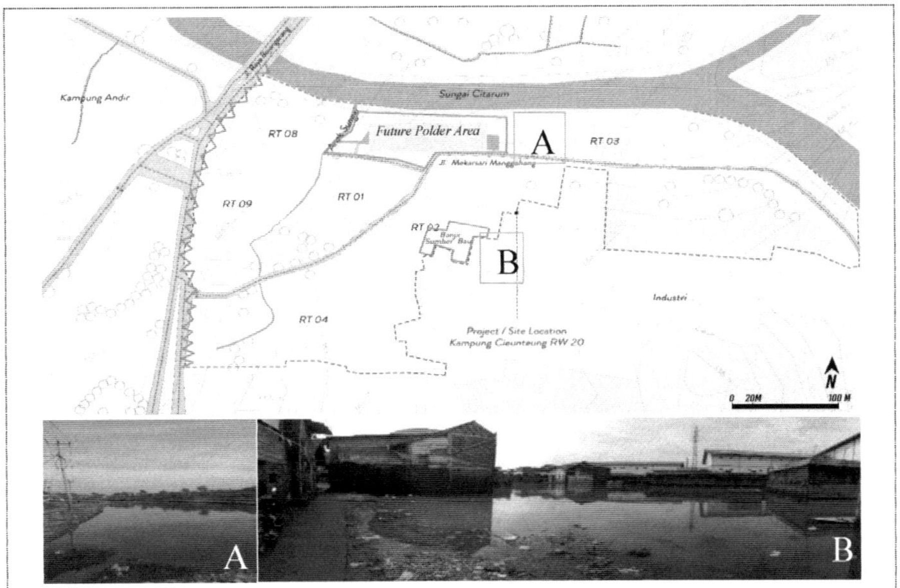

Figure 1. Illustration of Kampong Cieunteung location
(Source: Researcher documentation)

2.2. *Contingent Valuation Method as a Study Approach*

Contingent valuation method (CVM) is a valuation method of a product based on user preferences. This method is used to assess commodities that do not have an approximate market price. The user or community is perceived as having aspirations or preference for the commodity and transforming it to value. An example of this method application is the assessment of goods such as the beauty of a natural environment or the quality of the built environment. This method is widely used in the assessment of built environment quality because of the difficulty of determining the value of its products through market mechanisms (Tresnadi [5], Edwin Chan [6], and Timmermans [7]).

CVM generally uses questionnaires and interview respondents directly about their willingness to pay (WTP) about specific built environment design or product. In this case study, the value of housing that adaptive to flood condition is highly dependent on the respondent's aspiration. In general, the CVM method consists of three stages: the identification of design options for changes in the quality of occupancy, assessing the economic ability of respondents to know the affordability level and the socioeconomic information of respondents. With CVM method, the views of the residents as respondents are more sought out to accept a specific housing design, as well as reduce the value of compensation (material/goods) from outsiders, to improve the quality of their living environment with their own participation.

2.3. *The Research Phases*

The data collected from the residents consisted of social, economic, existing housing condition, assessment of current occupancy condition and aspiration to adaptive flood housing design model. Fifty two (52) respondents were interviewed in Kampong Cieunteung.

The question technique used is bidding choice, which is in first phase, the respondent was ask to choose the preference of housing model according to his wish and in the second phase, the respondent was ask to choose the housing model according to his ability to pay (Manalang [8], Christine [9]). When some adaptive flood housing model was offered, respondents were asked for their preference and reason for the selection of the model. At the initial offer, the respondent did not know the value or the cost of the model, so it was expected to collect the ultimate preference data from the respondent without being affected by his or her personal ability. Then, the question continues with the respondent was offered or informed the price (construction / production) of each housing model. At this stage, respondents would re-select the housing model based on the price and knowledge they know from the initial question.

3. Analysis and Discussion

This study offers a flood adaptive housing design in Kampong Cieunteung that is affected by flooding. The latest conditions, the remaining floods in the form of hardened mud have damaged the houses of the residents as high as 2m from the height of the road and became the basis for the development of adaptive flood design. The two main housing design types offered. The first type, the conventional two-story house design with the ground floor is used for public activities (called as the 1st type). The second one, the amphibious house designs that can float up during floods. The amphibious house is divided into two variants: the design is equipped with exterior and interior walls (called as the 2nd type design) and a wall-less design that only columns, beams, and roofs structure (called as the 3rd type design). The second variants is offered to reduce development costs and the community can develop their own home layouts.

It is assumed that both types need land consolidation as a pre-requisites to rearrange total area and infrastructure. The main differences principles between the conventional house and the amphibious house is that the amphibious house need a co-operational and modular housing structure to optimize the buoyant (amphibious) foundation, utilities and basic infrastructure performance. The three alternatives of housing design model can be seen in Figure 2.

The size of the building and the area of land for the building is assumed to be standardize in accordance with the average residential unit in this Cieunteung Kampong. The 1st type is a conventional house building with two-story in concrete construction. The ground floor can be used for flexible needs. The land area is approximately 36m2 and the building area is 22m2 for each household. Estimated price of this building is Rp. 2,500,000/m2.

The 2nd type is a house that can float with steel construction, equipped with the wall. Total building are is about 21 m2. The house will be lifted if water goes through the drainage due to the pontoon deck that will make the house become floating. The use of steel in the column of buildings together with the foundation to make the house can rise and fall with the roll system on the sides of the floor plate of the building. This model is estimated to cost around Rp. 4,000,000- Rp. 5,000,000/m2. The 3rd type is a floating house with steel construction but not equipped by the walls. The walls or home lay out could be incrementally modified by the owner. Similar with the 2nd type, this type uses a pontoon deck that will make the building floating. It is estimated to cost around Rp. 2,500,000/m2.

At the first phase, respondents were given all three options without being given pricing information. From 52 respondents, 50% of respondents chose the 1st type because the building is considered more secure, more stable, the ground floor that can be used, and easier maintenance. While the other 46% of respondents chose the 2nd type because it more comfortable, flexible, practical, and attractive design. While the 3rd type is the least selected type (4%) because the respondent considers the building is incomplete as a house.

At the next phase, the respondent is given information about the estimated construction price of the three types offered, and is given the opportunity to change the choice. There was a considerable change in selection of respondents who initially chose the 2nd type to switch to the 1st type (become 66%) for reasons of the price was too expensive. Eight percent of respondents switched to the 3rd type due to cheaper building costs (become 12%). Therefore, the remaining was 22% of respondents who still choose the 2nd type. Illustration is in Figure 3.

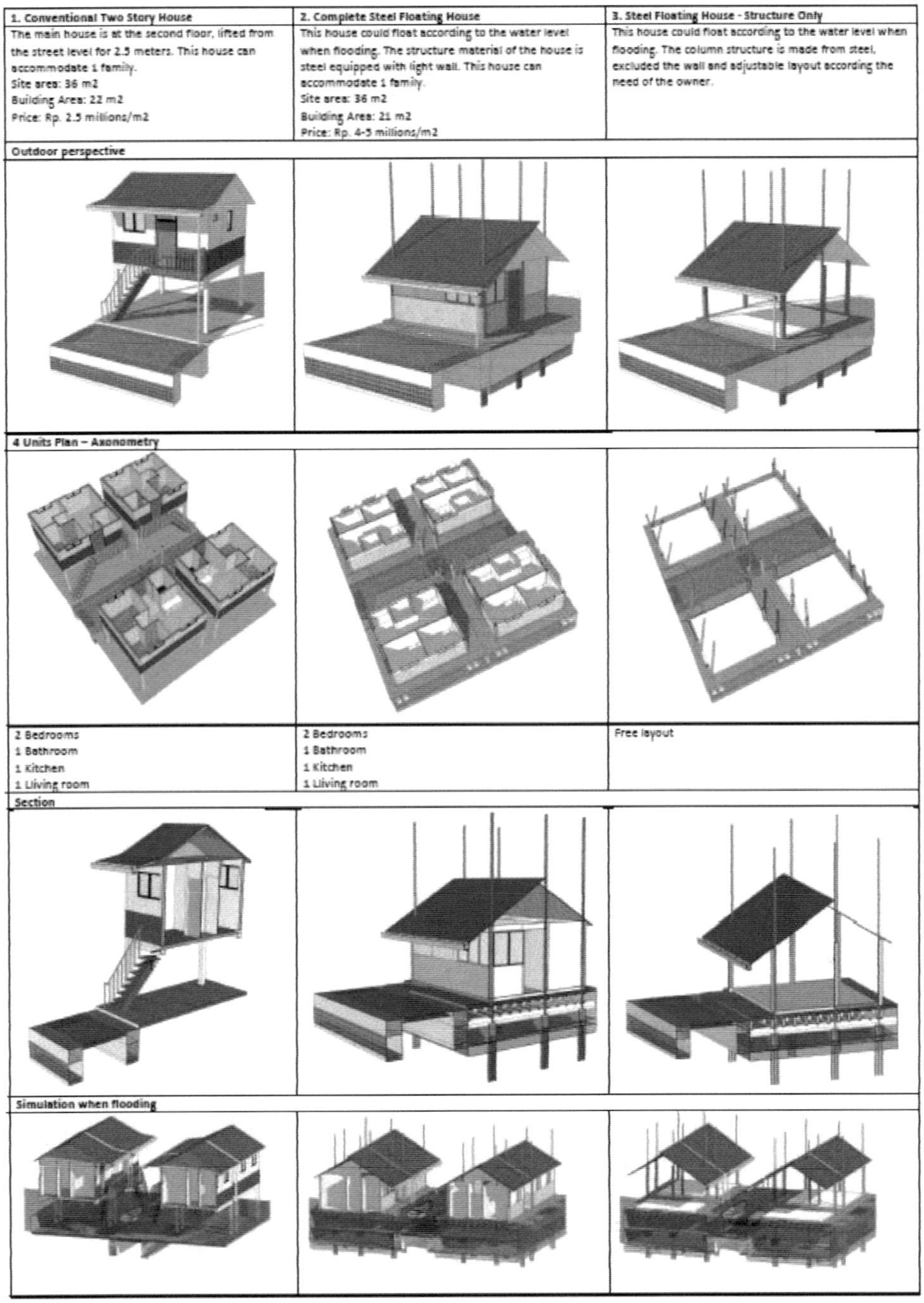

Figure 2. Alternatives model of flood adaptive housing design
(Source: Researcher analysis)

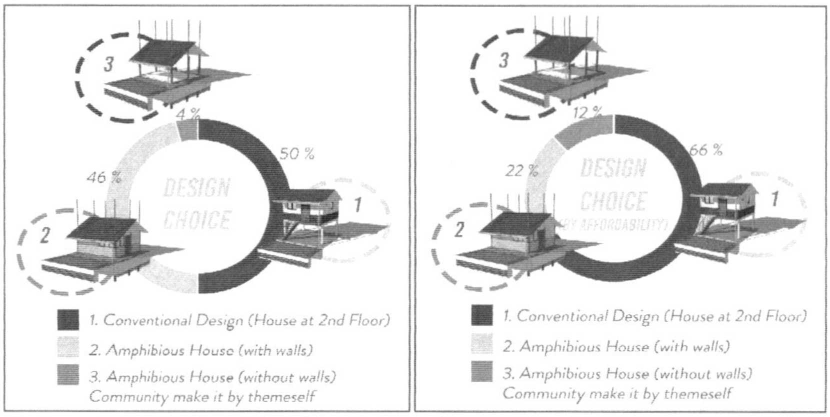

Figure 3. The resilient housing choice from the respondents
(Source: Researcher analysis)

This shift of preference is interesting to be analysed, because it is influenced by the land tenure, income levels and their knowledge of home building systems. The majority of the inhabitants of Kampung Cieunteng have occupied the land for 31-50 years (48%) make them difficult to leave long-established's social and economic networks. This condition could be seen also at the income level that the low income residents is major (90%) which their income under 2.5 million rupiah per month. It make them more difficult to build a life in a new location. The design choices are closely related to the resident motives of housing maintaining and their knowledge to improve the dwelling to be adaptive to floods. This data can be seen on Table 1.

Table 1. Percentage of respondent's year of occupancy, income, and willingness to maintain the house

Year of occupancy	Percentage
< 10 years	6%
10-30 years	29%
31-50 years	48%
> 50 years	17%
Income level	**Percentage**
< Rp. 2,500,000	90%
Rp. 2,500,000 - Rp. 5,000,000	4%
Uncertain	6%
Willingness to maintain the house	
Low	21%
Moderate	59%
High	16%
Very high	8%

(Source: Researcher analysis)

According to correspondences analysis, residents who categorized in informal job sector, the new resident (less than 10 years), the lowest income (like labourer, driver, or unemployment) and the lower education level tend to choose conventional design (the 1st type). The 1st type also reflect the common gradual/incremental housing. The 2nd type or "the complete unconventional type" was chosen by the respondent with more stable jobs, more length of stay (30-39 years), more stable income and more higher education level. The 3rd type or "the gradual unconventional type" was chosen by the entrepreneur job, the youth resident (about 20-29 years length of stay), the highest income, and relatively could be accepted by every education level of residents. This analysis can be seen on Table 2. It could be identified that the lower the income and the lower the education or knowledge about housing construction (reflected both on education level and length of stay in site), correspondence to the higher need of incremental housing design.

Table 2. Correspondence analysis

<table>
<tr><th>Correspondence analysis between job type and housing choice</th><th>Correspondence analysis between length of stay and housing choice</th><th>Correspondence analysis between income per month and housing choice</th><th>Correspondence analysis between education level and housing choice</th></tr>
<tr><td>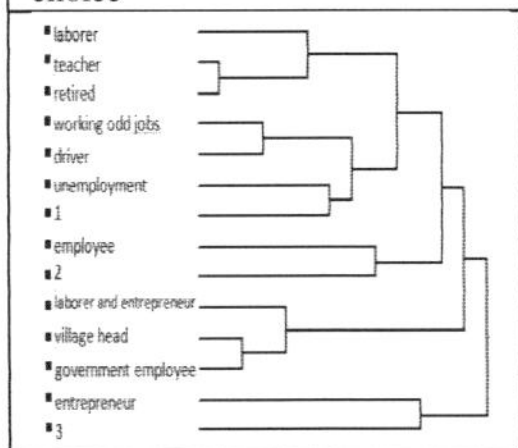
</td><td>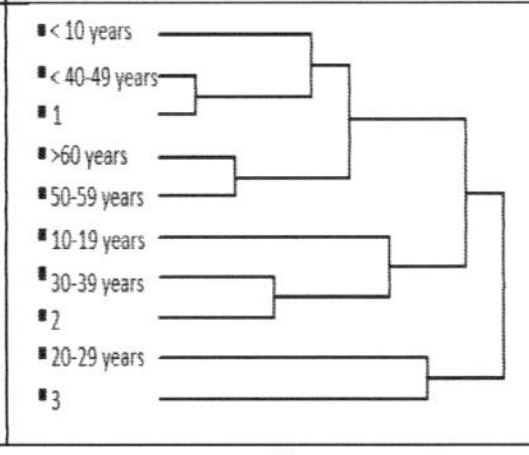
</td><td>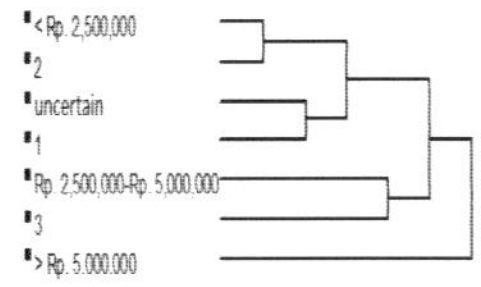
</td><td>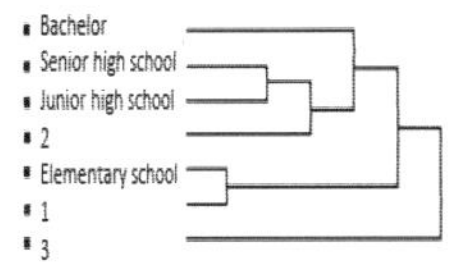
</td></tr>
</table>

(Source: Researcher analysis)

4. Conclusion

From this study, it is concluded about two important aspects: the community choice for resilient housing design and the affordability level of community for resilient housing design. Actually, the community have ability to choose the resilient housing design based on their aspiration and based on their ability to pay.

The community choices are determined by land ownership, income levels and knowledge of housing construction. Adaptive flood designs with amphibian systems (functioning both in normal and flood condition) are major selected when the residents evaluate the design as cost equivalent to their conventional construction. When the information of cost construction is issued, then the residents are divided into two options: the conventional two story house or the amphibious design without walls and free lay out in order to reduce the cost. This result indicates that there are actually open opportunities for various adaptive flood housing design schemes to be implemented involving partial funding participation from the communities themselves.

It is also concluded that the affordability level of community for resilient housing design is equal to their conventional cost construction of the existing house. To reduce the cost and to achieve affordability resilient housing for low income people, the design should consider some scheme like modularity of housing lay-out and land parcel, co-operational construction and maintenance, and also option for incremental/gradual house development.

References

[1] ARUP. 2014. City Resilience Framework. The Rockefeller Foundation, *City Resilience Index*.
[2] Folke, C. 2006. Resilience: The Emergence of a Perspective for Social-Ecological Systems Analyses. *Global Environmental Change.*
[3] Moench, Marcus. Tyler, Stephen. 2012. A Framework for Urban Climate Resilience. *Climate and Development.*
[4] Vale, Lawrence J., Shomon Shamsuddin, Annemarie Gray, Kassie Bertumen. 2014. What Affordable Housing Should Afford: Housing for Resilient Cities. *Cityscape: A Journal of Policy Development and Research*. Vol.12, Number 2. pp 21-49.
[5] Tresnadi, Hidir., 2000. Valuasi Komoditas Lingkungan Berdasarkan Contingent Valuation Method, *Jurnal Teknologi Lingkungan*, Vol.1, No. 1, pp.38-53.
[6] Edwin Chan, C. Y. 2008. Value of buildings with design features for healthy living: a contingent valuation approach, *Emerald*, pp. 229.

[7] Timmermans, H., Molin, E., & Noortwijk, L. V. 1994. Housing Choice Processes: Stated Versus Revealed Modelling Approaches. *Journal of Housing and the Built Environment*, Vol.9, pp. 215-227.

[8] Manalang, R. T., Munemoto, J., Yosida, T., & Espina, C. 2002. A Study on Residents' Self-built Improvements at MRB Dwelling Units in Metro Manila, JAABE, pp. 177.

[9] Christine, Heather. 2014. Amphibious Architecture Living with a Rising Bay. *Theses of The Faculty of California Polytechnic State University. San Luis Obispo*

Resident satisfaction on their residence and environment (case study of Srondol Bumi Indah Housing of Semarang City)

Paulus Hariyono

Pandanaran University, Semarang, Indonesia

Corresponding e-mail: phariyono@yahoo.com

Abstract. A piece of an architecture work will be meaningful if it meets the needs of the residents. Likewise, the design and natural environment of a residence will surely be meaningful if it is able to satisfy the residents. The degree of satisfaction can be referred to the theory of human need hierarchy proposed by Abraham Maslow. Although his theory is an old one, it is still a good one for a reference. Socio economic status (SES)also affects someone in understanding the comfort of his resident. This research has some purpose: 1) to know the satisfaction level of the residents, 2) to know the effects of socio economic status towards the residents, and 3) to know the natural environment aspect to resident satisfaction. The methode analysis used is qualitative analysis. The major finding are: 1)security factor is the main aspect of the human need residents; 2) upper and lowerclass residents have different knowledge and understanding regarding the natural environment satisfaction on the house they live.

Keywords: satisfaction, house, Maslow Theory and socio economic status, natural environment.

1. Introduction

A piece of an architecture work will be meaningful if it meets the needs of the residents. Likewise, the design and natural environment of a residence will surely be meaningful if it is able to satisfy the residents. The degree of satisfaction can be referred to the theory of human needs hierarchy proposed by Abraham Maslow. Although his theory is an old one, it is still a good one for a reference. Aside from that, the socioeconomic status affects someone in understanding the comfort of his resident.

1.1. Maslow theory

In his theory, Abraham Maslow [1] mentions that human needs continue to thrive. In 1954 Maslow suggested that there were five human needs that continue to thrive. However, in 1956 Maslow and his friends [2] added three other hierarchies of needs: the needs to understand, the needs for aesthetic and the needs for transcendence. Therefore, based on Maslows' theory [3], there are now eight hierarchies in human needs:

1) The basic needs of human being are physilogical needs that include the needs of shelter, food and clothing. Physiological needs should be met first.
2) Once a person's physiological needs are relatively satisfied, their safety needs will come after. Safety and security needs include the security of their residences and environment and job, safety against accidents and their adverse impacts. Therefore, the building where they live should be strong and safe from theft and interference.

3) After physiological and safety needs are fulfilled, the third level of human needs is the needs of affection. This need involves feelings of being attended, having friendships and intimacy. Therefore, a special room where every member of the family and their friends can express their affection is needed.
4) All human have a need to feel respected; this includes the need to have self-esteem and self-respect. On this level, the residents have already had proper social and economic status. Therefore they need to have their house provided with prestigious philosophy concepts such as styles from Ancient Greece, Rome and even the styles of local but complicated culture such as Joglo.
5) Cognitive Needs or the needs of knowledge and understanding. On this level, the residents are able to express their knowledge and understanding about particular culture with its symbols through their chosen styles. For example a house with an Ancient Greece style will not only have exotic Greece pillars, but also furnished with Ancient Greece antiques and statues.
6) Aesthetic needs. These needs are fulfilled by completing the residence with exterior and interior elements from a particular culture such as Ancient Greece or Ancient Javanese Art.
7) Self-actualization. On this level of need the residents will express their philosophy innovations through exterior and interior elements of a particular culture. For example is by combining Javanese philosophy and modern arts.
8) Self-transcendence. This need refers to an understanding of spirituality in which the residence will be decorated with religious nuances.

This reseach shows what level of satisfaction of residents in Srondol Bumi Indah Housing towards their house according to Maslow's theory above.

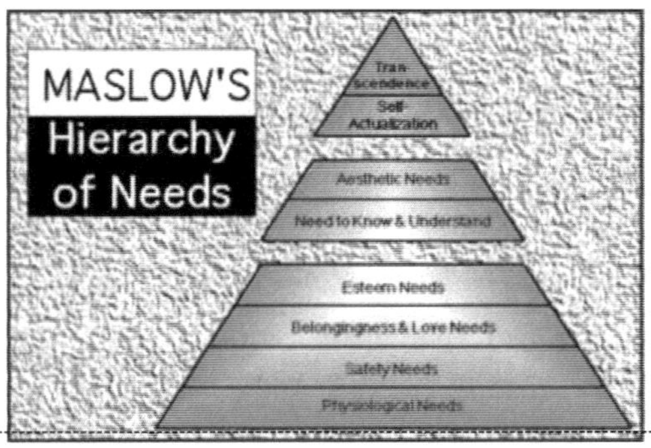

Figure 1. Maslow's Hirarchy of Human Needs [4]

1.2. Maslow theory

Everyone has different level of satisfactions. One of the most influential aspects which lead people to fulfill of demands is socio economic status. Those with low social status level tend to have more simple demands. According Mangkunegara [5] the low social economic status needs a thing that is cheap. But, according to Maslow's Pyramid of Need Hierarchy [6], the demand of those who have low social economic status might be very simple. They need shelter for them to live. They even ignore the needs of safety. They will take a simple house even though it is almost collapsed because what they need is only a place to shelter.

Those with middle socioeconomic status require safety needs to be fulfilled in their houses. The constructions should be strong enough to give them safety feeling. They also need special rooms for them and their friends to meet and chat.

People with higher social and economic status require meeting the needs of aesthetic and transcendent. Their residence will be furnished with beautiful exterior and interior decorations. Thus there is a tendency that the socio economic status affects the level of residents' satisfaction towards their houses.

1.3. Maslow theory

According to Chiras [7], Natural Environment is all factors which biologically affect organism. In order to give comfortable feelings for the residents, there are biological elements, called Healthy Residence, that need to be considered in designing a residence. Frick [8] says that the indicators of Healthy Residence include: air circulations, temperature, noise, air quality, light, security and humidity. Residents from different strata – upper and lower, will have different understanding on those indicators.

There is a tendency in which residents from upper strata will replace the comfort from natural environment with technology. For example, they replace the coolness of natural air with Air Conditioner (AC). However, residents from lower strata will keep the natural environment to give them comfort in their residence, such as air circulation and greening to enhance the quality of the air.

Referring to the above background, some problem can be formulated as: how much the resident's satisfaction towards the design of his residence?

The objectives of the research:

1) To know the level of satisfaction of residents towards their house according Maslows' theory.
2) To know the influence of socio economic status of the resident to the satisfaction of his house.
3) To know the influence of natural environment to the satisfaction of the residents.

2. Methods

2.1. Research description

A design and arrangement of residential space will give the occupants satisfaction if it is designed in certain ways. This study will look at the extent to which a residential design will give a certain satisfaction to its inhabitants. The notion of "satisfaction" according to Kamus Besar Bahasa Indonesia [9] is "fulfilled desire". Therefore, to be satisfied is when the needs and the desire are fulfilled. Abraham Maslow's theory about Hierarchy of Human Needs will be used as a benchmark to measure satisfaction [10], i.e., the first and basic need of human being is a physiological need (the needs of shelter, clothing, and food). When their standard of living increases they will try their best to fulfill the needs of safety and their needs of affection (love and social belongings), the needs of self-esteem, the needs of cognitive, aesthetic, self-actualization and self-transcendence as well. This study shows to what extent the occupants' satisfaction towards their house can be fulfilled based on Maslow's theory, the effects of socioeconomic status of the occupants, and natural environment towards satisfaction of their house.

2.2. Research sites

The level of socio economic status of the residents affects the level of satisfaction of the residents towards their house, therefore, the location of the research is chosen in the area which consists of different levels of residents from different socioeconomic status. Srondol Bumi Indah Estate fulfills the criteria. There are several types of houses located in Srondol Bumi Indah Estate such as: 1) small house (type 90/120); reflecting that the occupants are of low socioeconomic status; 2) medium house (type 150/250) reflecting that the occupants are of moderate socioeconomic status; and 3) large house (type 300/450) reflecting that the occupants are of high socioeconomic status.

2.3. Research subjects

The characteristics of human satisfaction level is determined by the condition of one's socioeconomic status indicated by the type of house they own, in order to obtain the characteristic differences. Different types of housing samples are required.

The group of housing type can be described with the three simplified Likert'scale: not large, sufficient, and large. A group of residences type 90/120 is taken as a type of not large one. A group of residences that is more spacious is taken as a type of house type 150/250. While a group of residences with larger area is taken as a house with type 300/450. These house type groups can be found in Srondol Bumi

Indah Semarang. The stratified purposive random sampling is used as the sampling technique which according to Sutrisno Hadi [11] it is the sampling which is based on a certain strata to achieve a particular goal.

In Srondol Bumi Indah Housing there are about 400 residential units with compositions as follow:

1) House type 90/120 (small type): about 100 units, it is assumed that the occupants earn less.
2) House type 150/250 (medium type): about 200 units, it is assumed that the occupants are from medium income class.
3) House type 300/450 (large type): about 100 units, it is assumed that the occupants are of high-income class.

Based on the explanation above, it can be concluded that the ration between the occupants of the 90/150 house type: 150/250: 300/450 is 1: 2: 1. The number of samples is taken proportionally according to the composition of the number of residents and types of house. Below are tehe propotion number of respondents.

1) Owner of small type of house	: 3 respondents
2) Owner of Medium type of house	: 6 respondents
3) Owner of Large type of house	: 3 respondents
Total	: 12 respondents

2.4. Data retrieval
Data were collected using questionnaires, interviews, and observations.

2.5. Analysis methods
This research uses analytical descriptive analysis method from questionnaire, interview, and observation of space and building condition.

3. Discussion

3.1. General description
The description includes ages and incomes of the residents who become the research respondents.

*3.1.1. Age.*There were three groups of respondents. The first group consisted of retired men who were over 61 years old (67%). All of them were the head of the family. The second group consisted of middle aged respondents, 51-60 years old (25%). The third group was a group of 41-50 years old respondents (8%). They were private employees and entrepreneurs.

3.1.2. Income. Incomes are sorted into low, medium, and high-income. Respondents that are classified as low incomes earners are those who earn less than Rp. 10,000,000, -. While respondents with moderate income are respondents who have income Rp. 10,000,000, - up to Rp. 20,000,000, -. Respondents with high income are respondents who earn more than Rp. 20,000,000, -. The majority of respondents' income is low (below Rp 10,000,000, -), 58.3%, because most of them have been retired from jobs, and over 61 years old. The moderate income - 25%- (Rp. 10,000,000, - to Rp. 20,000,000, -) comes after and followed by high-income respondents (more than Rp. 20,000,000) - 16.7%..

3.1.3. Income. Incomes are sorted into low, medium, and high-income. Respondents that are classified as low incomes earners are those who earn less than Rp. 10,000,000, -. While respondents with moderate income are respondents who have income Rp. 10,000,000, - up to Rp. 20,000,000, -. Respondents with high income are respondents who earn more than Rp. 20,000,000, -. The majority of respondents' income is low (below Rp 10,000,000, -), 58.3%, because most of them have been retired from jobs, and over 61 years old. The moderate income - 25%- (Rp. 10,000,000, - to Rp. 20,000,000, -) comes after and followed by high-income respondents (more than Rp. 20,000,000) - 16.7%.

3.2. Satisfaction of residents
The residents' satisfaction in Srondol Bumi Indah Residence can be measured by using Human Need Pyramid theory proposed by Abraham Maslow (Globe 1985). In his theory, Maslow mentions eight levels of human needs, namely: 1) Physiological Needs; 2) Safety and Security Needs; 3) The Needs of Affection; 4) The Needs to have Self-Esteem and Self-Respect; 5) the Needs of Knowledge and Understanding; 6) Self-Aesthetic Needs; 7) the Needs of Self Actualization; and 8) the Needs for Transcendence. These needs appear in hierarchies. First of all, people need to meet their first basic needs (physiological needs). Once a person's physiological needs are relatively satisfied, their safety needs take precedence and dominate behaviour. After physiological and safety needs are fulfilled, the next level to be fulfilled is the needs of affection followed by the fulfilments of needs for understanding and knowledge, the needs for aesthetics, the needs for self-actualization, and last but not least the needs for transcendence. Maslow's theory is called "Hierarchy of Needs Pyramid".

3.2.1. Physiological Needs. House conditions can be classified into three. The first one is a house with maximum condition (50%) in which the house is the best place for people to protect themselves from heat, cold, and rain. The temperature inside the house is good. The second is a house with air ventilation but the residents still need fans to make them feel more comfortable (41.7%). The last one is a minimum type of house (8.3%). This size of this type of house is very narrow. Referring to physiological needs, it can be concluded that the number of houses with maximum function as a place to shelter reaches 50%, while those that are considered sufficient reaches 41.7%. The fulfilment of residential house as a shelter cannot reach 100% because some of the respondents still feel that their house is less spacious and the temperature inside the house is still hot.

3.2.2. Safety and Security Needs. Safety and security needs include the security from theft and the safety of the house building. The security from theft has been 100% fulfilled because since the last 10 years there has been improvement in the security system. Previously there were only two security guard posts available. That condition was far from expectation. Since then another security guard post and some portals on the streets were built. This additional portals and security guards narrow the thieves' movement, and therefore the needs of safety and security of the residential area can be obtained. In terms of building design, the residential buildings are usually equipped with trellises to prevent thieves from entering. The safety of the building had not been able to meet the needs of safety since some of houses' walls were not sturdy enough; there were leakage and cracks on the roof because of termites. Only 50% is perceived to be safe after the standardized frame which had been eaten by termites was repaired. 33.3% considered the house condition is enough, while the rest, 16, 7% said that the house condition was not safe due to leakage and termites. Based on the analysis, it was found out that the common concerns about the needs of safety and security are about the security from theft and the safety of the building condition. 100% respondents said that the needs ofsecurity from theft are absolute whereas the needs of the safety of the building take 50%. The average of this fulfilment is (100% + 50%):2 = 75%

3.2.3. The needs of affection. 50% of houses fulfilled the needs of spaces where the residents can express their feelings with other members of the family, while the other 50% said that their houses were not spacious enough for them to have a gathering with other members of the family, in other words, the living room was less spacious. 33.3% of houses had spacious room for family gathering. 16.7% of houses did not have enough places for that activity. This fulfilment is very relative. There are residents who do not mind with their small family rooms because they seldom have time to meet their family due to their busy schedules. Parents go to work, children go to schools. There are also residents who are satisfied with their less spacious family rooms because for them as long as they can sit and have a chat together with other members of the family they are happy.

3.2.4. The needs of self-esteem. The needs of self-esteem are indicated with the sense of pride on the house they live in. The majority of residences (50%) give pride to the owners for some reasons; many people like the housing complex, the house was bought by their own, and that the designs of the houses are always up to date and everlasting. 41.67% of houses had given enough pride due to their up-to-date designs and their popularity among society. 8.33% of houses were not able to give pride due to the small size of the land and the building

3.2.5. Cognitive needs. The design of a residence reflects particular knowledge and understanding such as the knowledge of air circulation, lighting, and coolness. This kind of understanding found in Srondol Bumi Indah Housing is as follows: The knowledge given to the society by the majority of residences (50%) is about good air circulation, a house that is ecologically healthy and a house that does not need an Air Conditioner. 33,3% of houses are considered standard and have enough knowledge while 16.67% of other house building are not considered having enough knowledge.

3.2.6. A house as an aesthetic reflection. The majority of houses (50%) have been reflecting aesthetic sense of the residents. This is because the residents cannot design their own house freely. Additional designs should be coordinated with the environment with the same design. Just since few years ago, the residents were given freedom to design their own house.33.3% of the findings show that the house aesthetic is good due to the up to date designs while 16.7% of the houses have very minimum aesthetic values because the designs are made by the residents.

*3.2.7. Self actualization.*The majority of residents (50%) are able to actualize themselves because they feel comfortable, with good air circulation and fresh air. 33.33% residents also stated the same reasons. There are 16.67% of houses do not need to have self-actualization of the owners.

3.2.8. The need to transcedence. The majority of houses (50%) do not need to express the transcendence value of the owner as they think that they do not need to express their self-transcendence. Houses that are able to express their transcendent values are 25%. Those houses usually provide specialplaces for praying. The other 25% of houses are considered able to express their owners' transcendence becauseSrondol Bumi Indah has become a favourite residence. If the benchmark of respondents' opinion is more than 50%, it can be concluded that the needs of safety and security have been more than enough fulfilled (75%). The other needs such as physiological needs (a need of shelter), affections, self-esteem, self-actualization and transcendent are also sufficiently fulfilled (50%).The reasons why those levels of satisfactions seem low (not more than just the needs of the safety of the house buildingand the need of security from theifits) maybe because: first, the residents have been staying in Srondol Bumi Indah for a long time. Approximately, they have been there for more than 15 years. They have been bored with the housing conditions.Second, the financial condition of the house owner is static, they will not be able to renovate their house in terms of knowledge, aesthetics, pride, self-actualization and self-transcendence.Financial condition will affect their satisfaction.

3.3. The Impact of socioeconomicstatus and natural environtment to the resident's satisfaction

According to Mangkunegara, one's socio economic status affects them in choosing goods and services. Those who come from high socio economic status tend to choose branded products and long lasting ones. Therefore for the sake of their comfort, they prefer to have instant and constant facilities, such as the use of Air Conditioner (AC) for their stable and comfortable temperature and air circulation at home.

While those who are from lower socio economic status tend to choose cheaper goods to fulfill their daily needs. In order to safe their expenses. They manage the circulation of natural air in their homes in such a way so that they do not need Air conditioner (AC). They even grow plants in their garden to enhance the coolness of their homes.

Figure 2. Living and family rooms are set in the same room in a small type of residence so that the room will be more spacious and the air will be circulated more freely.

Figure 3. Family room is designed more spaciously than a living room in a small type of residence. The more spacious room enables the air circulates better.

Figure 4. Plants are parts of natural environment in a small type of residence. Natural fresh air from plants has become important element to make the house more comfortable to live in.

Figure 5. Hanging plants in a small type of residence. Natural fresh air from plants has become important element to make the house more comfortable to live in.

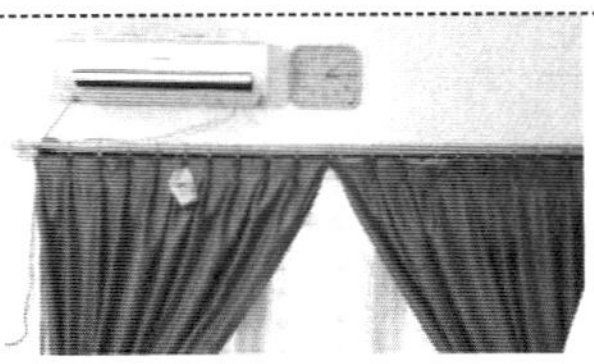

Figure 6. Technology, such as Air Conditioner (AC) is used in houses owned by upper class society to gain constant comfort and nice temperature.

There are differences in understanding about the comfort of residence in regards with socio economic status and natural environment, as follow:

1) As a shelter (physiological needs) for middle class society and lower society: they tend to be less satisfied because of size which is not spacious enough, and that the air is getting warmer.
2) In terms of design, a house is able to give more than enough particular knowledge (50%) for example knowledge about air circulation, room arrangements, and garden.
3) The resident of a small type of house tends to take air circulation and greening as considerations because it contains knowledge of an ideal house.

4) The resident of a big type tends use technology for fulfill their satisfaction.

4. Closing

With a benchmark of 50%, Srondol Bumi Indah Residence has been able to fulfil the needs of safety and security for its residents. The fulfilment is considered enough. But, the satisfaction of design and aestetic tend can not be fulfilled as they need, because changing the design and aestetic of the residence need high financing.

The small type of residence owned by lower strata group of society, fulfill the needs of shelter to the residents (physiological needs) due to the good air circulation and its greening program which creates cool and fresh environment.

Big houses, owned by higher strata group of society, do not always give such good environment as small houses do, but they use technology. Socio cultural factors and socio economic status of the residents, education background and knowledge of the residents affect how they fulfill the satisfaction of their residence.

The highest level of fulfilment given to the residents in Srondol Bumi Indah is the fulfilment of physiological needs, safety and security, pride and knowledge sharing, whereas, the needs of affection, aesthetics and self-actualization are sufficiently fulfilled. And the transendent need is fulfilled in low level. It is suggested that socio cultural aspects and socio economic status of the residents towards their residences need to be done in the next research.

Acknowledgments

Author would like to thank the respondents at Srondol Bumi Indah Housing who have made their time to provide explanation and data collection for the purposes of this research.

References

[1] Goble. 1985:17 Mazhab Ketiga Psikologi Humanistik. Yogyakarta: Kanisius.
[2] Goble. 1985:17 *Mazhab Ketiga Psikologi Humanistik*. Yogyakarta: Kanisius.
[3] Goble. 1985:17 *Mazhab Ketiga Psikologi Humanistik*. Yogyakarta: Kanisius.
[4] McLeod, S. A.2016. "Maslow's Hierarchy of Human Needs". Retrieved from www.simplypsychology.org/maslow.html
[5] Hariyono, Paulus. 2007. *Sosiologi Kota Untuk Arsitek*. Jakarta: Bumi Aksara. McLeod, S. A.2016. "Maslow's Hierarchy of Human Needs". Retrieved from www.simplypsychology.org/maslow.html
[6] McLeod, S. A.2016. "Maslow's Hierarchy of Human Needs". Retrieved from www.simplypsychology.org/maslow.html
[7] Hamzah, Syukri. 2013. Pendidikan Lingkungan. Bandung: Refika Aditama.
[8] Frick, Heinz & Mulyani, Tri Hesti. 2006. Arsitektur Ekologis. Yogyakarta: Kanisius.
[9] Kridaklaksana, Hari Murti, dkk. 2010:793. *Kamus Besar Bahasa Indonesia*. Jakarta:Departemen Pendidikan dan Balai Pustaka.
[10] McLeod, S. A.2016. "Maslow's Hierarchy of Human Needs". Retrieved from www.simplypsychology.org/maslow.html
[11] Hadi, Sutrisno. 1980:17. *Metode Penelitian Research*. Yogyakarta: F. Psikologi.

IOP Conf. Series: Earth and Environmental Science **99** (2017) 012015 doi:10.1088/1755-1315/99/1/012015

Exploring collective memory and place attachment using social media data

Sushardjanti Felasari[1*], Herybert Setyabudi[2], Djoko Budiyanto Setyohadi[3], Sinta Dewi[4]
[1,2,4] Department of Architecture, Universitas Atma Jaya Yogyakarta,Yogyakarta, Indonesia
[3]Department of Informatics, Universitas Atma Jaya Yogyakarta, Yogyakarta, Indonesia

E-mail: sfelasari@staff.uajy.ac.id

Abstract. This paper describes how collective memory and level of place attachment can be explored using social media data to develop a sustainable travel destination in the city of Yogyakarta. Yogyakarta is famous as a tourist destination city for tourism in Indonesia. One of the reasons why people visit an object at travel destination is to recall the memory of the place. Memory is important for creating memorable space and places as it differentiates one place to another. Memorable places could grow as a symbol and an identity of a district in the city. This paper explores the collective memories recorded as status in the social media. The study identifies the distribution of such locations or nodes representing the memory footprint of the city of Yogyakarta, which can be achieved by determining the quality level of nodes based on the level of place attachment. Analysis is done by looking at the number of caption status by location and time. Qualitative description is used to present the level of place attachment based on the content of the status captioned. The study shows that level of place attachment seems not to be influenced by the popularity of an object. However it affects how strong a travel destination might be sustainable in future development.

Keywords: memory, place attachment, nodes, social media, travel destination

1. Introduction

Yogyakarta is famous as one of the travel destinations for tourism in Indonesia. During the year 2015-2016 there was an increase in the number of visitors both domestic (16%) and international (21%) compared to 2014-2015. The support from central and local governments in developing tourism in Yogyakarta seems inline with the planned development of New Yogyakarta International Airport (NYIA) located in Kulon Progo, DIY and Trans Java toll road connecting NYIA to Borobudur Temple. There are many reasons why people visit a tourist destination. Besides good infrastructure, they might come to recall memory of particular places and their attractions.

Memory is the mental capacity of humans to maintain and revive facts, events and etc by remembering or recognizing previous experiences [1]. Memory perceived by tourists can be from their previous memory or a memory that has been experienced by other people. Memories attached to a tourist spot become a collection of meaningful memories that are collectively remembered by a group of people who share and engage in forming memory or it is called collective memory [2].

Sharing and engaging to form memory of particular spaces and places are greatly supported in modern times like today, where everyone is connected each other with sophisticated mobile devices and supported by a good internet network. Not surprisingly, the tourist attractions and new interesting places are rapidly shared by the existence of social media used by people who are visiting the location. Using mobile devices, people can easily upload images or status on social media telling about the condition of the sights. These practices encourage the formation of places as indirectly people have made memories about the sights with the uploaded images and status on social media.

The context in which people interacting synchronously either directly or indirectly through artefacts, devices, and storytelling, will have implications for the identity, memory, history, and meaning of the place [3]. Unfortunately, the lack of understanding of the mobility and the activities of tourists and the lack of information related to the urban context sometimes resulted in the creation of places and facilities that are less meaningful and unmemorable.

Therefore this paper addresses the issue of how collective memory and level of place attachment can be explored using social media data to develop sustainable tourism destinations in Yogyakarta.

It is expected that the tourist objects or nodes presenting the memory footprint of Yogyakarta city can be mapped.

2. Collective Memory and Place Attachment

The development of tourism is believed could affect the urban space, as the tourist attractions will form such node and a gathering point where people meet. The space would be a place if it has a meaning to the environment which is derived from its local culture [4].

The gathering people in a particular place like a tourist spot unconsciously contributes to the formation of a memory. Collective or individual memory forms identity, especially memory of events that have an important meaning to a community. Such meaningful events will thicken the boundaries of one's feelings about himself in relations to others and public society [5]. These memories can affect the creation of a space because memory distinguishes one place from another [6]. Memory is also born from the movement of people that occur in the urban space. Past memory attached to buildings and places in the urban space can be explored and can be associated with the needs of today's memory. As places have a past, the place also continues to grow in the future. Therefore memory is very important in a tourist location as it helps one to better understand the urban context [7].

Memory collected collectively at a tourist location can be described as an emotional bond between the individual and the location. The emotional connection between individual to a location/node forms a place attachment. Place attachment is a symbol of relationship with the place where it is generated by giving emotional and common sense meaning to a specific place or boundary and explaining how people perceive and relate to them [8]. Place attachment also refers to the emotional or affective bond between an individual and a particular place as well as the functional bonds that occur repeatedly in interactions between people and places [9].

There are 4 levels of place attachment according to Rubinstein & Parmelee [10]. The intensity of "Place Attachment" is described as (1) one only knows about a place and thinks about it without experiencing strong personal feelings/memories, (2) "personalized attachment", if one has a memory of a place from his personal experience, (3) "extention," when a place provides an emotional memory or psychologically involves the individual in various ways, and (4) "embodiment" / embodiment, when the boundary between the self and the environment becomes blur or even for some individuals their personal identity and place identity melted into one.

3. Methods

Using mixed methods, this paper explores collective memory recorded as status related to specific objects mentioned in the social media i.e. Twitter. In this study, Twitter data related to Yogyakarta tourist location shared from users' mobile devices were captured based on factors such as location and time. Using a filter called 'stop-word', data were separated from the meaningless words. Data is also classified based on the characteristics of tourism location (i.e. cultural and historical tourism, culinary tour, shopping tour, nature tourism, or spiritual tourism).

The level of popularity of a tourist location can be obtained by analyzing status based on the location and time of such objects. The study also look at the content of status to measure the level of place attachment. Word categories to filter the contents of the status have been predefined. Analysis is translated into intensity level so that the relationship between location and its collective memory can be seen.

4. Discussion

The study uses data sampling retrieved from November - December 2016 due to a consideration that the month includes holiday season, where almost all tourist sites are filled with domestic and international tourists. In order to see the space usage by the tourist and the popularity of a tourist location/object, data were divided into 4 groups of visitors based on time i.e. morning (06:00-10:00), day-time (10:01-14:00), evening 14:01-18:00), and night-time (18:01-24:00).

4.1. Cultural and Historical Tourism

There were 20 locations recorded as cultural and historical attractions in the city of Yogyakarta of which there are 13 tourist sites with number of visitors are over 100 people.

Table 1. Total number of visitors and level of place attachment for cultural and historical tourism

No	Location	Number of Visitors					Level of Place Attachment					
		Time					Level					
		Morning	Day time	Evening	Night time	Total	1	2	3	4	Total	Mean
1	Bank Indonesia Yogyakarta	0	0	0	15	15	11	4	6	0	21	1,40
2	Panggung Krapyak	0	0	0	15	15	12	0	9	0	21	1,40
3	Museum Sonobudoyo	36	4	5	9	44	87	0	18	0	105	1,13
4	Mandala Krida	39	1	32	43	115	106	8	15	0	129	1,12
5	Museum De Mata	131	25	35	65	256	237	18	30	0	285	1,11
6	Gembira Loka	75	66	33	33	207	202	0	15	0	217	1,05
7	Keraton Yogyakarta	142	471	45	302	960	928	36	42	0	1006	1,05
8	Tugu Yogyakarta	1	51	24	4	80	77	4	3	0	84	1,05
9	BNI 46 Nol Kilometer	134	40	297	70	541	532	8	15	0	555	1,03
10	Taman Budaya Yogyakarta	40	354	21	163	578	568	2	27	0	597	1,03
11	Balai Kota Yogyakarta	188	107	139	113	547	540	6	12	0	558	1,02
12	Bentara Budaya	78	32	69	137	316	310	10	3	0	323	1,02
13	Kantor Pos	10	7	7	39	63	62	2	0	0	64	1,02
14	Malioboro	27	45	1	413	486	477	12	9	0	498	1,02
15	Kotabaru	54	55	8	96	213	211	2	3	0	216	1,01
16	Stasiun Tugu	22	4	25	29	80	79	2	0	0	81	1,01
17	Taman Sari	35	11	94	36	176	175	0	3	0	178	1,01
18	Gedung Agung	66	68	19	44	197	197	0	0	0	197	1,00
19	Masjid Agung	41	26	29	24	120	120	0	0	0	120	1,00
20	Pasar Beringharjo	15	44	27	9	95	95	0	0	0	95	1,00

It can be seen from table 1 that Panggung Krapyak and Bank Indonesia Yogyakarta have the same level of place attachment. Nevertheless, Panggung Krapyak has more value at level 3 than Bank Indonesia Yogyakarta. It means the visitors has a deeper emotional bond to the place. Panggung Krapyak is part of a series of imaginary axis of Yogyakarta, so it is still interpreted to have a great influence in the life of the people of Yogyakarta. The location of Panggung Krapyak itself is a little bit far from the Yogyakarta Palace. However it is not influenced much by culture outside of Yogyakarta. The environment is still thick with past conditions that bring tourists back to ancient times. Tourists are invited to understand, live, and perceive the past events that occurred at that location.

The second rank of place attachment level is at Bank Indonesia Yogyakarta. The quality of the location is not only influenced by the bank's own image, but also by the character of surrounding environment such as the Monument Serangan Umum, Gedung Agung, Bank BNI and Post Office.

Urban space on the area has helped visitors to remember the memory related to past events that happened in that location.

On the other side, Yogyakarta Palace can attract visitors as nearly as 1,000 people. Yogyakarta Palace becomes a must-visit location for every tourist who travels to Yogyakarta city. The most visited time is in the afternoon, because there are certain operational times imposed by the Yogyakarta Palace.

4.2. Shopping Tour

Table 2 shows 10 popular locations for shopping in Yogyakarta. There are 4 tourist sites with more than100 visitors when data was taken.

Table 2. Total number of visitors and level of place attachment for shopping tour

No	Location	Number of Visitors					Level of Place Attachment					
		Time				Total	Level				Total	Mean
		Morning	Day time	Evening	Night time		1	2	3	4		
1	Gramedia Sudirman	4	6	6	2	18	16	4	0	0	20	1,11
2	Galeria Mall	12	22	24	7	65	62	0	9	0	71	1,09
3	Gardena	58	88	90	0	236	231	10	0	0	241	1,02
4	Malioboro	2339	1509	1872	1348	7068	6987	80	123	0	7190	1,02
5	Lippo Plaza	262	395	617	312	1586	1560	38	21	0	1619	1,02
6	Mirota Batik	13	19	25	9	66	66	0	0	0	66	1,00
7	Pasar Beringharjo	11	36	52	10	109	109	0	0	0	109	1,00
8	Pasar Kotagede	3	14	31	12	60	60	0	0	0	60	1,00
9	PASTY	16	16	19	9	60	60	0	0	0	60	1,00
10	Pasar Ngasem	25	2	20	6	53	53	0	0	0	53	1,00

Malioboro is an old shopping district. It became the most popular shopping location since established and has the highest number of visitors. Morning and evening seems to be the favourite time for tourist to visit Malioboro. Visits can be increased by improving the environmental conditions, such as planting more shade trees, planting trees as a barrier between pedestrian ways and vehicles road, restrictions on motor vehicles entering the Malioboro area, etc. Lippo Plaza comes out to be the next most popular location. This building represents today's shopping place with a modern concept favoured by young people. Visits in the afternoon are high, because the building is facilitated by air conditioners. Utilization of its space for public use with adequate facilities, holding big events monthly and annually will increase the number of visitors.

On the contrary, Gramedia Sudirman, a book store with a complete collection of books and building conditions that are comfortable to visit, has the highest level of place attachment than Malioboro despite the less visitors. Visitors are pampered with the atmosphere of a quiet shop, complete book collection, wide corridors so that visitors feel at home. Galeria Mall shows a high level of place attachment as well. In this place, visitors are pampered with the atmosphere of a luxurious store, and rental tenants. On the contrary, Malioboro is less organized, crowded, and overheat sunshine during the day might be the cause of why tourists has less memorable of Malioboro. If these conditions are not addressed in sustainable urban planning, gradually the meaning of Malioboro as a legendary tourist location for shopping will fade away, lost by the time and replaced with a new one.

4.3. Nature Tourism

The development of Yogyakarta city into densely populated residential areas makes there are only few nature tourism locations found and mentioned on social media. The location of nature tourism recorded as many as 5 locations as shown in Table 3.

Table 3. Total number of visitors and level of place attachment for nature tourism

No	Location	Number of Visitors					Level of Place Attachment					
		Time				Total	Level				Total	Mean
		Morning	Day time	Evening	Night time		1	2	3	4		
1	Gembira Loka (The Zoo)	69	41	26	19	155	150	4	9	0	163	1,05
2	Sungai Code (Code River)	21	23	49	55	148	142	8	6	0	156	1,05
3	Alun-alun Kidul (South Square)	96	43	58	318	515	511	4	6	0	521	1,01
4	Mangunan Imogiri (Orchard)	9	2	6	10	27	27	0	0	0	27	1,00
5	Sungai Elo (Elo River)	0	4	10	0	14	14	0	0	0	14	1,00

Alun-alun Kidul (South Square) seems to be the most popular tourist location compared to the four others. It is an open public space at the southern part of Jogja Palace where citizens can use it freely for relaxing, doing some sports like running, jogging, and any other outdoor activities. Morning and night times are the most favourite time when the atmosphere is still fresh and the place is shaded from the overheat sunshine. By planting more trees and providing more parking lots in the area may increase the number of tourist visits during the day and afternoon.

In terms of level of place attachment, Alun-alun Kidul has less value than Gembira Loka Zoo and Code River. Gembira Loka has existed for a long time as a place for recreation while Code River flows among housing settlements in the city of Yogyakarta. For some people, both places have more meaning and that can be developed to keep their sustainability.

4.4. Culinary Tourism

There are about 10 tourist locations for culinary tourism of which 6 of them have visitor more than 100 people.

Table 4. Total number of visitors and level of place attachment for culinary tourism

No	Location	Number of Visitors					Level of Place Attachment					
		Time				Total	Level				Total	Mean
		Morning	Day time	Evening	Night time		1	2	3	4		
1	Gudeg Bu Tjitro	0	0	12	4	16	14	4	0	0	18	1,13
2	Greenhost Hotel	11	8	3	15	37	35	4	0	0	39	1,05
3	Bakmi Kadin	12	13	16	55	96	94	0	6	0	100	1,04
4	House of Raminten	18	14	43	40	115	112	2	6	0	120	1,04
5	Galeria Mall	12	20	27	21	80	79	0	3	0	82	1,03
6	Kopi Joss	19	5	16	253	293	285	16	0	0	301	1,03
7	Tempo Gelato	14	46	33	65	158	156	2	3	0	161	1,02
8	Gudeg Yu Djum	96	91	65	159	411	407	8	0	0	415	1,01
9	Agenda Resto	95	55	61	90	301	298	6	0	0	304	1,01
10	Couple Couple	26	28	28	32	114	114	0	0	0	114	1,00

Gudeg, a traditional food made from young jackfruit, remains as a favourite culinary among tourists. Gudeg Bu Tjitro has the highest level of place attachment among many other culinaries of Yogyakarta city, since it has been around since 1925. The taste of its food as well as the atmosphere of the dining places offered to the visitors brings culinary lovers back to the past. Regardless of the taste presented, the atmosphere of the place formed from the building can affect the visitors' emotions in enjoying the food.

Meanwhile another gudeg shop i.e. Gudeg Yu Djum has the largest number of visitors which popular to be visited at night. However as this place is relatively new, the level of place attachment seems not as high as the Gudeg Bu Tjitro.

4.5. Spiritual Tourism

Table 5 shows that there are 5 locations mentioned as a spiritual tourism spot in Yogyakarta of which 1 location has visitors more than 100 people.

Table 5. Total number of visitors and level of place attachment for spiritual tourism

No	Location	Number of Visitors					Level of Place Attachment					
		Time				Total	Level				Total	Mean
		Morning	Day time	Evening	Night time		1	2	3	4		
1	Klenteng Poncowinatan (Shrine)	0	21	0	0	21	14	14	0	0	28	1,33
2	Gereja Katholik Baciro (Church)	23	16	6	11	56	54	2	3	0	59	1,05
3	Masjid Agung (Mosque)	28	17	19	45	109	109	0	0	0	109	1,00
4	Masjid Sumur Gumuling (Mosque)	2	0	0	0	2	2	0	0	0	2	1,00
5	Gereja Katholik Kotabaru (Church)	23	8	6	34	71	71	0	0	0	71	1,00

Poncowinatan shrine has the highest attachment level among the other five locations. The shrine is located behind Kranggan Market - another chinese district besides Malioboro area. People around the shrine still perform ritual prayers with various offerings such as fruits and cakes as a form of their gratitude. Tourists visiting the shrine can still see this ritual and might be able to interpret what they see.

The Great Mosque is part of the Yogyakarta Palace and it has the largest number of visitors. It is popular to be visited at the evening and night which might be due to the overheat conditions at the site surrounding the mosque during the day and in the afternoon. Adding more trees and replacing the pavement cover of the mosque's yard with grass is expected to decrease the heat and to invite more visitors to come.

Figure 1 shows the popularity of all tourist attraction in the city of Yogyakarta and their level of place attachment.

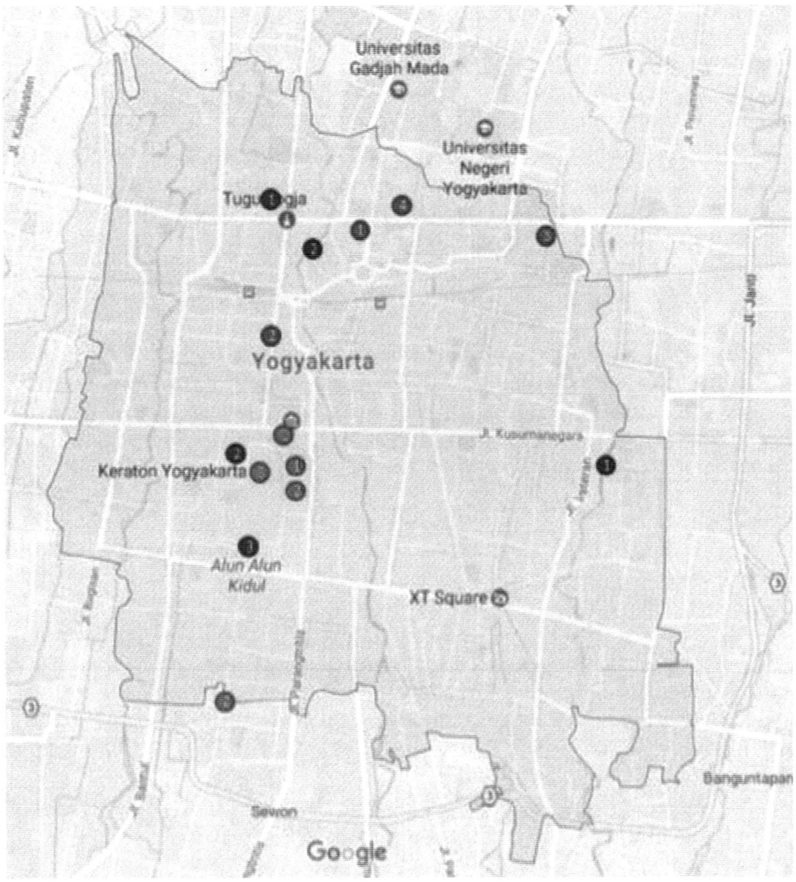

a. Cultural and heritage tourism
   1. Bank Indonesia Yogyakarta (1)
   2. Panggung Krapyak (2)
   3. Keraton Yogyakarta (3)
b. Shopping tour
   1. Gramedia Sudirman (1)
   2. Malioboro (2)
   3. Lippo Plaza (3)
   4. Galeria Mall (4)
c. Nature tourism
   1. Gembira Loka Zoo (1)
   2. Code River (2)
   3. South Square (3)
d. Culinary tourism
   1. Gudeg Bu Tjitro (1)
   2. Gudeg Yu Djum (2)
e. Spiritual tourism
   1. Klenteng Poncowinatan (1)
   2. Masjid Agung (2)

Figure 1. Map of Tourist Attractions and Their Level of Place Attachment

5. Conclusion

From the analysis of the number of visitors and the level of place attachment presented above, it seems that the popularity of tourist location is not necessarily correlated with the level of place attachment. This condition can occur as the more tourists visit a location, the more difficult the tourists to observe, interpret, and reflect the events that happened at that location. The less tourists to visit such objects, the more flexible they are to explore the objects, feel the atmosphere, and give meaning to the events existed in the location.

From the data presented above, it is found that almost all tourist sites are only in the first level of place attachment. Although some locations have above level one yet have not reached level two. This indicates that certain locations have been memorized and tied emotionally and psychologically by the visitors although they are still in the level of "personalized attachment".

A city can succeed in its social sustainability, if historical values and the memory, which are lived in the city are preserved and maintained by its citizens. In this case, Malioboro has a slightly higher level of place attachment than other locations for cultural and historical tourism categories and shopping tours, because people still respect and live on historical values which performed in their every day life and transmitting to the next generations.

This study shows that the degree of place attachment influences the strength of the purpose of the journey, which may be sustainable in the future development. Thus, the locations of travel destination will not lose their original identity, which will make them unmemorable for the next generation.

Tourist sites with low attachment level can be improved by architectural intervention, which is suitable to the character of the location. Location with large number of visitors is potential to increase their level of place attachment. By increasing the level of place attachment, the level of emotional bonded of the visitors to a particular location will also increased. This will support the sustainability of the tourist attractions in the future.

Acknowledgments
This paper is supported by Ministry of Research, Technology and Higher Education under the applied product decentralization research scheme.

References

[1]. Ardakani MK, Oloonabadi SSA. *Collective memory as an efficient agent in sustainable urban conservation*. Procedia Engineering. 2011;21:985–8.
[2]. Misztal B. Theories of Social Remembering. Open University Press; 2003.
[3]. Coyne R. The Tuning of Place: Sociable Spaces and Pervasive Digital Media. 1 edition. Cambridge, Mass: MIT Press; 2010. 344 p.
[4]. Ramstedt M, Thufail F. Kegalauan Identitas: Agama, Etnisitas dan Kewarganegaraan pada Masa Pasca-Orde Baru. Jakarta: Pusat Penelitian Sumber Daya Regional LIPI dan Max Planck Institute for Social Anthropology dan PT Grasindo; 2011.
[5]. Borden I, Kerr J, Rendell J, Pivaro A. The Unknown City: Contesting Architecture and Social Space. Reprint edition. The MIT Press; 2002. 562 p.
[6]. Felasari S, Peng C. The Role of a City's Collective Memory in Supporting Cooperative Urban Design Learning. In: Luo Y, editor. Cooperative Design, Visualization, and Engineering, Springer Berlin Heidelberg; 2012. p. 143–50. (Lecture Notes in Computer Science). Available from: http://link.springer.com/chapter/10.1007/978-3-642-32609-7_19
[7]. Trancik R. Finding Lost Space: Theories of Urban Design. John Wiley & Sons; 1986.
[8]. Altman I, Low S. Human Behaviour and Environments: Advances in Theory and Research. New York: Plenum Press; 1992.
[9]. Oh C., Lyu S., Hammitt WE. *Predictive Linkage between Recreation Specialization and Place Attachment.* Journal of Leisure Research. 2012;44(I):70–87.
[10]. Rubinstein RL, Parmelee PA. Place Attachment: Attachment to Place and The Representation of the Life Course by The Eldery. New York: Springer; 1984.

Ecological wisdom of Hindu-Javanese community settlement in Cetho Hamlet, Lawu Mountains, Central Java, Indonesia

Fauzan Ali Ikhsan[1*], Bambang Setioko[2], Atiek Suprapti[3]
[1]Architecture and Urbanism Doctoral Program , Diponegoro University, Semarang, Indonesia and Member of Urban-Rural Design and Conservation Laboratory, Architecture Department, Sebelas Maret University, Surakarta, Indonesia
[23]Architecture and Urbanism Doctoral Program, Diponegoro University, Semarang, Indonesia

E-mail: fauzan@ft.uns.ac.id,

Abstract. Dwelling culture of Java community has a long history. In Javanese community point of view, Javanese culture is not a homogeneous entity. There is a diversity characteristic of a regional nature of Javanese culture, which is caused by differences in the natural environment on the island of Java. Dwelling Culture settled in the mountains is a variant form of Javanese culture that has its own uniqueness. This study aimed to describe the local values of the rural settlement structure of in Lawu mountains with Hindu-Javanese culture background. This study used a qualitative approach with case study strategy. The locus of research is in the Cetho hamlet on the slopes of Mount Lawu Karanganyar, Central Java. The results showed that local values in the neighborhoods of Cetho hamlet is based on the Memayu Hayuning Bawana philosophy. These concepts are abstracted into various aspects of settlements in Cetho hamlet such as aspects of dwelling house setting, water management, and ritual activities related to environmental management. Memayu Hayuning Bawana philosophy becomes a cultural space coloring the ecological action of the population of Cetho hamlet. The comprehensive motivation of ecological action ultimately leads to one final goal, namely the desire to maintain a harmonic atmosphere with the surroundings.

Keywords: Local wisdom, settlements, Mountains, Hindu-Javanese

1. Introduction

Indonesia is an archipelagic country that has natural landscape and various ethnic groups. Based on the existing records, there are 250 languages and more than 300 tribes scattered throughout Indonesia [7]. The diversity of ethnic backgrounds supported by the diverse landscape of Indonesia, forms the diversity of the traditions of a particular group of people/tribes. Space to live is not an abstract concept of the traditional society. All the space contains the power of life that needs to be managed to maximize luck and avoid evil. For the traditional society, ignoring the power of space life gives bad influence to its inhabitants. The diversity of natural and cultural conditions creates spaces with local meanings. The perspective of each community group in managing space produces local values that distinguish between one place and another. The management of living space for the traditional Indonesian society largely covers the layout aspect, the division of the activity space and the appearance of the building [10].

Javanese tribe is one of the tribes in Indonesia which has a history of culture that is quite old, Javanese cultural region covers Central Java to the eastern part. In the history of its development, there are various differences in attitude and behavior of the Javanese in understanding the culture. This difference is due to differences in geographical conditions and the intensity of interaction with other religious and cultural values. This distinction of understanding leads to various variants of the Javanese cultural domain. According to Koentjaraningrat in Ronald [8] Javanese cultural region is divided into several sub-culture namely *Negarigung* (Yogyakarta, Surakarta and surrounding areas, Banyumas, Bagelen), the center of Javanese culture. *Pesisir Kilen* covers Cirebon to Kudus. *Pesisir wetan* include Bojonegoro to Gresik., *Mancanegari* cover Madiun until Malang. *Tanah Sabrang Wetan* covers the eastern parts of Java, Surabaya and Madura. Each sub-cultural region has a distinctive characteristic in implementing Javanese cultural philosophy into everyday life.

Studies of rural settlements in the mountains of Java are still relevant today, as the mountainous areas of Java comprise one-third of the total area of Java [14]. Some researchers have conducted researchs on rural settlements in the mountains of Java. Rejeki Research [15] took a locus in one villages on the slopes of Mount Sindoro, Wonosobo, Central Java. Rejeki research reveals the values of local wisdom in mountainous settlements on the slopes of Mount Sindoro. The determination of the orientation of the dwelling house facing the north-south, the concept of mancapat on the macro scale and the order of punden became the main findings of Rejeki research. Research on settlement of rural hamlet with Hindu-Javanese cultural background is done by Anggrayani [13] in Jenglong Hamlet located in Kawi Mountains, Malang. Anggrayani research findings indicated that the spatial arrangement of settlements Jenglong Hamlet partly adopted the arrangement concept of the area or environment of Hindu-Bali. The concept of space in Jenglong Hamlet is modified from the Hindu-Bali space concept, which is to adopt the concept of Tri Angga

Settlement research in Dusun Cetho has the same context with Anggrayani research, that is rural settlement in mountains with Javanese Hindu cultural background, yet different region of culture sub culture. The geographical conditions, technological developments, cultural environment differences may creates differences and adjustments in the application of the settlement arrangement concept. This study aims to locate traditional local knowledge systems and the ways in which they use such knowledge to adapt to their environment.

The term ecology was first introduced by Ernst Haeckel, a veterinarian there in 1869 as the science of interaction between all kinds of living things with their environment. Ecology comes from Greek, oikos (household/residence) and logos (scientific/scientific). So ecology is the science of the home or living place for human being [2]. According to Triyoga [12] Ecological wisdom is a system of beliefs, attitudes, behaviors and actions of local people in making their lives oriented in seeking harmony with the natural environment.

In the Javanese view of life, human life can not be separated from the environment in which he lives. He will always depend on and interact with his environment continuously. Through their experience and observation of the environment, the Javanese built their knowledge on how to live and ethically treat nature, ultimately giving birth to local wisdom values passed down from generation to generation. In regulating human relationships with nature, Javanese recognize the philosophy of Memayu Hayuning Bawana. Memayu comes from the word hayu (beautiful, beautiful or safe) by getting the prefix ma, to be mamayu (beautify, beautify or improve safety). The word Hayuning comes from the word hayu by getting the pronoun of belonging ning (his) which means beautiful, beauty or salvation. Free translation from memayu hayuning: striving for safety, happiness and prosperity. The word Bawana means the world. The world can be interpreted either physically or spiritually. The understanding of Memayu Hayuning Bawono is to strive for the salvation of the world and to keep the world and all its contents in order to remain preserved and harmonic [11]. This study aims to locate traditional local knowledge systems and the ways in which they use that knowledge to adapt to their environment on the settlements of the Hindu Javanese community on the slopes of Mount Lawu.

2.Methods

The research used qualitative approach with case study method. The research area is located in Dusun Cetho, Gumeng Village, Kabupten Karanganyar, Central Java, Indonesia. The data were collected by observation and interview. The selection of informants was done purposively after the grandtour observation. Data analysis technique used Spradley model data analysis techniques [9], the first stage is domain analysis to obtain a general overview and comprehensive about the social situation / research object under study. The second stage is a taxonomic and compositional analysis to obtain the domains or categories of the social situation under study. The last stage is the analysis of cultural themes to look for linkages between domains in order to obtain a complete picture of the object of research so that the results can be understood and become the basis for the conclusion.

3.Discussion

3.1. Site and Climate Conditions of Cetho Hamlet

Settlement of the Hindu community in the area of Java Cetho is a hamlet located on the western slopes of Mount Lawu, adjacent to the site Cetho. It is located on an altitude of approximately 1413 m above sea level, at the coordinates 7°35'43" latitude and 111°9'21" longitude. Climatic Character of Cetho hamlet is generally similar to humid tropical climate character of mountainous areas in Indonesia. The air temperature is relatively cool, the average temperature of the night ranges from 24-29°C and 29-34°C during the day. The topographic condition of Cetho hamlet located on the slope of Mount Lawu generally has a varied contour, with a land slope of about 10-35 degrees[see figure 2&3]. The soil in this area is relatively fertile, pine and pine trees dominate the vegetation located in the forest area of Mount Lawu, while land around the hamlet is cultivated for horticultural crops such as carrot, cabagge, onion, mustard and jar. Cetho Hamlet is administratively located in the administrative area of Gumeng Village, Jenawi District, Karanganyar District, Central Java.[see figure 1]. The location is about 35 km from Karangnyar with approximate area about 70Ha. Cetho hamlet is relatively isolated when compared to other hamlets located in Gumeng Village. The area of Cetho Hamlet is adjacent to the Perhutani Forest and the peak of Mount Lawu on the east and north sides and the agricultural fields on the south and west. The composition of the last population in the Cetho hamlet is 304 adherents of Hinduism, 134 Islam and 6 Christians

Figure 1. Location of Cetho Hamlet at Lawu Mountain, Karanganyar Distric, Central Java, Indonesia
Source : Bappeda Karanganyar (2015)

Cetho Hamlet can not be separated from the presence Cetho temple. Based on the clues point at two places in the Cetho temple complex, *Sengkalan Memet* sclupture (1451 AD) and *Ruwatan* building

(1475 AD), indicate that the temple was built on the clear end of the Majapahit kingdom in King Brawijaya V period. Cetho, a Hindu temple built *'ruwatan'* purpose, attempted to rescue from disaster and the pressures arising from the turmoil at the time. At the time, Majapahit kingdom is undergoing social unrest, political, cultural and even religious practices. Experts suspect this is one of the desolation causes, as well as from the influence of the Islamization of the West [5]. According to Lombard [4], at this time, the power to regulate all Majapahit territory began to weaken. In areas far from power, the ideology of Hinduism and Buddhism began to be abandoned and began streaming a new ritual, which glorifies figures savior. This trend is dominated elements of the original Java.

Figure 2.
Settlement Area of Cetho Hamlet
Source : Google Map(2016)

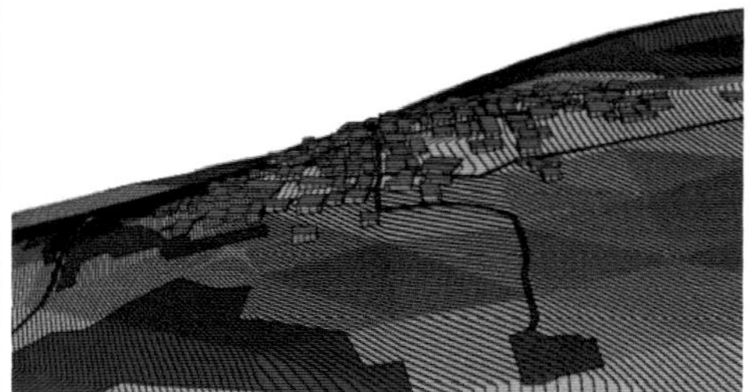

Figure 3.
Section View of Cetho Hamlet (West-East)
Source: Field Observation, 2016

3.2. Dwelling House Setting in the Cetho Hamlet

The topographic condition of Dusun Cetho which has a fairly steep slope of the land makes the settlement pattern of Dusun Cetho quite unique[see figure 6]. Differences in land altitude among the sites in certain locations are quite extreme, this requires the creativity of site management. The cetho population will level the land before it is constructed in a *"batur"* manner (cut and fill method).[see figure 5]. There is an unwritten agreement in which the landowners in the above positions are obliged to build a stone talud as a barrier to the land and at the same time resist landslide.

From sample of houses studied, Javanese house in rural settlement have an average two groups of building, *griya(*building function*) wingking* with *dhapur(*roof shape*) limasan sinom* and *griya gandhok* is also with *dhapur limasan or kampung. Griya wingking* of the Cetho Hamlet community is better known as *ndopo* and *senthong* (left, center, right*)*. While the *gandhok* was set up as *pawon* (a kitchen). Support facilities such as stables and bathroom facilities are not seriously equipped. Although located on the wide field, it generally shows two groups of building (*griya wingking and griya gandhok (pawon).* When referring to the guidelines for Javanese building, Kawruh griya Titika Wisma [6], griya wingking in the hamlet of Cetho have different conditions. Griya wingking description in Kawruh Griya script requires a high degree of privacy, otherwise the condition is different from griya wingking in the hamlet of Cetho. In front of senthong room (Cetho residents often call it *Ndopo*) the owners used it for various activities such as dining room and a couch to sleep on. This room is a family residence for all activities. It became one of the central features of houses in the Javaneese rural settlement, where a building accommodates various functions/activities.

Javanese houses in Cetho Hamlet mostly face in north and south direction. If forced by access then the house will be faced west, but in some cases the house remains cultivated facing north / south though home access from the west. In Cetho Hamlet there is a trust to avoid facing east, because the house facing the east is believed not to bring happiness and prosperity for its inhabitants.

Roof cover construction has used zinc as a roof covering material. The humid rainy climate conditions of Cetho hamlet cause people use roof covers that are capable to respond the climate change. However, based on the elder story, house roof in the Cetho hamlet used reed materials. When there is a clay roof tile material, residents had switched to tile. The use of tile material does not last

IOP Conf. Series: Earth and Environmental Science **99** (2017) 012016 doi:10.1088/1755-1315/99/1/012016

long, given the climatic conditions of the Cetho Hamlet is always wet. It causes the tile wet and also damage the wood batten roof. Since zinc material is known, the citizens then gradually replace the roof material to zinc. According to informant story, zinc material more waterproof and able to dispel rain water especially when there is rain accompanied by strong winds. Zinc material make the house warmer, and able to drive the cold mountain air.[see figure 4]

The house wall initially uses *gedhek* (bamboo weaving) and partly uses natural stone construction. The stone is arranged using limestone adhesive. Stone and sand is obtained not far from Cetho Hamlet. But nowadays, as the transportation out of the sub-village is easy, new materials such as brick are already being used by some residents to build houses.

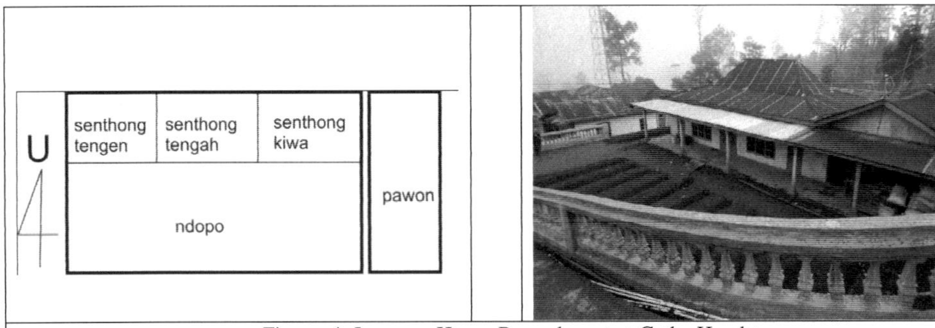

Figure. 4. Javanese House Room layout at Cetho Hamlet
Source: Field Observation, 2016

Figure 5. Stone embankment between building
Source: Field Observation, 2016

Figure 6. Topography of residential areas on a steep slope. Source: Field Observation, 2016

3.3. Water Management

The settlements of Cetho Hamlet are located on the western slopes of Lawu mountain, at an altitude of approximately 1400m above sea level. The soil contains many natural stones and deep ground water level. Such geomorphological character causes no well as a source of clean water supply. The Cetho people traditionally rely on mountain water springs that appear in some slopes of Lawu Mountain. The first source of water used as a source of clean water supply is the source of *Sendang Pundisari* water springs. The location of this water spring is right above the highest terrace of Cetho Temple (east of the temple), like the Hindu temples in other areas. In general, temples are built in locations that have a good source of water, because water for Hindus is part of the worship ritual.

Based on the stories of the informants, the first generation of people take water from *Sendang Pundisari* water springs. From *Sendang Pundisari* water springs, people used pipes to drain the water to the distribution tank.. Location of Cetho hamlet which is under the cetho Temple (West cetho Temple) simplifies the process of distribution of water without the use of pumps (relying on the force of gravity). From a distribution tank (illustrated in the figure 8), water flowed into houses and several public toilet in the middle of the settlement. Public toilet at Cetho Hamlet commonly called the *Kali Baku / Kali Umum*. Early generations of residents in Cetho Hamlet do not have bathrooms in their homes. Supplies of water used the resources in the *Kali Baku*. In certain moments, *Kali Baku* was used for bathing the baby and bathing the villagers who died. Along with population growth at Cetho hamlet, water supply from the *Sendang Pundisari* water springs is insufficient to meet the need of clean water of Cetho Hamlet. Finally, the Cetho villager find the greater water springs. The source of water springs is located at *Sendang Macan*, about one kilometer above Cetho temple. Water from *Sendang Macan* water springs is distributed to a distribution tank that is not far from the water springs. From a distribution tank , water flowed into houses and several public toilet in the middle of the settlement.

Along with the improvement of living standards and welfare of the population, at this time every house built bathroom each. Residents are able to drain water from the distribution tank directly to their house and respectively. To facilitate supervision, care, and equitable distribution of water, it was agreed that each family can only have one water pipe directly connected to the distribution tank. In addition to being used for daily primary needs such as drinking, bathing and laundry, a source of clean water supply from water springs, used for watering agricultural crops in the moor. Water is needed to improve the productivity of the moor land, given the character of the agricultural system in Dusun Cetho is rain-fed agriculture. Water is flowed from the water spring of Sendang Macan through the pipe to the moor of the population. In the moor, the water pipe is broken apart through the dividing line to fulfil the need of the land area. Usually at the end of the pipe installed a water vane (water diffuser outlet) so the plants watering can be evenly distributed.[Figure 7].

Figure 7. Water management at agriculture area
Source: Field Observation, 2016

Figure 8. Illustration of water distribution tank Source: Field Observation, 2016

3.4. Ritual activities related to environmental management

3.4.1. Mondosiyo Ritual

The name *mondosiyo* is taken from one of the names of *wuku Mandhasiya* in the order of 14 of 30 *wuku* based on Javanese calendar. The Mondosiyo ritual falls on Tuesdays *kliwon* wuku *mandhasiya*. This ritual will be repeated every day, considering that one wuku lasts seven days. The Ritual of Mondosiyo in Cetho hamlet is a tradition of "clean hamlet" conducted by local people.[figure 9 &10] Mondosiyo Tradition is an activity of the citizens of Cetho hamlet expressing gratitude for the social harmony and nature sustainability that has been given by the omnipotent to all citizens of Cetho Village. According to the beliefs of the villagers of Cetho, respect for nature is an attempt to maintain harmony between macrocosmic natural and microcosmic realm. Clean hamlet ritual is an effort to *meruwat* (regenerate) the orderliness of the cosmos to create harmony and tranquility of village's life. This tradition is also at the same time as the tribute of Eyang Krincing Wesi birthday believed to be *Dhanyang* / embryo of the village that became protector and guard harmony village life order. According to the known Myths story in Dusun Cetho, Eyang Krincing Wesi is the caretaker of Cetho

Temple during the time of King Brawijaya reign (the last King of Majapahit ruling around the 15th century). He always served the needs of the king every time doing activities at Candi Cetho. Eyang Krincing Wesi is believed to have been died and was buried at the *Krincing Wesi Punden* which is currently in Cetho temple area. The Spirit of Eyang Krincing Wesi is believed to always keep the environment and always be respected by all of Cetho Hamlet citizens.

Dhanhyang or danyang which according to Geertz [3] is considered as the spirit of the deceased historical figures (the founder of the local village, the first person who opened the land). *Danyang* while still lived as a man came to the village while still a jungle, opened the land and divided the land to his followers, family, friends and he became the first village leader. After his death he was buried near the village and his tomb became *punden*. But in some places punden is not always related to the location of the tomb. *Punden* is the pioneer / founder of the village who is considered sacred and their spirits are believed to still keep an eye on the village and can provide assistance.

Figure 9. Mondosiyo ritual

Source: Field Observation, 2016

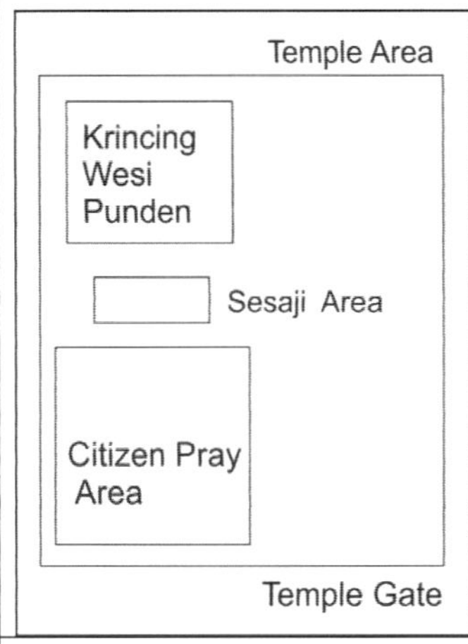

Figure 10. Setting of Mondosiyo ritual

Source: Field Observation, 2016

Mondosiyo ritual is believed to have been conducted from hundreds years ago until now and never broken up. The Mondosiyo ritual begins one day before (Monday-Wage), the residents prepare traditional offerings of special food (*tumpeng panggang*). Then, on a Monday night *tirakatan* (ceremony) is held in front of the Krincing Wesi Punden. On The Day (Tuesday Kliwon), residents go together to Krincing Wesi Punden around 10:00 am and offerings to carry out a pray together. In the evening, the usually made a Puppet show. The story was selected by the elders. For some residents, the story plays at the time of the leather puppet show is believed to give luck and serenity to the next Mondosiyo celebration.

3.4.2. Dawuhan Ritual

Dawuhan ritual is done every six months on Saturday *kliwon*. Dawuhan Ritual is an expression of gratitude for the citizens of Cetho hamlet on the availability of water and at the same time a form of homage to the water god of the water source in the village of Cetho. Dawuhan ritual series is carried out by water channeling the source center and then ending with a pray at *Sendang Pundisari*. *Sendang Pundisari* is an ancient water springs located above Candi Cetho. Through the ritual, Cetho villagers want *meruwat* the existing water sources in order to always provide adequate water needs for all citizens. For the residents, the water source is a holy place, because water is used for all the life needs and water serves as a complement of worship rituals. A set of *roasted tumpeng* (rice, roast chicken, side dishes, fruit, flowers) was prepared during the Dawuhan ritual. Events usually start around 10:00 and finish at noon. The activity is led by *Mangku*(the customary stakeholders) in the hamlet area of cetho.

3.5. Ecological Wisdom of Settlement in the Cetho hamlet
The results of observations on the values of ecological wisdom in Settlements in Cetho Hamlet can be grouped into two dimensions[figure 11], the value of space culture and the value of spiritual culture wisdom. Space culture is formed from the relationship among humans as well as that of humans and their environment. Spiritual culture is formed from human relationship with god. The space culture dimension is shaped by the values of local wisdom in Dusun Cetho such as consensus construction of embankment, cut and fill concept, building material selection, determination of building orientation, consensus of water management and agricultural land. Meanwhile, the spiritual culture is formed by rituals and customs that still run in Cetho Hamlet such as Mondosiyo ritual, Dawuhan Ritual and Ritual in Kali Baku. For the people of Cetho Hamlet, the balance between the dimensions of Space culture and the Spiritual Culture Dimension became the basis of the realization of the *Memayu Hayuning Bawono* philosophy, an ideal condition that is expected by every Javanese people in arranging the environment of his birth and mind to the tranquility of life.

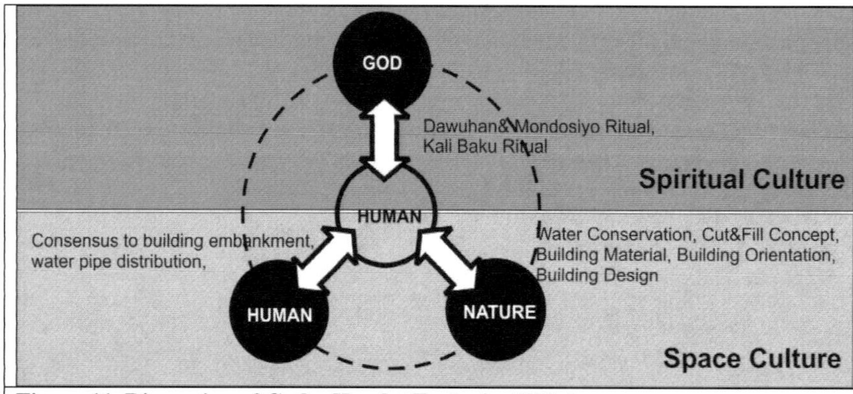

Figure.11. Dimension of Cetho Hamlet Ecological Wisdom

4. Conclusion

The mindset, attitude and actions of managing the environment for the people of Cetho hamlet can not be separated from the knowledge built through the empirical experience inherited from their ancestors. Geographical and climatic conditions with a rural Javanese cultural background have formed a distinctive adaptation pattern for the inhabitants of Cetho. Ecological wisdom that is manifested in the physical / outward and non physical / inner levels shows the characteristics of *Kejawen* (Javanese beliefs) that still strongly color the life of the people of Cetho hamlet. *Memayu Hayuning Bawana* philosophy becomes a cultural space coloring the ecological action of the population in Cetho Hamlet. The comprehensice motivation of ecological action ultimately leads to one final goal, the desire to maintain a harmonic atmosphere with the surroundings. The harmony of living with the environment contains two meanings, namely First, the harmonic relationship between humans and the surroundings. Second, the harmonic relationship between humans and God.

Acknowledgments
The author would like to thank the informan (persons) of Cetho hamlet who have been willing to accept the author to discuss and retrieve research data.

References
[1]Bappeda Karanganyar, 2015, *Geografi Kabupaten Karangnyar, Pemerintah Daerah Kab. Karangnyar*. http://www.karanganyarkab.go.id/20110119/candi-cetho/

[2]Frick, H and Suskiyatno, B.*1998. Dasar-Dasar Eko Arsitektur*. Yogyakarta: Kanisius. **1**

[3]Geertz, Clifford.2014. *Agama Jawa: Abangan, Sanri, Priyayi dalam Kebudayaan Jawa*, Depok: Komunitas Bambu. **23**

[4]Lombard, D, 2000, *Nusa Jawa: Silang Budaya Kajian Sejarah Terpadu; Bagian III:Warisan Kerajaan-Kerajaan Konsentri*s: Jakarta : PT. Gramedia Pustaka **25**

[5]Perpustakaan Nasional Republik Indonesia (Perpusnas), 2014, *Kepustakaan Candi Jawa Tengah-Candi Cetha*. http://candi.perpusnas.go.id/temples/deskripsi-jawa_tengah-candi_cetha

[6]Prijotomo, J, 2006, *(Re)konstruksi Arsitektur Jawa: Griya Jawa dalam tradisi Tanpa tulisan*.Surabaya: PT. Wastu Lanas Grafika **173**

[7]Rigg, Jonathan, 2002, *Indonesian Heritage: Manusia dan Lingkungan*, Jakarta: Grolier International.**40**

[8]Ronald, A, 1990, *Ciri-Ciri Karya Budaya Di Balik Tabir Keagungan Rumah Jawa*, Yogyakarta: Penerbitan Atma Jaya University **183**

[9]Sugiyono, 2016, *Metode Penelitian Kuantitatif, Kualitatif dan Kombinasi (Mixed Methods)*, Bandung:CV. Alfabeta. **344**

[10]Tjahyono, Gunawan.2009. Sejarah Kebudayan Indonesia. Jakarta: PT. Rajagrafindo Persada **29**

[11]Endraswara, Suwardi.2016.*Memayu Hayuning Bawono: Laku Menuju Keselamatan dan Kebahagiaan Hidup Orang Jawa*, Yogyakarta: Narasi **18**

[12]Triyoga, Lucas Sasongko. 2010 *Merapi dan Orang Jawa : Persepsi dan Kepercayaannya*. Jakarta: PT. Gramedia Widiasarana Indonesia. **151**

[13]Anggrayani, Irma.2010. *Tata Ruang Permukiman Komunitas Hindu di Dusun Jenglong Desa Sukadadi, Kecamatan wagir, Kab.Malang*. Tesis .Program PascaSarjana Universitas Gadjah Mada.Yogyakarta

[14]Hefner, Robert W., 1999, *Geger Tengger: Perubahan Sosial dan Perkelahian Politik*, Yogyakarta: Lkis Yogyakarta dan The Asia Fondation. **25**

[15]Rejeki,VG Sri. 2010. *Nilai Kosmologi Pada Tata Spasial Permukiman Desa Kapencar Lereng Gunung Sindoro Wonosobo*, Forum Teknik Vol 33 No3 September 2010 p.**140-148**

Thermal evaluation for exposed stone house with quantitative and qualitative approach in mountainous area, Wonosobo, Indonesia

Hermawan Hermawan[1*], **Eddy Prianto**[2]
[1]*Architecture Department, Faculty of Engineering and Computer Science, Qur'anic Science University, Wonosobo, Indonesia
[2]Department of Architecture, Faculty of Engineering, Univeritas Diponegoro, Semarang, Indonesia

E-mails: hermawanarsit@gmail.com

Abstract. A building can be considered as having a good thermal performance if it can make the occupant comfortable. Thermal comfort can be seen from the occupant's respond toward the architectural elements and the environment, such as lighting, the room crowding, air temperature, humidity, oxygen level, and occupant's behaviours. The objective of this research is to analyse the thermal performance of four different orientation houses in mountainous area. The research was conducted on the four expose stone houses with four different orientations in the slope of Sindoro Mountain which has relative cool temperature, about 26°C. The measurement of the elements above was done quantitatively and qualitatively for 24 hours. The results are as follows. First, the most comfortable house is west-orientation house. Second, based on the quantitative and qualitative observation, there is no significant difference (±5 %). Third, the occupant's behaviours (caring and genen) also become factors influencing occupant's comfort.

Keywords : orientation, thermal performance, occupant's behaviours, qualitative, quantitative

1.Introduction

In architecture, research in thermal and performance comfort has a close relation. Thermal comfort is the expression of occupants' satisfaction related to the building thermal. Thermal performance is the ability of building in managing thermal environment. Both of them should be integrated to make a thermal condition so the occupants can do activities in their house well. Thermal comfort related to the occupants' feeling is known as active thermal. Basically, researches in active thermal use the qualitative approach, and building thermal (passive thermal) use the quantitative one. However, people can also use quantitative or mixed approach to know active thermal, and use qualitative or mixed one to know passive thermal.

Most of thermal comfort measurements are based on PMV (Predicted Mean Vote) and PPD (Predicted Precented Dissatisfied) theory. Although the theory of adaptive thermal comfort has been developed, it still cannot be separated from PMV and PPD theory. Adaptive thermal comfort is based on three main things: Thermal Comfort Vote (CV), Preference vote (PV), and Thermal Acceptance and Neutrality. Adaptive thermal comfort can also predict thermal comfort through statistic tests [1]. Occupants' adaptation is a factor influencing the building's thermal comfort. Adaptation is an important factor in adaptive thermal comfort. Adaptive thermal comfort also refers to the people's perception toward hope and choice related to comfort. Adaptive thermal comfort orientate to occupants' comfort, while thermal comfort related to the building is known as passive thermal comfort.

Passive thermal comfort can be known by analysing the building envelope. It is considered as an important aspect in creating passive thermal comfort. Design of building envelope must be adaptive and interactive to be able to respond climate surrounding the building. Design of building envelope has relation to aspects of the building's facade in controlling solar radiation. One method to improve the performance of building envelope is simulation. One of the simulations is CAD (Computer Aided Design). It can also save energy [2].

There are a lot of building's cover researches related to energy saving. Researchers in Malaysia conclude that energy waste of buildings reaches 19% of total energy waste. The temperature in Malaysia's buildings is 25–30 [3]. In addition to temperature, there are some climate factors related to building envelope: temperature, humidity, wind speed, rain, and solar radiation [4]. Design of building envelope which is influenced by climate variables will influence the energy consumption [5]. Materials of building envelope also influence the thermal comfort. Materials of building envelope also influence the thermal comfort. The previous researches have shown that building envelope made of stone and exposed stone have different result related to thermal comfort in the coastal areas [6]. In addition, houses with wooden wall have different thermal comfort if the location have different climate [7]. The part of building has a big role for achieving building thermal performance including balcony and windows [8]. Optimized design of windows can make a good thermal performance [9].

Heat is the greatest factor influencing the building envelope [10]. Heat is closely related to solar radiation, and orientation of building has big impact in creating thermal comfort for the occupants. Thermal comfort of the occupants and buildings should be researched at the same time. Aspects of occupants and building can be observed quantitatively and qualitatively. Qualitative approach should be focused by the researchers so when they conduct research related to thermal comfort, they can mix both quantitative and qualitative method.

In architecture, it needs to see qualitative and quantitative aspect of the elements. Elements which can create thermal comfort are lighting, air temperature, air humidity, room level of oxygen, room tightness, and occupant's behaviours. Researches which see thermal comfort from qualitative and quantitative aspects was done in Peru. It analysed the building elements qualitatively and analysed the thermal qualitatively. It was done in a city which has height 3200 meters. The result shows that the occupant feels uncomfortable when the elements and design of buildings did not meet the requirement of good buildings [11].

Indonesia is located in a tropical area with warm and humid climate. It also has low and high land. The high land has cold temperature but there is still sun light. The people sometimes feel cold and they create a traditional heater. The heater is in form of stove using wood fuel which produces smoke. For long term it makes the house condition unhealthy because the occupants feel breathless. The characteristics of housing in high land are that the walls are made of wood, stone, brick, and bamboo. However, people begin to leave stone-walled house and replace by modern one although sometimes it is not appropriate. Some researches showed that traditional houses could adapt to the local climate [12]. Therefore, it needs to know the characteristics of stone-wall traditional house in facing climate so it can find the strategies to get thermal comfort. Research objectives are to analyse thermal performance of exposed stone-walled houses in mountainous area, to analyse thermal receipt of the occupant of exposed stone-walled houses in mountainous area and to analyse characteristics of exposed stone-walled house which meet thermal comfort.

2. Methodology

The research was done in houses with river stone wall and four different orientations. They are located Gunung Alang Village in Sindoro slope and Tambi Village, Kejajar District, Wonosobo. The research used four climate variables and two personal variables. The climate variables were air temperature, solar radiation temperature average, wind speed and humidity. The personal variables were clothes and occupant's activities. Occupant's responds were ranked using ASHRAE (American Society of Heating, Refrigerating, and Air-Conditioning Engineer) which covers seven indicators: very hot, hot, warm, comfortable, cool, cold, and very cold. The result was displayed in form of graphs and pictures. Analysis was conducted by calculating average, maximum and minimum data. The difference of climate variables between outside and inside room would be known. From the occupants side, their respond toward micro environment condition was analyzed. Graph analysis was used to the quantitative approach. Habit and occupant's response used qualitative approach. Analysis related to buildings condition and elements was linked to graph and interview results. The research findings were building strategy to create thermal comfort in tropical high lands.

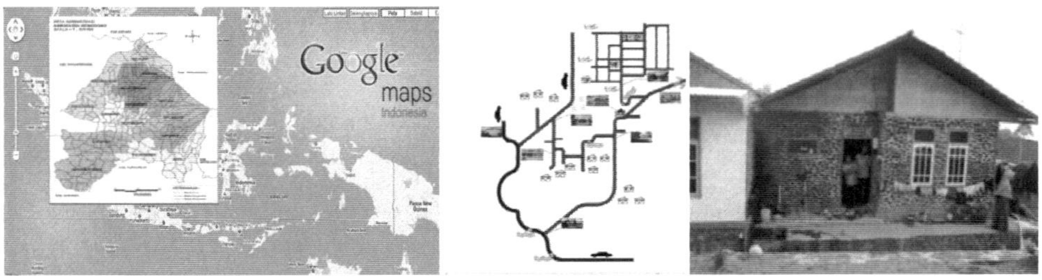

Figure 1a)Indonesia and Wonosobo Map (Source : Google Maps); b)Location of sample research; c)Type of research sample

One research which used quantitative and qualitative approach was thermal evaluation in historical building. This research used decision support system (DSS). DSS can be used as a solution to determine the criteria of buildings renovation. The elements of criteria were determined by using Graphical Analysis for Interactive Aid (GAIA) and DEPHI. The result is the suggestion to make a policy to reduce energy consumption on masonry buildings. Some data used here were building information and climate zone, building facade plan, organization, building volume, construction method, building aperture, and energy consumption. The research data were collected by measurement (quantitative) and interview (qualitative). Measurement was done to know the levels of energy consumption and investment cost, and interview was done to know the risk of the loss of historical aesthetic features and spoilage fabric [13].

3. Results and Discussions

Descriptive discussions are combined with graph and three-dimensional pictures. People's job and habits, as well as culture in building a house are also analyzed here. The measurements of house elements related to thermal performance discussed are lighting, air movement, room crowding, and house environment. The discussion on climate variables is linked to energy saving, so the strategies of a house to save energy would be known.

The main job of people in Gunung Alang is tobacco farmers. It influences the design of their house rooms. Their house consists of wide living room for keeping tobacco crops, bed rooms, bath room, kitchen, and family room for watching TV. The roof is covered by zinc. The walls are made of stone or wood, these materials are easy to get in the surroundings. In constructing their house, people still keep the ancestor culture tightly. For example, it is forbidden to build houses facing or oppositing mountain since it can cause negative atmosphere, and it is bad for the occupants. Therefore most of houses face to north or south although there are also people who don't believe.

The process of house building is started by collecting stones or woods. People collect stones from the river under the village by breaking the giant stones into small sizes. Then they put the stones on the main road and carry them to village by car. When materials to build foundation and wall are adequate, the people plant banana trees on the location in which the houses will be built. After the trees are forty days old, people start to build foundation and wall. In building house, people employ masons and get aid from their neighbors. In the process of setting nok (munggah molo), people hold a ceremony to show their thanks to God and hope to get salvation and prosperity. They put rice, corn, and red white flag as symbols of wealth and prosperity which will cover the house. Some people also put Raja banana as the symbols that the house will be strong and doesn't need many renovations. The flag symbolizes independence and red fabric also functions to repel pests that can attack wood on the roof structure.

3.1. Natural Lighting

The observed house have different orientations: north, south, west, and east. It creates a difference difference in getting sun light. The house of Subardi's, orienting north, gets sun light from the front windows and western side of the house. There is no room which can get sun light continuously. The living room and family room more often get sun light from western windows.

The house of Dipon's has east orientation. Sun light comes through the front side and hole in the bath room which has function as the entrance of rain. The other sides are challenged by neighbour's houses. Meanwhile in the Supardi's house, sun light can enter freely since it faces north and the windows are 180 x 80 cm and main door is 220 x 200. In the west side, there is 120 x 60 cm^2 window. The sun light can enter the living room but the light is not dazzling. According to the occupant, the lighting in Mubasir's house is good enough to do activities. It has two windows; 180 x 120 cm^2 and 60 x 40 cm^2 and one main door, 200 x 220 cm^2.

Of the four houses, most of natural lighting just can illuminate the main room (living/family rooms). Other rooms such as kitchen, bath room and bed rooms don't get adequate natural lighting and need artificial lighting. One house has roof because the wall is so close to the neighbor's houses.

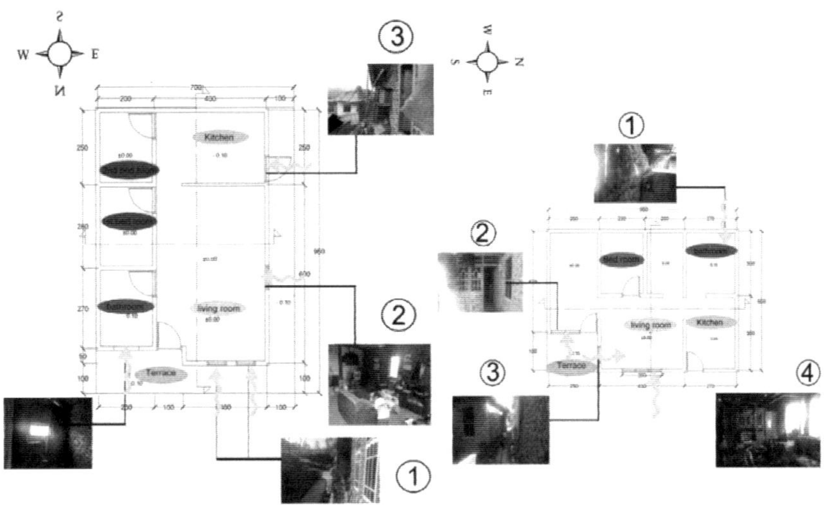

Figure 2 Natural lighting analysis, a)Darto's house, b)Dipon's house

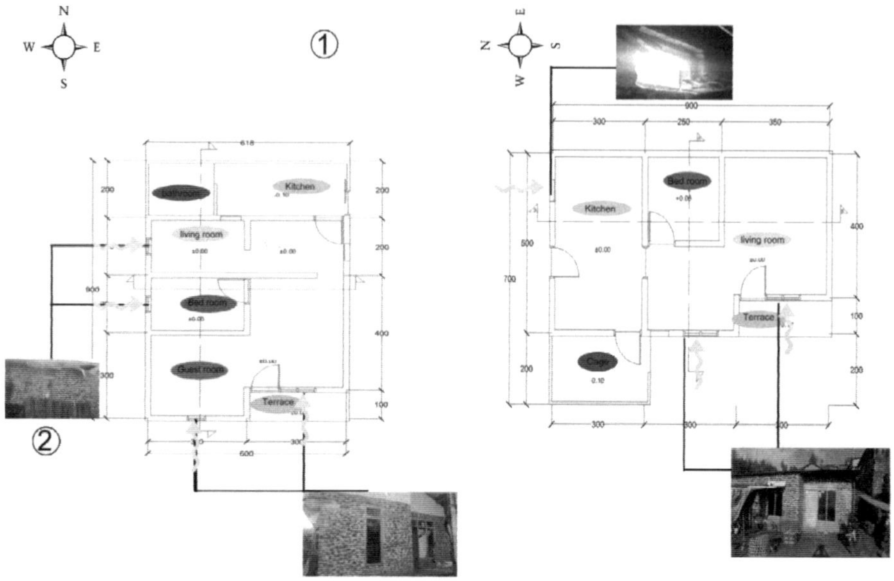

Figure 3 Natural lighting analysis, a)Supardi's house, b)Mubasir's house

3.2.Natural Wind Movement

Wind movement determines the comfort of inside rooms. In the tropical areas the wind blows fast enough. Wind from valley and mountain blow from east to west and enter Darto's house through side windows. In this house, the wind enters through living room, then family room, and kitchen at last. Bed rooms don't get wind movement directly. West wooden window, 70 x 120 cm^2, opens at 06.00–17.00. The wall of Darto's house is higher than the neighbors so that the wind can enter easily. Meanwhile wind movement at Dipon's house is challenged by neighbor's house. Wind moves to living room and family room through front ventilation and windows. Bed rooms don't have ventilation so that they don't get wind movement.

The absence of the ceiling in the Supardi's house makes the wind enter through the roof but cannot come inside the rooms. The wind also enters through side windows. Basir's house whose west orientation makes the wind enters the house fast but stopped at living room and storage. Ventilation in the kitchen, 60 x 40 cm^2, is not closed tightly, closed by wire net only and makes the wind enter the house fastly. In the Supardi's and Mubasir's house, at 11.00 am and 3.00 pm, the wind blow fast, 7.3 m/s, and causes rumble when the wind hits pine trees located behind the house.

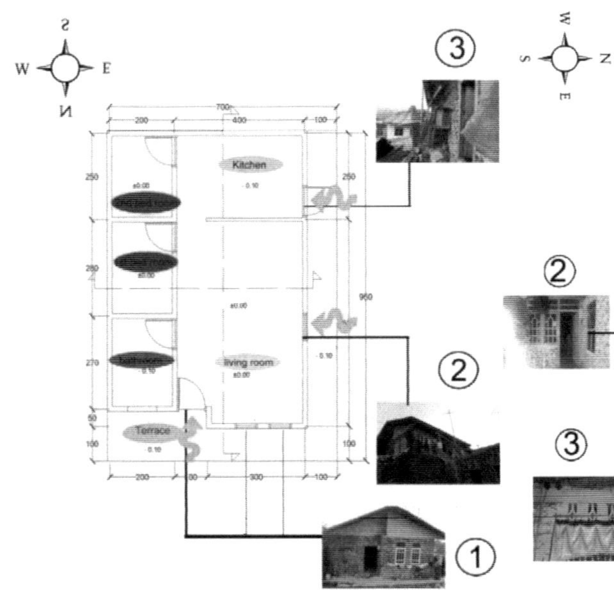

Figure 4 Natural wind analysis, a)Darto's house, b) Dipon's house

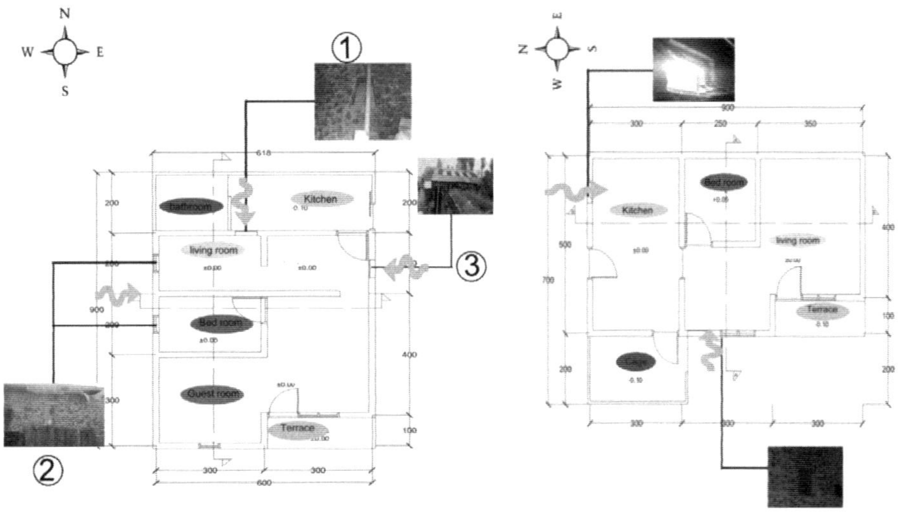

Figure 5 Natural wind analysis, a)Supardi's house, b)Mubasir's house

3.3.Room Crowding

Air volume in a room is influenced by the crowd of the room. The crowd can be known from the content of the house and the volume of the room. In Darto's house, the most crowded room is at the room number 2. It is only 2.5 x 2 m but it contains 1,8 x 2 m bed, cupboard, and a television. On the contrary, the less crowded room is living room containing chairs, tables, big cupboard, small cupboard, TV set, and sound system. The width is 24 m. In Supardi's house, the crowded situation happens at the living room, 6.5 x 9.5 m^2, containing long chair, tables, and sound system. The furniture is made of wood and it should make the room warm. However, there is no sun light entering the room and makes it cold.

In the family room of Dipon'shouse, there are some furnitures, such as chairs, cupboard, aquarium, television and table. Here the family members usually meet and there is enough sun light. It makes the room warm and comfortable for the occupants. The most crowded room is family room, 3 x 2 m^2. In this room, there is big wooden cupboard for television. In the Mubasir's house, the most crowded room is bed room. It contains big bed, big wooden cupboard, television table and the television. Actually these furniture can make the room hot, however the absence of window which can function as entrance for the wind make the room cold.

Figure 6 Room crowding, a)Darto's house, b)Dipon's house

Figure 7 Room crowding, a)Supardi's house, b)Mubasir's house

3.4.Patterns of Occupants' behaviours

Natural lighting, air movement, and room tightness have a close relationship to the patterns of occupants' behaviors. The three factors influence the thermal comfort. A research in China showed that patterns of occupants' behaviors have close relation to the reach of thermal comfort. Courtyards and shade spaces influence the thermal comfort of the occupants [14]. In addition, individual and social environment also influence occupant's perception to thermal comfort. Individual environments are gender, age group, exposure to sun, level of activity and clothing insulation, skin color, while social environments cover position, companionship and cultural background [15]. Both factors are relevant with the research which has been done. Pattern of occupant behaviors in this research is to do natural warming by making fire and sitting near it (*genen*) at night and sitting under the sun light in the morning (*karing*). Both activities influence the reach of thermal comfort. In addition, people have sustainable concept in building house. They take stone from the river little by little. It protects the nature from damage.

3.5.Oxygen Level, Temperature and Humidity

By measuring the oxygen level, it can be known that the family room which contains the lowest level of oxygen is Supardi's. The level is 5,1°C at 12.00 am. It happens because of the location and the width of the room, 3 x 2 m^2, and the less effectiveness in using windows in west side. The windows are blocked by the cupboard from inside and there are many fire woods outside the windows so the air is difficult to enter.

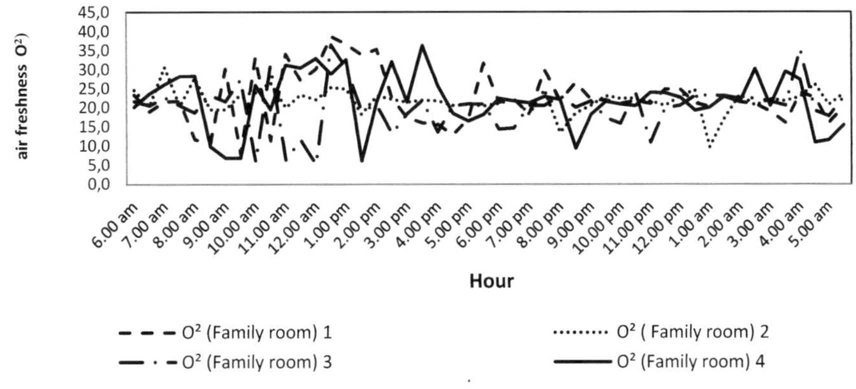

Figure 8 Result of Oxygen measurement in living room

Each kitchen observed has both traditional stove, which uses fire wood, and gas stove. Both stoves are used for cooking and heating body, geni. The highest level of the temperature is in Mubasir's house. In the afternoon the temperature reaches 28,6 °C and the humidity is 48%. Although each kitchen has similarity in function and the temperature, Mubasir's is warmer than others because there are windows in the north side and transparent zinc on the roof so that the sun light is easy to enter the kitchen. The roof is also lower than others so that the heat is easy to spread and makes the temperature increase.

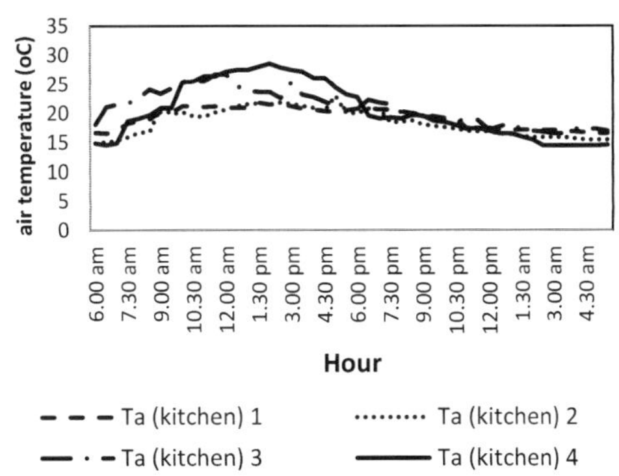

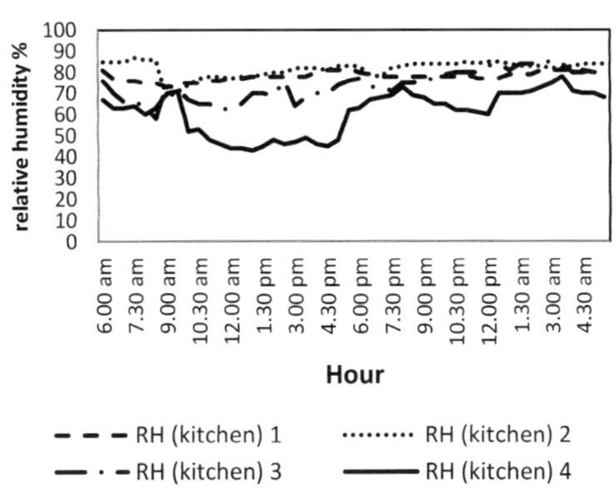

Figure 9 a)Result of air temperature (Ta) in kitchen, b)Result of relative humidity (RH) in kitchen

The four houses which become the objects of research have same characteristics; there is no ventilation and the ceiling is made of board, bamboo or thick fabric which functions to slow the warm air up. The difference is that Supardi's has glass window. The lowest level of temperature is in Dipon's house, 15°C and 2°C. It may be influenced by the bedrooms location which is in the west side and directly linked with the neighbor's wall so that the air in the room becomes humid. In addition, the solar radiation is so little because it only gets sun light from 09.00 am–02.00 pm and after that it will be blocked by neighbor's roof.

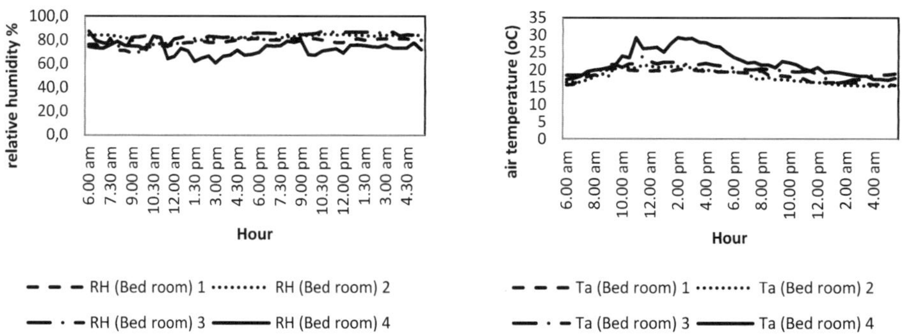

Figure 10 Result of bed room measurement's air temperature (Ta), b)Result of relative humidity (RH) in the bedroom

Temperature is one factor in thermal comfort both in hot and cold areas. Research in China showed that temperature became the parameter to see thermal comfort in cold areas. It shows that average temperature in living room is 13.5°C and in the bed rooms is 12.7°C. These conditions make the occupant feel uncomfortable and need artificial heating [16]. In addition, air humidity and CO_2 level influence the occupants' thermal comfort. Research in Singapore used variables of temperature, air humidity, humidity and CO_2 level to know thermal comfort in the house, restaurant, and work places. The result shows that those three variables can predict thermal comfort of the occupant [17]. Both researches confirmed that temperature, air humadity and CO_2 level can be a basic to measure thermal comfort. In my research, CO_2 was replaced by O_2 by argument that the freshness of the air in the house is related to O_2 and there is no much smoke which can make the occupant in danger.

4.Conclusions

Some people's belief that the houses should not face or opposite to the mountain is something which can be explained scientifically. It is due to the fast mountain wind which will make the house temperature cold. The Raja banana trees planted on the location of building is believed to make the materials durable because they can reduce pest which attacks the materials. The people also believe that on the roof frames red white flag should be put. The red color can avoid beetles eating wood because the red color dazzles battles. The local people use traditional equipment to collect stones and it will not destroy nature.

House orientation, ventilation, shape and size have effect on lighting and air condition. Room crowding and the furniture placement influence the amount of entering. Comparison between the room size and the number of furniture influence the amount of fresh oxygen. The great amount of oxygen will create thermal comfort for the occupants. Vegetation surrounding the house will provide oxygen and protect it from the wind. Adaptive strategies which are done accidentally are proper house orientation karing (heating body under the sun light) and genen (heating body near traditional stove and boiling water for tea or coffee)

References

[1] E. Kuchen, Variable Thermal Comfort Index for Indoor Work Space in Office Buildings: A Study in Germany, Open Journal of Civil Engineering, 2016, 6, 670-684.

[2] Cherif Ben bacha, F. Bourbia, Effect of kinetic façades on energy efficiency in office buildings - hot dry climates, 11th Conference on Advanced Building Skins, pp.458-468.
[3] Seyedehzahra Mirrahimi, Mohd Farid Mohamed, Lim Chin Haw, Nik Lukman Nik Ibrahim, Wardah Fatimah Mohammad Yusoff, Ardalan Aflaki, The effect of building envelope on the thermal comfort and energy saving for high-rise buildings in hot–humid climate Renewable and Sustainable Energy Reviews 53 (2016) 1508–1519.
[4] Sadineni SB, Madala S, Boehm RF. Passive building energy savings: a review of building envelope components. Renew Sustain Energy Rev 2011;15 (8) : 3617–31.
[5] Al-Saadi S, Budaiwi I. Performance-based envelope design for residential buildings in hot climates. In: Proceedings of building simulation; 2007.
[6] Hermawan, Prianto, E., Setyowati, E., (2015), The difference of thermal performance between houses with wooden walls and exposed brick walls in tropical coasts, Procedia Environmental Sciences (23) 168 – 174.
[7] Hermawan, Prianto, E., Setyowati, E., (2015). Thermal comfort of wood-wall house in coastal and mountainous region in tropical area. Procedia Engineering 125 725 – 731.
[8] Prianto, E & Depecker, P. (2003). Optimization of architectural design elemens in tropical humid region with thermal comfort approach. Energy and Buildings 35, 273-280.
[9] Prianto, E & Depecker, P. (2002). Characteristic of airflow as the effect of balcony, opening design and internal division on indoor velocity: A case study of traditional dwelling in urban living quarter in tropical humid region. Energi and Buildings, Volume 34, Issue 4, 401-409.
[10] Chan K, Chow W. Energy impact of commercial-building envelopes in the sub-tropical climate. Appl Energy 1998; 60 (1) : 21–39.
[11] Houmam MELIANI, Jacques TELLER, Shady ATTIA, Architectural and environmental housing typology analysis in Huamachuco, Peru, PLEA2016 Los Angeles – 32nd International Conference on Passive and Low Energy Architecture Cities, Buildings, People: Towards Regenerative Environments, 11-13 July, 2016.
[12] Prianto, E., Bonneaud, F., Depecker.P., Peneaud, J-P. (2000). Tropical-Humid Architecture in Natural Ventilation Effecient Point of View-A Reference of Traditional Architecture in Indonesia. International Journal on Architectural Science, 80-95.
[13] Seddiki, M., Anouche, K., Bennadji, A., & Boateng, P. (2016). A multi-criteria group decision-making method for the thermal renovation of masonry buildings: The case of Algeria. Energy and Buildings, 129, 471–483. http://doi.org/10.1016/j.enbuild.2016.08.023.
[14] HaiyanYan, LiuYang, Wuxing Zheng, Wenfang He, Daoyi Li, Analysis of behaviour patterns and thermal responses to a hot–arid climate in rural China, Journal of Thermal Biology 59 (2016) 92–102.
[15] Salman Shooshtarian and Ian Ridley, The effect of individual and social environments on the users thermalperceptions of educational urban precincts, Sustainable Cities and Society 26 (2016) 119–133.
[16] Borong Lin, Zhe Wang, Yanchen Liu, Yingxin Zhu, Qin Ouyang, Investigation of winter indoor thermal environment and heating demand of urban residential buildings in China's hot summer - Cold winter climate region, Building and Environment 101 (2016) 9-18.
[17] Toby C.T. Cheung, Stefano Schiavo, Elliott T. Gall, Ming Jin, William W Nazaroff, Longitudinal assessment of thermal and perceived air quality acceptability in relation to temperature, humidity, and CO2 exposure in Singapore, Building and Environment 115 (2017) 80-90.

Community participation towards the value of traditional architecture resilience, on the settlements' patters in Tenganan village, Amlapura

Listyana Febriani[1*] **I Gede Wyana Lokantara**[2]
[1]Policy Science, Ritsumeikan University, Ibaraki, Japan
[2]Urban and Regional Planning, Selamat Sri University, Semarang, Indonesia

E-mail: listyanafebriani@gmail.com

Abstract. Ecological conditions such as landslide, flood, and the global warming issues are the disasters that should be anticipated. The value of traditional architecture resilience has a role towards a city as cultural heritage. Based on that influence, the role of architecture is needed in fostering the environment to be able to survive and sustain just as an architectural concept that considers human needs and natural balancing. The purpose of this study is to analyse the concept of traditional architecture and community participation in maintaining this condition, so it would be able to have a value of sustainability. The research method used is mix method that is start from observation and macro analysis element (main building and public facility) and micro element (house of resident) to analyse community participation in realizing traditional architectural resilience in Tenganan Village, Amlapura. The results of this study found that the traditional settlements in Amlapura, Karangasem, Bali is a form of urban architecture that can survive in a sustainable way of macro elements and micro elements oriented to environmental ecological conditions. This condition happensbecause the community has a high enough participation to maintain it in the form of custom rules.

Keywords: traditional architecture, community resilience

1. Introduction

Global warming is a phenomenon that is currently a big topic for discussion, due to the fact that it is a crucial issue that related to the environmental problems. As an impact, natural disasters such as landslides, floods, earthquakes, and others happened. Therefore, it requires a way to survive from such natural disasters. One way to survive is in a form of architectural design. Indeed, developments and advancements in the field of architecture have contributed to the global warming and the changes in ecological conditions. Based on this effect, the role of architecture is needed to foster the environment and to adapt to the degradation of the ecological in order to survive and sustain. It is just as the form of architectural concept that considers human needs and natural balancing. The existence of the concept of sustainability architecture put forward the concept of conservation and maintenance the ecological element itself and the used of environmental friendly technology are a breakthrough that is widely used.
The concept of sustainable architecture is a design that needs to be implemented as an effort to reduce the effects of global warming and the prevention of natural disasters [1]. Traditional architecture is an

architecture form passed down from one generation to the next which puts forward natural and local cultural aspects. It is a tradition to maintain the value of the culture and the ecological or environmental aspects in which the resident resides. Amlapura is one of the cities in Bali that have many traditional architecture based on nature and until now it is still survive in the middle of technological development and architectural design form. Amlapura residents who uphold the customs and beliefs set forth in daily life, as well as on their settlements. The traditional architecture in Amlapura is able to apply the concept of sustainability in architecture, in the utilization of energy, land, materials, and waste management. This study was conducted to identify, analyse, and describe the concept of sustainable architecture in Amlapura, using cases in Tenganan village. Tenganan village is one of the villages located not far from Amlapura. The village has the concept of traditional architecture that maintains aspects of local culture through tradition to maintain the original design of the settlement and ecological conditions. Therefore, it is important to study typology and elements of traditional architecture in the Tenganan village, Amlapura and participation in maintaining the value of the traditional architecture.

2. Methods

This traditional architecture study concept of the settlements in Tenganan Village, Amlapura was done by direct observation on the environment and buildings condition and interviewing the residents using mixed method. To get the concept of sustainability environment on traditional architecture Tenganan Village, Amlapura, descriptive analysis method was used. Identification and analysis based on the sustainable environmental aspects. It used a parameter to assess the state of the city in macro element (public facilities) and micro element (house or traditional architecture). The aspects used as sustainable environmental parameters on traditional architecture in Amlapura are as follows:

- The ability to adapt to the climate, and utilization of local building material and traditional architecture typology.
- Community participation in preserving the sustainability value of traditional architecture will be analysed using quantitative method with 40 random respondents (random sampling).

The parameters are chosen based on tangible elements because the overall research is qualitative.

3. Discussion

3.1. Macro architectural concept of Amlapura

The area of Amlapura city center is formed by the existence of the former kingdom and the surrounding settlement design (royal hamlet). Activities undertaken in this area were related to the culture and in the context of efforts to empower the objects of cultural heritage building and trade and service activities. The effort to develop the image of the glory of this region in the past is by making the old city as a cultural tourism (heritage), a city centre, or a landmark. Amlapura city center can be developed as the center of cultural-related activities, while the supporting area is expected to be developed in line with the center of the region and in accordance with the characteristics of each cultural heritage objects in both tangible and intangible. The goal is none other than the creation of a living urban area, accommodating the activities and needs of urban space for the time to come as needed. The concept is used to maintain the existing condition of the building but revitalize the supported parts. Thus, it will be beneficial for the city of Amlapura itself as a tourist area of the old city culture that has special landmark characteristics and has a high ecological value to be an iconic form to maintain the traditional architectural design.

In reality, the preservation of historic urban areas, cultural heritage, traditional architecture is not just an effort to maintain and create a harmonious visual relationship between old and new forms. Utilization of old buildings in the historical area and traditional architecture in Amlapura, as an important contribution in the efforts to improve the quality of the area and to face the changes of environmental quality due to natural disasters. Therefore, the creation of a sense of place is more than merely restoring the authenticity of the city area. New in-fill developments should be made more meaningful and responsive, for the sake of ensuring social, cultural, and environmental sustainability. It means that preservation through physical intervention should also be understood as an effort or continuous activity in shaping the environment

3.2. Architectural traditional concept in Tenganan village ability to adapt with the tropical climate and disaster

The concept of traditional architecture in Tenganan village, Amlapura is a form of architecture that is made as a way to adapt with the hot tropical climatic conditions and contouring landscape. Tenganan village is located on the mountainous area with sloping contoured land conditions. The condition is tricked by cultivating staircase land following the slope. The village design is staircase lengthwise higher as it goes to the North. The settlements are contoured with cut and fill system. The foundation of the building used stone to adapt to the humid tropical climate on the mountainous area. The small and closed building is used to adapt cold mountainous weather. Each settlement is divided into smaller clusters so it has many open spaces with air holes for natural circulation.

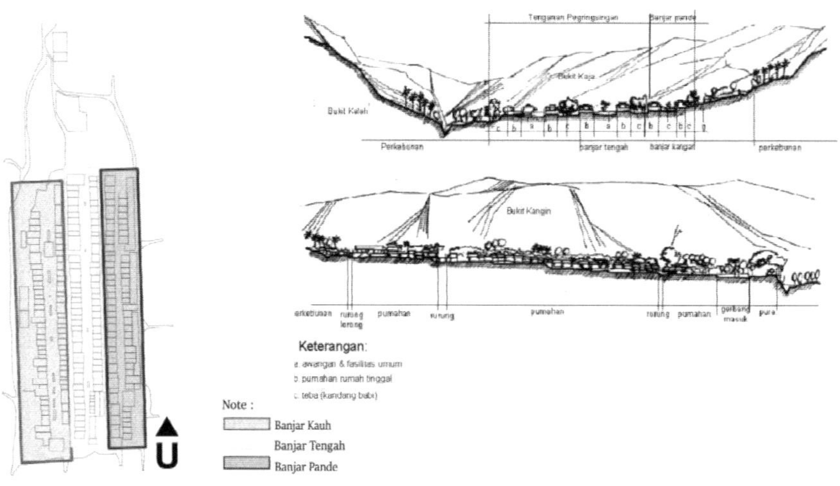

Figure 1. Settlements space pattern in Tenganan village, Amlapura[2].

The sunlight can expose the insight of the building due to a lot of holes and impermanent partition. The roof is made of the mix of *ijuk*, tile, and sago palm (*rumbia*). *Ijuk* and sago palm (*rumbia*) are light materials on top of the roof. It creates cool shade and feels integrated to the nature. Many households nowadays use tile roof due to supply shortage of *ijuk* and sago palm (*rumbia*). The settlements are adjacent in series, facing each other with the face of the building head to the *awangan*. The settlements in Tanganan village is made with traditional architecture concept representing the community to preserve the environment and able to survive from natural disaster such as landslide, flood, as well as impacts of global warming.

3.2.1. Utilization of local building materials as a concept of traditional architecture design.

In general, the building materials for traditional settlements in Tanganan village, Amlapura are taken from local area, including both public buildings (the buildings placed in the front), and private buildings (the settlements). Most of the materials are taken from forest garden around the community area and approved by customary chairman. There are custom rules, which regulate the utilization of forest garden as building materials called *awig-awig,* because Tenganan village uphold preservation of the environment. Lintang and Arum (2013), mentioned some of those rules (*awig-awig)* are [3]:

- The tree is allowed to cut down if it is already dead and it is should also be reported and check by the village.

- The tree that is still alive is allowed to be cut down as building materials if it is their own, as for a house of just married family and should be approved by the village. It is called *tumapung*.
- Tree logging is allowed for village needs without any consideration of the tree condition and its owner because of *rampangan,* for instance to build temple.

The value of forest preservation as the basis of their rules is the main point that Tenganan village is able to survive from of natural disasters or the effects of global warming that occur today. Therefore, the village of Tenganan is called as a pleasing village, cold, and is being a cultural heritage of Amlapura city due to its natural conditions and architectural concept of its own characteristics. More clearly the form of traditional settlements and environmental ecological conditions found in Tenganan village, Amlapura can be seen in figure 2.

Figure 2. Utilization of local construction materials in Tenganan village, Amlapura city [7].

Such rules are highly obeyed by the community, thus it is really support sustainable environment in the village. In addition, eventhough this Tenganan village is located in the mountainous area, it was very rarely affected by the natural disasters which can destroyed the local community.

3.2.2. Element structure of traditional architecture in Tenganan village.
Traditional architecture form in Tenganan Vilage, Amlapura is micro observation by looking at its elements. Then, make a map of function and value of each part of its architecture. Based on Lintang and Arum (2013) [2], figure 4 below showed the space of building typology in Tenganan village that used by the public and some examples of local building materials.

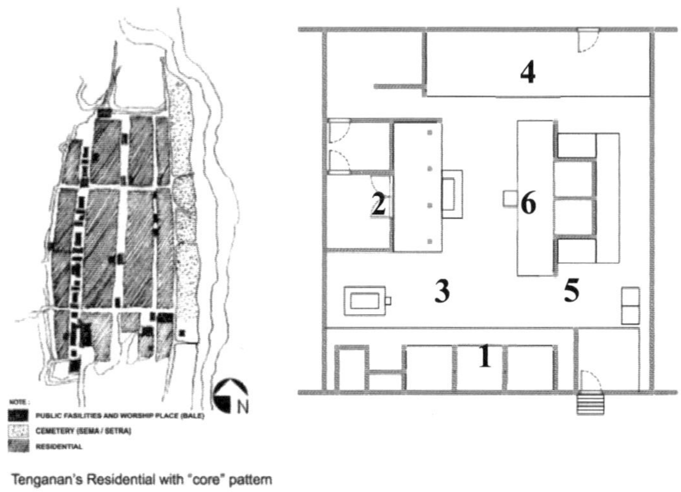

Figure 3. Elements structure of architectural design of Tenganan village, Amlapura city[10].

- Bale Agung

Bale Agung is located in *awangan* from the North to South, in which the building is rectangle. The foundation is made of stone stack of 50x5 meter and 1 meter height. The body is 47x3 meter and 1.5 meter height. The body is made of jackfruit wood (*ketewel*) with 28 pillars. The head is made of coconut leaves. The building materials of Bale Agung are taken from village's forest garden.

- Bale Banjar

Bale Banjar is a sacred building, which is also a village's reserve. The body is 13.5x4 meter and 50-centimetre height made of river stone stack glued together by clay. The body is 12x2.5 meter and about 1-meter height. The column made of wood, while the roof frame made of wood and bamboo with cover made from coconut leaves (*palpalan*).

- Bale Kenca

Bale Kenca or also called Pekenca is used as a place to discuss important case. The building base is square with 4x3 meter and 70-centimetre height. The foundation is made of stone stack glued together by clay. The body is 3x2 meter and 65-centimetre height made of wood and bamboo in 4 (four) columns. The roof cover is made of *ijuk.*

- Bale Petemu (Patemon)

Bale Petemu is used as a place for youth organization meeting (*seka teruna*). It consists of 3 (three) rooms, Bale Petemu Kaja, Bale Petemu Tengah, and Bale Petemu Kelod. The foundation is made of river stone, except for Bale Petemu Kaja which is made of brick. The column of all the buildings is made of wood. The roof is made of wood and bamboo with cover is made from *ijuk.*

Based on some explanation from the sample of housing or buildings in Tenganan village, Amlapura, it can be concluded that the building in Tenganan village, Amlapura uses traditional local building materials which is available around the area such as river stone, bricks and clay as the foundation, and wood and bamboo as the head. Meanwhile, the roof uses i*juk* and coconut leaves. These have been appropriate with the concept of sustainability environment architectural with low cost and low technology, which minimizes the negative impacts toward the building itself and the environment. Therefore, traditional architecture in Tenganan village becomes cultural heritage of Amlapura as the preservation and conservation of the traditional architecture in East Bali.

IOP Conf. Series: Earth and Environmental Science **99** (2017) 012018 doi:10.1088/1755-1315/99/1/012018

3.2.3. Community participation in sustainable conservation value of traditional architecture in Tenganan village, Amlapura

Their attention to the environment used to manifest their involvement to participate in maintaining their traditional architecture sustainability. The data used in this study is the primart data obtained by a questionnaire. The indicators used in determining the level of community participation are:

Tabel 3.1 Community participation parameters on the conservation of traditional architecture design in Tenganan village, Amlapura. (a)

Indicator	Parameters
Involvement in community event	• Interest to be involved in conservation efforts. • Provide opinions or solutions or ideas, etc.
Provide helps	• Provide helps in form of energy or thoughts or money, etc.
Attendance frequency at the meeting	• Eager to be present and often involved in every activity.
Obey the local rules (*awig-awig*)	• Never break the rules. • Build houses by using traditional architectural concepts.

Table 3.2 Grade criteria for the respondents.

	Score
Very Agree	5
Agree	4
No Opinion	3
Disgree	2
Strongly Disagree	1

This analysis model explains the respondent's statement by describing it via tables, and the measurement using Likert scale. Table 3.3 below showed the total score of all respondents are:

Table 3.3 Total score of the respondents.

Information	Score
Maximal	40 (respondents) x 5 = 200
Minimal	40 (respondents) x 1 = 40
Median	40 (respondents) x 3 = 120
Quartile I	40 (respondents) x 2 = 80
Quartile II	40 (respondents) x 4 = 160

Based on Sugiyono (2008), the scores was analyzed using several approaches to determine the level of community participation:

- If Quartile III < Score < Maximal; means very positive. (active community participation).
- If Median < Score < Quartile III; means positive. (community participation is considered quite active).
- If Quartile I < Score < Median; means negative (Community participation is considered less active).

- If Minimal < Score < Quartile I; means very negative (Community participation is considered inactive).

If projected, the level of community participation can be calculated based on the following formula:

$$\text{Participation Rate} = \frac{\textit{Obtained Scores}}{\textit{Maximal Score}} \text{ x } 100 \qquad (1)$$

This section will measure the level of community participation based on the benchmarks that exist in each indicator. Obtained data then processed using Microsoft Excel, to facilitate data processing. Table 3.4 below will explain it.

Table 3.4 Final score by the respondents based on the parameters.

Parameters	Score	Median
Interest to be involved in conservation efforts.	145	120
Provide opinions / solutions / ideas, etc.	146	120
Provide helps in form like energy / thoughts / money, etc.	129	120
Eager to be present and often involved in every activity.	137	120
Yery abide by applicable rules with never breaking .	149	120
Build houses by using traditional architectural concepts.	137	120
Total	140,5	120

To be able to know the level of community participation can be seen from the overall score obtained. In table 3.4 it is shown that the overall score obtained is 140.5. The score is between the median area and the third quartile, or in the positive area. So it can be said that the level of community participation supporting the preservation and resilience of traditional architecture in Tenganan village, Amlapura is active.

$$\text{Participation Rate} = \frac{\textit{Obtained Score}}{\textit{Maximal Score}} \text{ x } 100 \qquad (2)$$

$$= 140{,}5/200 \text{ x } 100\% \qquad (3)$$

$$= 70{,}25\ \%$$

4. Conclusions

The Tenganan Village architecture uses traditional architectural concepts based on nature, climate, and material availability are taken to the account. The location of the settlements in the mountainous area is combined with the shape of the village that has contour site, with the adaptation of the building form to the tropical climate of the mountains. Building materials use natural materials, with village (*awig-awig*) rules as the main guidelines, which can balance the needs of the community with existing natural resources. The use of traditional architecture concept that consider the needs and availability of present and future made Tenganan village being a cultural heritage of Amlapura and able to survive in tropical climate and slope contour. The concept of balance between human and nature should be able to be applied in every architectural design. The concept of sustainability in traditional architecture is able to maintain natural conditions, thus reducing the impact of global warming and natural disasters. Community participation through obeying the rules of *awig-awig* is quite pleasant in supporting the preservation and resilience of traditional architectural design in Tenganan village, Amlapura city.

Acknowledgements
Many thanks to the Tourism Office for the data and participation and also for our relatives who are involved to help this from the beginning. We wish to give away the benefits of this study for the heritage tourism and ourselves, either nowadays or in the future.

References
[1] Tanuwidjaja, Gunawan 2011 Sustainable Architectural Design in Indonesia: *Responding the Current Environmental Challenges*. SENVAR 12th University of Brawijaya: Departement of Architecture Faculty of Engineering
[2] Veronica and Setia 2009 Pola Perumahan dan Pemukiman Desa Tenganan Bali. 1. 3-5
[3] Lintang and Arum 2013 *Arsitektur Lingkungan Berkelanjutan pada Permukiman (Sutudi Kasus Desa Tenganan, Bali).* 3-9
[4] EngineeringDepartment of Infrastructures Region, Directorate General of Rural and Urban Governance 2004 Technical guidelines for structuring and Revitalization Area. Bali
[5] Suharjanto, Gatot 2011 *Membandingkan Istilah Arsitektur Tradisional Versus Arsitektur Vernakular: Studi Kasus Bangunan Minangkabau dan Bangunan Bali.* **2** 592-602
[6] Satrio, Janus 2009 Preservation Of The Ancient Region Between Concept And Reality. Spatial bulletin ISSN: 1978 – 1571 Edition of November-December 200 The Ancient Relic Departement Directorate Culture And Tourism
[7] Department of Tourism and Culture of Karangasem Regency of Bali 2010-2029 2010 Spatial Plan Area of Bali, Bali.
[8] Strategic Plan Culture and Tourism of Karangasem Regency in 2011-2015 2011 Department of Tourism and Culture of Karangasem Regency
[9] Triwardhani, Ayudia 2010 Study of Elements City Space Improvement Image of Area Value of History, Old Town Case Study Benteng Kuto Besak Palembang (Master's degree course Planning Regions and Cities). Bandung Institute of Technology. Bandung
[10] Budiharjo E 1986 Arsitektur dan Kota Indonesia
[11] Law No. 11 2010 about Heritage
[12] Zahnd, Mark 1999. Designing and Integrated City. Kanisius. Yogyakarta

Mural art as a media on making urban kampung's public space

Dalhar Susanto* , Widyarko Widyarko, Ajeng Nadia Ilmiani
Department of Architecture, Faculty of Engineering, Universitas Indonesia, Depok, Indonesia

Corresponding E-mail: dalhar3001@yahoo.com

Abstract. The lack of public space is one of the main problems in the big cities in Indonesia. Urban kampungas part of the city is also no exception. Rapid growth on population sparks uncontrollable physical development that erode open space inside urbankampung . Sometimes, what is left is just neglected space which don't 'live' and far from the definition of public space. Mural art has been existed since the beginning of human civilization. Now, it has evolved into one of the popular urban art. The previous research has proven that the process of urban art making through participatory approach could trigger community interaction in a space. Interaction itself is a main factor that may trigger the establishmentof a public space. With the same method, this research attempts to build mural in a neglected space inside urbankampung named Palsigunung. After all of the process done, the space still haven't changed from the previous condition,which is still a neglected space. Together with facilitator,kampung's residentsneedto be involved identifying the problem and also the solution to the lack of public space in their kampung. Particularly for urban kampungPalsigunung, the needed solution might not be mural.

Keywords: urban kampung, public space, participatory, mural art

1. Introduction

In 2011, a new history has been created in Indonesia where more than half of the people live in urban areas. The yearly growth of occupant in urban areas of Indonesia has been drastically decreased from its climax of 5.71% in 1982 to 2.59% in 2015. However, the number is still twice the national population growth of 1.2% in 2015 [1]. This rapid growth started in 1960 which is marked with the change of focus in national economy development that was initially based on agriculture and changed to industry [2]. It triggers major urbanization from regions to big cities in Indonesia. Later on, informal settlement arise taking both the newcomers and the city'snative inhabitants which could not be accommodated in the formal housing system at the time. Urbankampung is a generic word which can be used to represent the informal housing inside the city. To this day, the word kampung is not enough to be pictured with only physical description such as: high-density, informal, or low-cost housing, but it also describe the whole community who lives in it [2]. As community, kampung residents has a very close social relationship. "Neighbours are more valuable, special and important than relatives who live in other places", This traditional phrase is well known among Indonesians. It describes the concept of relationships among people in kampung [3].

Human beings are social creatures and the balance between the public and the private life appears to contribute to a healthy psychological and physical life. Public life as opposed to a private life provides opportunities for engagement and social contact [4]. Currently, public space as space for public life has become a rare place in urban kampung. While the space of urban kampung have not changed, its population keeps increasing, thus the area for its public space is getting shrinked. What is left sometimes is just a narrow alley or just a'left-over'space. This public space could decrease more, or even gone as a result of horizontal development of housing within kampung. When that happens, people willforced to utilize other open space which previously neglected for their new public space. The problem is that the concept of public space can be interpreted differently by communities according to each culture [5]. To change public space is not an easy matter because the neglected space that is forced to be made as public space may be not relatable to the culture of the community itself.

Mural art is a painting which is done in a wall surface or ceiling.In the history of modern civilization in America, this kind of art form was once obstructed because the development of 'modern art'. The art in that era is not 'owned' by the public and doesn't have any social role. The art that was approved was exclusively owned by artist. This changed in 1960 when the exclusivity of art was turned to the public, marked by the booming of urban art such as statue, mural, or other art form in almost every big city in America [6]. Now the use of mural has been spreading in the world as a 'filler' for public space in which not only limited as esthetic element but also as media for spreading information, marketing, and political movement. Chicago, USA is an example of the city using mural to show their expression of political views; meanwhile, Penang, Malaysia government sets up regulation to protect their heritage area as canvasses for the city creativity. Along with the advance of technology, mural making can also be developed not only by using paint but also by using printed wallpaper.Today, the presence of mural could also have side-effect. In physical context, several studies have proven that placement of mural art in the wall will decrease vandalism such as graffiti attack [4][6] [7]. In non-physical context, its presence could be used as space revitalization tool – mural that is presented in aneglected space could re-activate the places and its community [4]. If the mural is created based on local issue or theme, it will be worked as communal self-expression that could shape the sense of pride and sense of ownership for the community in which its present [4]. Although several literatures have agreed on the positive effect of mural, the presence of mural could not be appeared instantly without any process.

Figure 1 Dr. Siddha Webber and team create a political themed mural in Bronzeville, Chicago. Pictured by Max Herman

Figure 2 Mural as tourist attraction at George Town, Penang, MY. Pictured by K.E. Ooi

A place like public space is not a space that could be shaped by one-sided intervention, the planning should involve active participation from the community or in other word, public space is not something that could be created from the 'outside' [8]. According to classical design process,community/user only worked as a passive study object. This process could not guarantee the implementation of successful public space, but active participation which previously mentioned is co-design approach, which is a collaborative planning process between user and planner by viewing both in the same level of mastery. For this reason, the user has to be involved from the beginning of planning process [9]. Participatory approach through co-design process is almost fully guaranteed that the final product will be suitable with the values embraced by the user and also the history and time the product is used [10]. Mural as a final product is expected to fit the values comprised by community in which its present. However, several literatures mentioned that the emphasis of participatory practice is in the process. Collaboration experience in the practice encourages people who never interacted before to start communicating. After the practice, the interaction among community member is proven to be increased [4][10]. Therefore, mural construction through co-design participatory approach with community is expected to achieve two things; forming the sense of pride and sense of ownership of community to the mural and to enrich the interaction between people. This aspect is expected to be a catalyst of people interaction occurred the public space where the mural is present.

2. Research Methods

The general purpose of this research is try to understand the effect of mural which created through participatory approach in a neglected space inside urbankampung. Literature mentioned that the inclusion of public art such as mural must involve community as much as possible. Thus, the ideal research method for this case is action research method. By using this method, the research process is functioning democratically by the involvement of participant (kampung's residents) in most of the stages. Action research is a research method in which the researcher describes, interprets and explains particular social situation at the same time by doing changes or intervention with the purpose of fixing [11]. The researcher in this study has a role of facilitator while the community is the participant. The knowledge of facilitator in this process is constrained as much as possible, resulting that the participant will do the most important role in the mural making. Facilitator is also a 'adviser' in the co-design process which only give knowledge of a good technical workmanship for mural making. Observation-evaluation practice in action research is not only held in the end of practice (post-mural built) but linear since the beginning (mural planning).

3. Research Context: Urban Palsigunung

Palsigunung Kampung is one of the examples of an urban kampung, which is located at Depok City, in the province of West Java, Indonesia. This urban kampung has a population of 929 inhabitants with a percentage of 71% adult population (> 20 years) and 29% children-adolescent. Thekampung's location is very strategic, because it is located at the northern border of the city and adjacent to the East Jakarta municipality. This kampung is located between two main roads, which are Jalan Raya Bogor (Raya Bogor Street) on the eastern side and Jalan M. Jasin (M. Jasin Street) on the northern side. Both main roads are very crowded with vehicles and serve as a connection between cities or as one of the arterial roads of Depok City. On Jalan Raya Bogor, particulary on the area which intersects with the kampung, Traditional Tugu Market, supermarket or other economy facilities are located. Space for the public life of the kampung members has been through a dynamic process since 2007 until 2017. According to Mr. Budi, one of the kampung'spublic figure, formerly the kampung had a volleyball field, which they usually use. Because the field was surrounded by chairs and also had a small stall located nearby, it not only functions as a sport facility, but it became a place where kampungresidents socialize in their everyday lifes. At certain moments, this field is often used as a place to celebrate certain occasions such as Indonesia's Independence Day event, Election Day, or even weddings. The users are all of the kampung members from elderly to children, and from men to women. These descriptions attest that the volley field is the only public space for the kampung at that time. Unfortunately in 2011, this public space was no longer usable because the legal owner decided to sell it to newcomer, which then build residential building upon it.

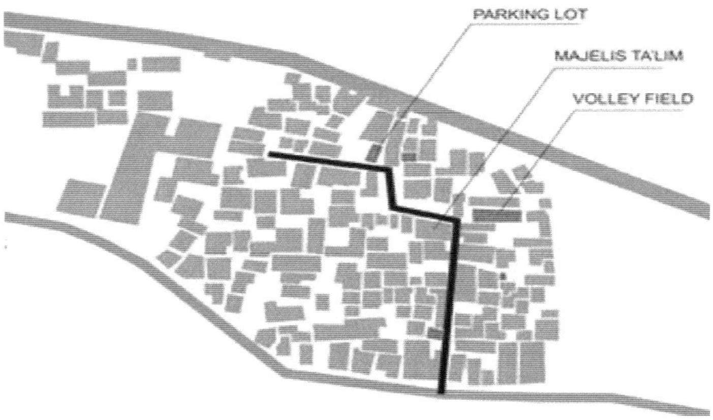

Figure 3 Figure and ground plan of KampungPalsigunung

Meanwhile for the education facility, in a radius less than 500 meters there are various kinds of school from various levels (kindergarten to high-school). Arround this kampung there are also several health facilities such as TuguIbu Hospital and Puskesmas (Government's community clinic). Therefore, the information above proves that PalsigunungKampung, like other urban kampungs, is surrounded by various public facilities. However, a different condition occurs inside the kampung. In this kampung, there is only one road with 100 meters in length that can be passed by single four-wheel vehicles. Other than that, the kampung circulation is mostly dominated by 'narrow' alleys with a width of 1 to 2 meters, which can only be accessed by pedestrians or two-wheel vehicles. Based on the satellite imaging analysis, in a period of 2015-2016 there are at least 25 new residential buildings that built on the area that used to be open spaces. This condition surely affects the public life of the kampung itself.

Figure 4 Cramped situation in 24 sqm public space in front of Mr.Budi and Mr.Agus house

Figure 5 Parking space that temporarily converted into Independence Day celebrating area.

At this time, the public life for the kampung members is no longer concentrated in one area, but it is spread to several areas, where one of it is at the MajelisTa'lim Building. With a dominant user of adult and elderly, this building serves its formal function as a place for religious activities like reading Al-Quran and praying together. Despite its formal function, this building is also use for social activities between kampung members or as a meeting space for important discussion. At first, this building was being donated by one of the kampung's resident for the communitiy. Nevertheless, after the donors was deceased, one of his heirs sued his ownership status of the building to the court. Like the previous volleyball field, the existence of this building is in danger of being lost.

After 2011, kampung members celebrate big public events like the Independence Day celebration at an open space with a size of 150 sqm. This open space can be declared as the largest empty open space that the kampungs have at this time. When that day takes place, community can build a performing stage and arrange other competitions. However, this open space is still far from the definition of a kampung's public space, because the use of it by the community is still categorized as temporarily. Other than the celebration days, this open space only serves as a parking area for kampungmembers who doesn't have cars. In everyday life this space is filled with four-wheels vehicles, which leave no rooms for interaction between residents.

As for now, the active open space, which becomes the place for kampung communities to gather is a place that wasn't expected before. This open space is actually a joint terrace from two houses that is own by Mr. Budi and Mr. Agus. With an area of about 24 sqm (4 m x 6 m), this open space can be defined as the last public space that the kampung has because it is actively used by all community elements. Children use it in the mornings to the late afternoons as their play space, meanwhile from late afternoons to the evening this area is 'owned' by the adults. Adults that usually use this space are mostly men, using it as a place to relax. Besides relaxing and having conversations, common activities that these men do in this 'narrow' space is: watching soccer games on television, playing chess, and playing cards. Based on interviews with the kampung members, this space just became active in 2015 after some institutions from outside the kampung donated a non-permanent installation in this space. The installation that is made of wood material was initially meant to be use for a mini library. Instead, its existence became a catalisator for kampung's residents to start gathering and interacting with each other. The issue of this space is that the human capacity is very limited. During one of a visit by researchers, it is observed that this space can befilled by 20 human at once. The overall explanation about this kampung indicates that eventhough the public life of the kampung is still going along, the space for it has decreased from time to time. This urban kampung could be a stereotype condition that occurs to all urbankampung in Indonesia.

4. Process and Result

4.1. The Neglected Space of KampungPalsigunung.

The chosen neglected space for the mural making, is located at an open space that can be accessed from an alley of 2 meters in width. Like all neglected space, this empty space with an area of 40 sqm is always far from human activities. Other than because it is the largest unused space in this village, this space was chosen because of its strategic location (almost at the center of the kampung). At first, this was a place for collecting unused construction materials, such as rock, sand, even concrete. The owner then supports to allocate his space as public facility rather than not be managed. Moreover, he initiated to clean the area and made the concrete cover for the flooring. The neglected space given became one of the resident participation to the community.

Figure 6 The chosen neglected space before the drawing started. The left pictures the first condition, the right shows the condition after concrete covering.

4.2. Socialization and Discussion

This phase invited all the community member to gather and discuss for the plan proposed. Located at the MajilisTa'lim building, around 50 villagers consisting of women and children came to join the socialization and discussion phase.This phase is done to give awareness to the community about the aim of the activity, and also listening to their expectations about both the mural and the space.

Figure 7 Situation inside MajelisTa'lim building during socialization and discussion phase

Figure 8 Children as participants drawing concept for mural during co-design phase

4.3. Mural Co-design
At this phase, village members acting as participant will be acoompanied by facilitators in designing the mural that will be created. Most participants are the children although the invitations were addressed to all kampung's residents. The facilitators triggered the children to draw anything related to the theme used, green kampung. Green kampung theme came from the dweller willing to make their kampung more livable and comfortable. The drawings made then was used as the mural design's basic idea.
4.4. Implementation
Village members, represented by children and teenagers created the mural while accompanied by facilitators. Unfortunately, the adults which were expected to join the process didn't come along the implementation; whereas, their technical skills would support more.Besides drawing mural, used rubber tirewere also re-furbished as chairs and placed at this space as a media for sitting and also playing for children. These additional elements are results of the discussion from the previous phase.
4.5. Result
Based on the observation done by researchers, this open space is only used by children several days after the mural had been created. Afterwards, this space went back to being a neglected space.

Figure 9. Children working on rubber tire that will works as chair

Figure 10. Children and facilitator working on mural on the wall

Figure 11. Facilitators and children taking photo after mural completion

Figure 12. Situation in the neglected space after the mural has been constructed.

5. Discussion

The construction of mural and its participatory process have failed to be a catalyst for creating a public life in a neglected space inside the urban kampung. This failure is questioned for the researcher because the literaturs and previous studies mentioned otherwise. However, it must be acknowledged that the practice process has several important notes to be discussed. The great purpose of this practice is to create a public space. The problem is 'public' in the word public space is incoherent, not a singular group but an entity that could be split to several groups based on the aspect of socio-economy, age, and gender. Each group will relate itself to the public space in different approach [12]. In participatory process which was held in Kampung Palsigunung, the group that followed the entire process is only the children. Other groups only present in the beginning process (socialization and discussion). This shows that the presence of 'public' in this process is still incomplete, therefore the public life in neglected place could not be created.

The general stages of participatory process held in the practice is divided to five major phases; socialization, discussion, co-design, implementation, and evaluation. When each phase is completed, the next phase will begin by ignoring the problems found in the previous phase. Sanders said that participatory is not simpy a method or set of methodologies, it is a mindset and an attitude about people [13].The major evaluation note taken from the first phase (socialization) shows that even though all of the community were invited, only the group of mothers who dominate the guests. A good participatory process is involving the proper analysis to solve a conflict, consensus building, and also decision making in the community. In other word, this practice focuses on the dynamic process that happens within, not on the result [14]. Hence the implementation of participatory practice in this was not supposed to be working linearly but circularly so the spreading of information and consensus could reach all of the elements inside the community.

Another issue in this research lies in the 'mural as solution' itself. Before the practice begins, mural has become a 'goal' that is wanted to be achieved. The researcher hoped that the 'process' for finishing themural could be the catalyst of public life. This research placed the researcher or facilitator as the party who did the problem identification and making the solution to solve the problem, which is mural. The presence of participant orcommunity was only involved after the first stage, which is the constructing mural with participatory approach. According to Masters, action research method with social-economy problems could only be solved by placing the power on the group, not to the individual either the facilitator/researcher or participant [11]. In other word, the identification of problem and solution should be decided together amongst the researcher andcommunity, not only one-sided like what happened in this study. Therefore, the solution needed by the people of urban kampung Palsigunungfor creating new public space might not be mural.

Moreover, there is a possibility that the mural is still seen as a way of vandalism. The residents in Kampung Palsigunung are familiar with the unwanted graffiti which easily found in some spots. Those graffiti express the identity of individual or group of people, and sometimes the drawings are not showing the proper morals. Although the purpose of the mural is more to reactivate the neglected space, the residents still accept it in the opposite opinion. The participation which only comes from the segmented group -the children and teenagers- still cannot change the adult perspective about the mural.

6. Conclusion

Carmona said that there will never be a one-size-fits-all universal model of public space, and critiques and celebrations of public spaces alike will always require questioning and interpreting in the light of localcircumstances [12]. The statement is relevant to this research because the process and final product that has been succeed in other place could not create any effect in KampungPalsigunung. This conclusion is not suggested that the participatory approach of making mural will not able to develop public life in urbankampung. Nevertheless, urbankampung's community together with the facilitator, need to identify the problem and seek the solution themshelves for the issue of their decreasing public space.With the heterogenousnature of communities around Jakarta and its surrounding areas, the possibility of both problem and solution can be different on each urban kampung.

References

[1] World Bank. (2017, May) http://data.worldbank.org. [Online]. http://data.worldbank.org/indicator/SP.URB.GROW?end=2015&locations=ID&start=1960&view=chart

[2] Happy Ratna Santosa, Purwanita Setijanti Rita Ernawati, "Facing urban vulnerability through development," Humanities and Social Sciences, pp. 1-6, 2013.

[3] Antony Sihombing, "Living in the s: a Firsthand Account of Experiences in Jakarta's s," in FORUM International Journal of Postgraduate Studies Architecture, Planning and Landscape University of Newcastle, Newcastle, UK, 2002, pp. 15-22.

[4] Tony Matthews Deanna Grant-Smith, "Cork as canvas: Exploring intersections of citizenship and collective memory in the Shandon," Community Development Journal, pp. 138-152, 2015.

[5] Dasimah Omar, Rozyah Mohd Yunus, Zalina Samadi Maimunah Ramlee, "Revitalization of Urban Public Space: An Overview," in Asian Conference on Environment-Behavior Studies, Teheran, 2015, pp. 360-367.

[6] Anusapati, "Murals in Yogyakarta's Public Places.," download from: http://urp.fib.ugm.ac.id/images/download/1901121350Anusapati.pdf.

[7] Craw Loius S, Leland Michelle G, Bussel Simon J, Munday Karen W Penelope J, "The Mural as Graffiti Deterrence ," Environment and Behavior , vol. 38, no. 3, pp. 422-434, May 2006.

[8] Karen Puren Weesel Strydom, "A Participatory Approach to Public Space Design as Informative for Place-Making," in Challenges of Modern Technology , Warsaw, Poland, 2013, pp. 33-40.

[9] Elizabeth B.-N.Sanders & Pieter Jan Stappers, "Co-creation and the new landscapes of design," International Journal of CoCreation in Design and the Arts, vol. 4, pp. 5-18, June 2008.

[10] Lois Frankel Gabriela Maria Sanches, "Co-Design in Public Spaces: an Interdisciplinary Approach to Street Furniture Develpoment," in drs2010 Unmontreal, 2010.

[11] Janet Master, "The History of Action Research," I. Hughes (ed) Action Research Electronic Reader, The University of Sydney, 1995.

[12] Matthew Carmona, "Re-Theorising Contemporary Public Space: a New Narrative and New Normative," Journal of Urbanism , pp. 373-405, 2014.

[13] Elizabeth B.-N. Sanders &, "From user-centered to participatory design approaches," Design and the social sciences, 2002.

[14] Frances Cleaver, "Paradoxes of Participation: Questioning Participatory Approaches to Development," Journal of International Development, pp. 597-612, 199.

The impact of green building approach to office property value

Yosephine Sitanggang[1*], **Dalhar Susanto**[1]
[1]Universitas Indonesia, Depok, Indonesia

E-mail: yosephinesitanggang@yahoo.com

Abstract. A real estate development often produces negative impacts towards the environment such as the reduction of the ecological capacity in the site and its surroundings, energy exploitation, and excessive pollutant emission. To overcome these issues, the green building concept or approach has been adapted by several real estate businesses in Indonesia especially in the office sector. According to the data provided by GBCI in 2017, there are 17 buildings listed as a certified green building office in various levels. As what has been known, the green building approach results in the increase of price in the planning, construction and the building's maintenance. This paper will discuss about the research results regarding the effect of the green building approach towards the property value of office buildings especially in Jakarta. The research will be executed through the comparison method, which is the process of comparing office building that have already adapted the green building concept with the one that have not, or in other words, the conventional office buildings. Data gathering is done through observation and interviews with developers and building managers. The research results show that by adapting the green building approach for office buildings in Jakarta, the property value regarding the utility, scarcity, effective demands, and transferability aspect can increase.

Keywords: real estate, green building approach, office, property value

1.Introduction

A real estate development activity that proceeds excessive construction can harm the environment. The building sector are the biggest energy consumer and primary contributor of gas emission in global warming [1]. Human activities that dominantly contribute in terms or the main energy consumption happen in the building sector [2]. To lessen buildings' negative impact towards the environment, a certain approach has been developed to improve environmental awareness, which is also known as the green building concept.

Green building constructions are more costly compared to those conventional buildings, especially its soft-cost expenses due to the addition of design, analysis, technical, efficiency calculation, and green building certification cost [3]. As quoted by [4] according to CalRecycle, green building approach might need a bigger capital expenditure, but by applying its purposed concept, green buildings can save operational costs [4]. Therefore, an important question rose: will the amount of capital expenditure due to the application of green building concept on a certain real estate development project be proportional with the kind of benefits green buildings can provide so that it can increase the value of a property, in this case office buildings.

Green Building Council Indonesia [5] explained how the green living or green building approach is an innovation towards creating a residential area that fulfils its performance in terms of land use, build to save water and energy, having facilities that lessen waste, and maintain the air quality inside a room. According to [3], the green building approach is a technological innovation. This approach is not new to the real estate development. Woodson [6] explained that this approach began to be adapted and developed in several countries around the world since the 1980s [6].

Research regarding the green building approach towards property value has been conducted in several countries. Yudelson [3] explained that the benefits obtained from applying the green building concept on a building is: lessen operational costs; lessen maintenance costs; increases productivity of users; management risks and relationship between stakeholder; environment management; **increases property value**; and produces a product that is more competitive in the market in where the green building has a lower operational cost and possess a better environmental quality will overshadow conventional buildings [3].

Quoted by Kubba [4], studies by "Landmark International" (2008) showed that the green building approach can increase value (can be seen from the increased sales and leases, increased occupancy, and decreased changes of ownership) [4]. According to Fachrudin [13], based on the research by "Research of Institute for Building Efficiency" (2010), it is explained that the green building approach has advantages such as the increase of occupancy rate, sale and leasing value, decrease of capitalization, and the increase of productivity [13]. Popescu [1] also confirmed that green buildings can decrease operational costs, increase reputation and image, decrease negative impacts towards the environment, and decrease the use of natural resources [1].

A real estate development is defined as an activity that develops a project through building parcels on a land with the goal of increasing its real estate value in order to gain profit. According to [11] property's product can be classified into four categories which are residential, commercial, agricultural and special purpose. Shilling [11] stated that office buildings are considered as a sub component of a commercial property. Shilling [11] explained that residential and commercial properties are the primary type of developments that are being exchanged due to the high value. According to Schmitz & Brett [10], office buildings can be differentiated through several factors, which are: class which that covers age, location, quality, building's system, available facilities, rent price, and renters' profile; location; size and flexibility; function and ownership; along with its features and facilities [10].

A real estate development's nature is an activity that creates and increases the property value. [9] stated that a property value has four aspects, which are: (1) Utility, which is the function or utility of a property in fulfilling the user's needs; (2) Scarcity, the property's level of availability; (3) Effective demand, the level of needs and purchasing power; (4) Transferability, its ability to transfer ownership [9]. According to Arnold [9], a property's value is determined by those four aspects. A property is said to have value if it contains those four aspects.

2.Methods

A comparison was operated towards the property value of an office buildings who adapted the green building concept with one that does not (conventional office building). Its purpose is to determine the effect of the green building approach towards its property value. This research used the mixed method approach. According to [12] a mixed method is the combination of qualitative and quantitative method in where the analysis is in the form of text (qualitative analysis) and numeric data (quantitative analysis). To measure the value of an office building, the researcher has designed an instrument in the form of property value table assessment that covers four aspects which are utility, scarcity, effective demand, and transferability.

This assessment instrument was designed by conducting a theoretical review that relates to the quality aspects of an office building and aspects that relate to the green building approach. The assessment instrument was designed though incorporating valuation aspects from [7] research results which identified several factors that become indicators which affects the quality and utility of a general office building. Ho, Newell & Walker's [7] research results produced 27 indicators to assess quality and utility of a general office building. The indicators are classified into six categories which are functionality, services, access and circulation, presentation, management, and amenities.

The six assessment indicators are then modified with the addition of green building's assessment indicators. The researcher has included assessments from the green building approach according to worldwide institutions who has released a rating system such as GREENSHIP Indonesia, BREEAM, GBTool Canada, IISBE, LEED, EcoProfile, CASBEE Japan, Green Star, Green Globes, and BCA Green Mark Singapore. The resulted indicators from each assessment system has been processed by the researcher and is adapted with [7] assessment. The green building approach assessment indicator in this research consists of 29 indicators. In the designed assessment instrument, the researcher has found 5 indicators that consists of general conventional office buildings and green office buildings aspect which is called mixed aspects.

Yudelson [3] explained that the benefits of a green building are classified into two, which are measurable benefits and non-measurable benefits (which covers its own prestige). Therefore, the researcher has added 5 indicators as an assessment tool which relates to the level of prestige that is generated due to the application of the green building concept which refers to [3] opinion, brand image, public relations, enhanced marketing capability, market positioning, and"doing the right thing".

Assessment aspects were processed so that it can be used as assessment tools which can determine the value of an office building. The assessment instrument used 69 indicators to provide valuation towards an office building. The indicators that are used for the assessment instrument are classified into three general office aspects, which are general aspect indicators, green aspect indicators, and mixed aspect indicators. The applied assessment system for each indicator is presented through scoring with the range 1 (lowest) to 5 (highest). Each research object was valuated based on the assessment criteria that was previously determined by the researcher.

The total score that can be obtained from each research object will be transformed into a value that relates with four aspects which are utility, scarcity, effective demand, and transferability, in where these aspects contribute evenly in determining property value. In assessing a research object, the total score will be divided by the total score received from each indicator from every aspect and multiplied by twenty-five (the amount of value for each aspect will add to a hundred).

There will be two research objects. The first is an office building that has already applied the green building approach and is certified as a gold rated by GBCI, the Alamanda Tower green office. The second building is a conventional office building, the Talavera Office. The two buildings are located in Jl. TB Simatupang, South Jakarta. These buildings were chosen because both possess the same qualifications: both are office grade A, located alongside each other, more or less the same size, and possess an identical function.

Data gatherings was done in several steps which are, interviews with developers and direct observation to give out an assessment regarding its level of utility. To gather data for its scarcity, the researcher has collected data on the amount of office buildings that are located in the previously determined area. The amount of office types that were analyzed was then compared to the amount of office buildings located in the designated area. This was done to receive a scarcity ratio on the researched office building. The lesser the ratio received, the lower the scarcity of the office becomes. In gathering data for effective demand, the researcher collected data on the occupancy level of each research object. By doing so, the researcher can determine how much the office is demanded by the market. The bigger the level of needs, the bigger the value of the office becomes. To determine its transferability level, the researcher conducted interviews towards the development party. Through it, the mechanism of the ownership transferability can be determined. The more various the mechanisms are, the bigger the value becomes.

3.Discussion

Based on the assessment results on the two researched office buildings, it can be seen that the green office (Alamanda Tower) has a total score of 90.5, which is far greater than the conventional building (Talavera Office) with a total score of 47, for all assessment aspects which are utility, scarcity, effective demand, and transferability.

IOP Conf. Series: Earth and Environmental Science **99** (2017) 012020 doi:10.1088/1755-1315/99/1/012020

Based on the total points of the four aspects and the transformation in the form of value, the value that is received by the green office is bigger compared to the conventional office. According to analyzed research data, the green office's value on utility is bigger than the conventional building. From 66 indicators in the research instrument it is stated that the green office is more superior than the conventional building. This means that green elements in the green building approach is able to increase property value in terms of utility.

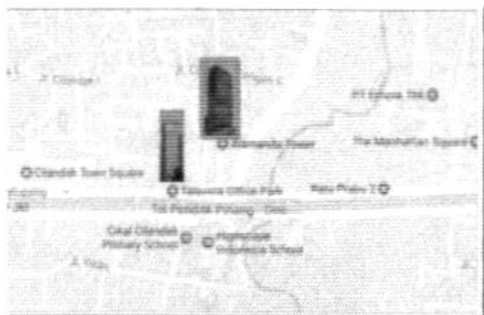	 ALAMANDA TOWER	 TALAVERA OFFICE
UTILITY 25% *66 indicators*	points: 285 **20.5**	points: 230 **17**
SCARCITY 25% *1 indicator*	points: 5 **25**	points: 1 **5**
EFFECTIVE DEMAND 25% *1 indicator*	points: 4 **20**	points: 4 **20**
TRANSFERABILITY 25% *1 indicator*	points: 5 **25**	points: 1 **5**
TOTAL	**90.5**	**47**

Table 1. Research Result Data
Personal documentation

In assessing the scarcity aspect, a scarcity ratio is needed from each research object which is the green office (an office building that inherits the green building concept) and the conventional office. In order to determine the scarcity ratio of the research object, the researcher has collected data on office buildings in the TB Simatupang area. Based on the office building list in TB Simatupang area, it was obtained that in the TB Simatupang area, office buildings that applied the green building concept (green office) and is already certified GREENSHIP from GBCI (Green Building Council Indonesia) is only covered by one building only, which is the researched building, Alamanda Tower (green office). Office buildings with grade A (Talavera Office, which is also the research object, is a grade A conventional office building) is covered by 8 buildings where in the same area is covered by 35 buildings. To determine the scarcity ratio of the research object, the researcher has applied this formula:

$$\text{Scarcity of Alamanda Tower} = \frac{\text{Total amount of green office in TB Simatupang area}}{\text{Total ammount of office in TB Simatupang area}}$$

$$\text{Scarcity of Talavera Office} = \frac{\text{Total amount of grade A office in TB Simatupang area}}{\text{Total office in TB Simatupang area}}$$

Scarcity ration of the research object is obtained by knowing the total amount of office types, in where if the calculation result is lessening (closer to zero), then that specific office type has a greater scarcity value. If a specific office building type is rarer, then its property value is higher. By using this formula, thus the result regarding the scarcity aspect of the two research object are as follow:

$$\text{Scarcity of Alamanda Tower} = \frac{1}{35} = 0.03$$

$$\text{Scarcity of Talavera Office} = \frac{8}{35} = 0.23$$

Based on the research analysis, green office is rarer than conventional office. This means that the green aspect that green offices inherit causes the building to be unique. This is caused by the lack of green office supply compared to the convention office buildings.

To determine the property value of the effective demand aspect, the researcher has gather data regarding the level of occupancy of each research object which is obtained by interviews with its developers. Based on the interview result, the level of occupancy of the green office is 95% while it is 90% for the conventional office. This means that for effective demand aspect, the green factor is not a big determinant for its property value, it can be seen from the level of occupancy of both research object (green office and conventional office building) that are relatively similar.

Transferability aspect relates with the type and form of ownership of the office space. The more types of offered ownership, the higher the property value will become. Based on the interview result from both the green office and the conventional office, it is proven that the transferability of both buildings have differences in where the green building offers two types of ownership of its office space, which is leased or strata titled. In contrary, the conventional building only has one type of ownership through leasing.

The analysis result of the office's property value stated that the application of green building approach can increase the property value of offices regarding the utility, scarcity, effective demand, and transferability. Based on the research result of the performed case studies, the researcher discovered that the application of the green building approach provides an influence on the property value related to utility and scarcity, but regarding the effective demand and transferability, the green building approach has not yet created a significant impact or influence on its property value.

This fact is similar to the statement of [3] about how the application of green building approach creates a product that is more competitive in the market with all its utilities and ability to create a better environmental quality. This relates to the utility aspect assessment in where the application of green buildings can increase its utility value.

The research result regarding the scarcity proves that office buildings that inherit the green building concept is rarer compared to those who do not (conventional office building), thus the value received on its scarcity is bigger than conventional office buildings. This is in line with [8] opinion on the awareness of the society on an environmental friendly construction thus increasing the need or demand of green labeled properties. The imbalance of high demand on green buildings and its lack of supply causes an increase in property value.

Regarding to its effective demand, according to the study by "Landmark International" (2008) and "Research of Institute for Building Efficiency" (2010), the application of green building concept can increase the occupancy rate. However, based on the research result of the study cases, the level of demand that is portrayed by the occupancy rate between the green office and the conventional one is relatively the same, thus the green building approach does not really affect the effective demand aspect for a property's value.

As for its transferability which is based on the offered ownership of office spaces, the green office provided more varieties of ownerships compared to the conventional one. However, it was not influenced by the green concept of the office.

4.Conclusion

The obtained research results that were operated to prove the hypothesis that "green building approach can increase the property value of office types" was confirmed by the comparison of two types of properties (office buildings who inherit the green building approach and conventional offices) case study, in where the green aspect in the green building approach is able to increase the property value of green offices. The green building approach can influence the property value of office buildings through the relation of its utility and scarcity. However, green buildings do not significantly impact property values regarding its effective demand and transferability aspect.

References

[1] Popescu, D. dkk. (2012). Impact of Energy Efficiency Measures on The Economic Value of Buildings. *Applied Energy*, *89*, 454-463.

[2] Morrissey, J., & Horne, R. (2011). Life Cycle Cost Implications of Energy Efficiency Measures in New Residential Buildings. *Energy and Buildings*, *43*, 915-924.

[3] Yudelson, J. (2006). *Marketing Green Buildings: Guide for Engineering, Construction, and Architecture*. London: Taylor & Francis Ltd.

[4] Kubba, S. (2012). *Handbook of Green Building Design and Construction*. Amerika Serikat: Elsevier, Inc.

[5] Green Building Council Indonesia. (2016, July 27). *Green Building Council Indonesia*. Retrieved September 25, 2016, from Menangkap Peluang Bisnis Properti Konsep Rumah Hijau Dengan Program EDGE: http://blog.gbcindonesia.org/menangkap-peluang-bisnis-properti-konsep-rumah-hijau-dengan-program-edge.html.

[6] Woodson, R. D. (2009). *Be A Successful Green Builder*. Amerika Serikat: The McGraw-Hill Companies, Inc.

[7] Ho, D., Newell, G., & Walker, A. (2005). The Importance of Property-Specific Attributes in Assessing CBD Office Building Quality. *Journal of Property Investment & Finance*, *vol.23, Iss 5*, 424-444.

[8] Alexander, H.B. (2014, September 29). *Kompas*. Retrieved December 16, 2016, from Label "Green" Efektif Mendongkrak Penjualan Properti: http://properti.kompas.com/read/2014/09/29/09261 2221/Label.Green.Efektif.Mendongkrak.Penjualan.Properti.

[9] Arnold, Alvin L. dkk. (1980). *Modern Real Estate*. Boston: Warren, Gorham & Lamont, Inc.

[10] Schmitz, A., & Brett, D. L. (2001). *Real Estate Market Analysis: A Case Study Approach*. Amerika Serikat: Urban Land Institute.

[11] Shilling, J. D. (2002). *Real Estate (13th edition)*. Ohio: Souther-West Thomson Learning University of Wisconsin.

[12] Cresswell, J. W. (2003). "Mixed Methods Procedures" dalam *Research Design: Qualitative, Quantitative, and Mixed Method Approaches*. California: Sage Publications, Inc. pp. 208-227.

[13] Fachrudin, K. A., Fachrudin, H. T. (2015). The study of Investment Portfolio Management and Sustainability of Property and Real Estate Companies in Indonesia Stock Exchange. *Procedia - Social and Behavioral Sciences*, *211*, 177-183.

Green Building Council Indonesia. (2016, July 27). *Green Building Council Indonesia*. Retrieved February 3, 2017, from Daftar Gedung Tersertifikasi GREENSHIP: http://www.gbcindonesia. org/greenship/2012-08-02-03-21-25/greenship-certified.

Local Material as a Character of Contemporary Interior Design in Indonesia

Dalhar Susanto[*], **Dini Puti Angelia, Tria Amalia Ningsih**

Department of Architecture at Universitas Indonesia, Depok, Indonesia

Corresponding e-mail: dalhar3001@yahoo.com

Abstract. Excellent design needs to fulfill universal requirements (utility, aesthetic, ergonomic, durability, and safe). Besides of all the requirements, an excellent design has to be shown its distinctiveness, uniqueness, and identity. To create an excellent design, we can use one of locality approach, it means local material utilization. From time to time, the material is linking each other in unity with environment context, human, knowledge, culture, social, economy, user needs and material availability. The aspects are the important part to get the reflective identity and local values in architecture and interior design work in Indonesia. It can be proofed by some of the architecture and interior work precedent, like traditional or vernacular in Nusantara or contemporary interior design work from Indonesian designer who has recognized to promote the locality value. However interior design works in Indonesia cannot be shown the characteristic of Indonesia identity and locality currently, it is different than another country work, like Japan, Italy, or Scandinavia. Interior design work from these countries can be easily known with accentuating of characteristic their places, such as material, color, detail, or geometry pattern in the product that has been produced. Meanwhile, some of the region in Indonesia are tropical climate and brought about much of local material and it has potential to make a unique work which has the local identity. This paper will discuss the result of a searching potential of local material usefulness as interior design identity in Indonesia. This research is done by typology method, which means discover the presence of some of the architecture elements appears to be related material. The elements are the pattern, color, craftsmanship, building element, object, and type of material in some of the contemporary interior design work in Indonesia were considered superior and capable of lifting elements recognized locality.

Keywords: local material, locality, identity, contemporary interior design in Indonesia

1.Introduction

The interior works were invented for human needs in order to help men's daily activities. Moreover, each human needs to make the interior works reflect in both individual and social life. According to Hidayatun (2013) in the architecture practice, an excellent design is a product that presents identity. Identity is an element of a set which if it is combined with another element by a specified binary operation, it leaves the element unchanged [1]. Through the material, a design can be identified by the quality of its design.

The great example why the major design from Japan can easily recognize is because the usage of the material that can be realized through the ornament patterns, color, etc. The pattern cannot be separated from the style of Japan architecture. It is the identity which gives the sensation for this country [2]. Historically, the pattern is associated with ornaments and decorations that create aesthetic value in Japan architecture building. Therefore, concept and pattern theory are channeled through spatial theory. They are reflected by structure, hierarchy, program of space, system, scale, symmetry proportion, balance, complexity, function, representation, imagination, and creativity. The concept of pattern is affected by other disciplines which are driven by the country's history.

On the other hand, Indonesia has a cultural diversity that shows interesting architectural elements [1]. It is colouring traditional architecture with environment and culture potential that create attention for some of the designers to develop. When this concept is used in a work of design, the materials will be presented with local wisdom from the knowledge and the ideas. Indonesia has various types of materials that can be used to make an interior work. However, the material does not accompany its application in works of the interior. Moreover, majority of the interior works that have been made, do not represent the identity of Indonesia.

As widely known, Indonesia has many ethnics and cultural diversity in Nusantara. With this different ethnics and tribes in Indonesia, all of which presents its own architectural form. The material usage is also influenced by the geographical location of Indonesia. This is affecting the material distribution in Nusantara. The tropical climate in Indonesia causes some areas in Indonesia to have fertile soil so that most of the forests in Indonesia are tropical rain forest that produce good quality plants. The forest products are used by the Indonesians in the making of their local products.

Material can naturally provide an identity for a product that is designed, because the material is the part of the architectural grammar. There are three things that exist in the architectural grammar, including modules, elements, and structure [3]. Each of these matters relates to mechanisms in architecture, one of all talks is about the material and how it is applied. According to Dplazes and Soffker (2005) "*All material has a shape, regardless of the existence of a forming will.*" Naturally, the material has its own form, regardless of the material is formed when the material is to be used. Even sometimes the existing form in the material is maintained, so the final form that comes out follows the material's pre-existing natural form.

However, the local potential is not limited only by traditional buildings in Indonesia. It includes all the unique properties, historicity, or role as a marker of area, city, and region (Roesmanto, 2007). To produce an architecture or interior work that have the uniqueness, it can be seen from the material used in the work, because each region employs a different material, but the difference makes the uniqueness of each region becomes visible.

In the local materials application in architectural or interior works in Indonesia, we can look back from the vernacular architecture. Back from the old days, people used to see the material around them and cultivated them to adapt. This can be a tool for the designer to innovate into the contemporary interior design. However, presenting local material elements is not as easy as imitating the use of existing materials. The local material is supposed to be used not only as an aesthetic element but also as a founder of interior design in Indonesia. This paper uses case studies on the work of contemporary interior design.

The major focus in this paper is to discuss how the local material is used by Indonesian designers to present the identity of Indonesian Interior design works. It is using the typology method to find the potential material in Indonesia. Moreover, this paper also discusses the main object that is often applied to be the identity of interior design works. The usage of typology conducts the interior elements such as object, surface, pattern and color. As the result, it can identify the local material as the identity of interior design works in Indonesia.

2.Methods

a. Methodology of typology for understanding the identity of interior works in Indonesia

Architectural works have unique characters. It shows the contextual response, as the design has the same purpose and the final form. However, the final form of the architecture product can be different. Although, the same function will have an item or a character which repeats for certain products. "*On the other hand, a work of architecture can also be seen as belonging to a class of repeated objects, characterized, like a class of tools or instruments, by some general attributes."(*Moneo, 1978). The work of architecture is an imitation of natural form, so that we will find similar elements appear on the certain architecture objects. All of that are the result of the human innovation with different representation [10]. Often we find similar elements in some designs that have the same purpose. For that reason, we can use the typology study to identify the similar characters that often appear at the architecture objects. The next step is grouping certain equations which are found on each object that has the same purpose. Because the architectural object is formed by type, it has similarities in certain architectural elements.

In order to see the type of issues by several designs, we need to understand architectural elements in certain usage of architectural and interior works. To understand the works, it needs to classify the type of elements into several categories. From these categories, it will be subdivided into several related derivatives. Moreover, we use secondary data such as architectural representation from the design for looking the major element appearance in interior design. section describes the rationale for the application of specific procedures or techniques used to identify, select, and analyze information applied to understanding the research problem, thereby, allowing the reader to critically evaluate a study's overall validity and reliability.

b.Methodology of applying interior elements be variable in typology to identify local material as the identity of interior works

The interior design is primarily concerned with the representation of space before the fact of its coming in to being, making it critical to convey the totality of the experiential qualities eventually comprised by the actual space to be lived in" (Power, 2014). There are several elements that are represented in interior design, object, surface, pattern and color. These elements were arranged in interior space to create the experiential qualities of the space.

One of building element is furniture or object as the space content in a room. It can have the function or just interior decoration. There are building elements which are named built-in furniture in interior space such as walk-in closet and TV panel. Then, the loose furniture is movable like the table, chair, armchair sofa, etc. Moreover, the objects that reflect as the decoration of interior space include frame painting, artworks, and sculpture give the ambiance of the interior space. Occasionally building elements have some of similarity in programming, like same dimension, scale, function, relation, or volume [8].

On the other hand, the majority of Indonesian objects of the interior has the pattern as the decoration of objects. Objects will have the form which has function because it will leave some of the impression in the room. "*It will make an impression on those who use it and they may well leave an impression on it.*" (Booth and Plunkett, 2014). There are two types of pattern that are mostly used in objects, geometrical and flora-fauna pattern which were inspired by forms the nature of Indonesia. These are usually combined in both traditional and contemporary design of interior Indonesia. The latter yet brings the identity of the interior place of Indonesia with the socio-culture context of the pattern.

Colour is related with the eye perception of the space seen. In certain Indonesian tribes have some colours often applied in their daily objects, for example the Chinese use red as their accent in their daily activity, from clothes until interior objects like furniture. Moreover, Jakarta's indigenous people called "Betawi's people" use green and brown from the woods in their daily objects like tables, chairs, and ornaments. In architecture and interior, colours can effect to symbolic and dimension that related with architectural form and space [8].

Meanwhile, Texture correspond the surface of interior works. For the example, a table uses the wood as material will produce different texture and qualities with the other table that use metals as the main material. It is because the wood has original texture which is coarse and little bit wavy, while the metal has slick and flat. Frequently, a designer chooses a material to accentuate sense of touch to show some of ambience in the design. Because of that, the designer shall choose the proper material that picture the identity of the interior design. Moreover, Indonesia has many types of indigenous material such as woods and stone. For that reason, most of interior designer used it to present atmosphere of traditional of Indonesia.

Each material has self-characters that we can treat according to materials types. The way we treat the materials to the interaction and ad justification with the initial design, all depends on the technique used. *"Such as density, transparency, reflection, erosion, stretching, bending, and cracking. These models do not attempt to represent a realistic representation of a design, but instead to suggest ways in which a design's components might act and interact."* (Simitch and Warke, 2014).

3.Study Cases

The study case uses the typology method to understand the patterns of materials usage that appear constantly in the interior design work, using the variable of interior element. There are two different hotels with two particular types of guest room from each hotels. Therefore, it has analysis of 4 rooms that were designed with Indonesian theme to create the traditional ambience in temporary design.

The first hotel that was designed by Jaya Ibrahim is Hotel Dharmawangsa Jakarta. It is one of the highly prestigious hotel that is exhibits Indonesian identity through the design. Additionally, each guest room is developed in one of five thematic schemes each only using traditional materials (The Dharmawangsa Jakarta, 2013). The first room, named Executive room, is a spacious King or twin bedded room with unit size of 66 square m2. It has large marble bathroom, an executive desk, sofa with coffee table, side tables and complementary chairs. The other guest room, Dharmawangsa Suite, has unit size of 107 m2 with a spacious king bedroom. It also has a separate living room, a large bathroom, an executive desk, and private terrace. The living room features a full sized sofa, dining table, and pantry. The rooms were beautifully decorated with regional Indonesian influence (The Dharmawangsa Jakarta, 2013).

The second hotel which is located in Seminyak, Bali, was designed by young designer of Indonesia, Alex Bayusaputro. It is named Four Point by Sheraton Bali. It has 3 different size of guestrooms with local batik theme surrounding the room. There are two particular rooms that are represented in this study. The first room which has size 60m2, named Junior Suite. It treats with bathroom, living area, and balcony. The second one, which is Vasanti Suite, has 80m2 sized provides with bathroom, king sized bed, living area, and outdoor Jacuzzi. This hotel has a modern interior with warmth of local touch (Four Point Bali Seminyak, 2016).

Dharmawangsa Hotel, design by Jaya Ibrahin

- *Executive room*

Figure 1. Executive room perspective.

Figure 2. Executive room perspective.

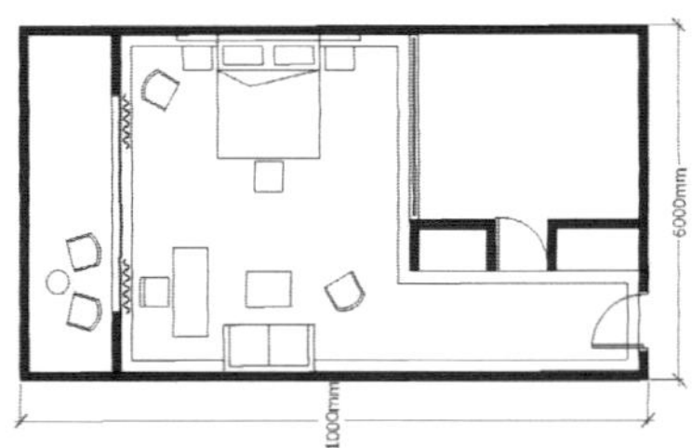

Figure 3. Executive room plan.

The some elements have find in this room are:

Table 1. The elements in Executive Room

The Elements in Executive Room													
NO	material	type of element		measurement	Surface		Pattern		Colour		Craftsmanship	
		Loose element	Building element	(surface area)	2D Flatten	3D Crafted	Geo metric	Organic	Natural	Artificial	Manual	Machine
1	wood	Console bed	-	0.72	flatten	-	-	-	Brown glossy	-	-	factory
2	wood	bedside	-	2	flatten	-	-	-	Brown glossy	-	-	factory
5	wood	bedside	-	2	flatten	-	-	-	Brown glossy	-	-	factory
6	wood	"betawi" armchair	-	1.67	flatten	(crafted at foot)	-	-	Brown glossy	-	craftsman	-

7	wood	bench chair	-	0.67	flatten	(crafted at foot)	-	-	Brown glossy		craftsman	-
8	wood	Working chair	-	1.82	flatten	(crafted at foot)	-	-	Brown glossy		craftsman	-
9	wood	Working table	-	4.188	flatten	(crafted at foot)	-	-	Brown glossy	-	-	factory
10	Wood	Coffee table	-	1.6	flatten	-	-	-	Brown glossy	-	-	factory
11	wood		Wall cornice	2.78	flatten	-	-	-	Brown glossy	-	-	factory
12	wood		Window cornice	1.07	flatten	(crafted at edge)	-	-	Brown glossy	-	-	factory
13	wood		wardrobe	9.3	flatten	(crafted at edge)	-	-	Brown glossy	-	-	factory
14	wood		Luggage rack	3	Flatten	(crafted at edge)	-	-	Brown glossy	-	-	factory
15	fabric	Artwork backdrop		2.88	-	Embossed	-	Songket floral	Red, purple, gold glossy	-	craftsman	-
16	fabric	bedcover	-	5.34	flatten	-	-	-	-	White matte	-	factory
17	fabric	Bed runner	-	0.96	flatten	-	-	-	-	Gold glossy	-	factory
18	fabric	4 Pillow cases	-	1.92	flatten	-	-	-	-	White matte	-	factory
19	fabric	4 Pillow cases accent	-	0.64	flatten	-	-	-	-	Gold glossy	-	factory
20	fabric	1 Pillow case accent	-	0.16		Embossed	-	floral	-	Red, Cream glossy	-	factory
21	fabric	Under bed	-	1.74	flatten	-	-	-	-	Cream matte	-	factory
22	fabric	Bench cushion	-	0.52	flatten	-	-	-	-	Cream matte	-	factory
23	fabric	Armchair cushion	-	0.52	flatten	-	-	-	-	Cream matte	-	factory
24	fabric	Working chair cushion	-	0.52	flatten	-	-	-	-	Cream matte	-	factory
25	Fabric	2 seat sofa	-	4.61	flatten	-	-	-	-	Grey matte	-	factory
26	fabric	-	Curtain blackout	15	flatten	-	-	-	-	White matte	-	factory
27	fabric	-	Curtain transparent	15	flatten	-	-	-	-	White matte	-	factory
28	fabric		carpet	35	-	embossed	-	-	-	Grey matte	-	factory
29	fabric		2 cups table lamp	1.44	flatten	-	-	-	-	Cream matte	-	Factory

- *Dharmawangsa suite*

Figure 4. Dharmawangsa suite perspective.

Figure 5. Dharmawangsa suite perspective.

Figure 6. Dharmawangsa suite plan.

The some elements have been find in this room are:

Table 2. The elements in Dharmawangsa suite.

The Elements in Dharmawangsa Suite												
NO	material	type of element		measurement (surface area)	Surface		Pattern		Colour		Craftsmanship	
		Loose element	Building element		2D Flatten	3D Crafted	Geometric	Organic	Natural	Artificial	Manual	Machine
1	wood	Console bed	-	0.72	flatten	-	-	-	Brown glossy	-	-	factory
2	wood	bedside	-	2	flatten	-	-	-	Brown glossy	-	-	factory
3	wood	bedside	-	2	flatten	-	-	-	Brown glossy	-	-	factory
4	wood	2 "betawi" armchair	-	3.34	flatten	(crafted at foot)	-	-	Brown glossy	-	craftsman	-
5	wood	bench chair	-	0.67	flatten	(crafted at foot)	-	-	Brown glossy		craftsman	-
6	wood	Working chair	-	1.82	flatten	(crafted at foot)	-	-	Brown glossy		craftsman	-
7	wood	Working table	-	4.188	flatten	(crafted at foot)	-	-	Brown glossy	-	-	factory
8	Wood	Coffee table	-	1.6	flatten	(crafted at	-	-	Brown glossy	-	-	factory

						foot)						
9	Wood	2 "betawi" dining table	-	3.34	flatten	(crafted at foot)	-	-	Brown glossy	-	craftsman	-
10	wood	Dining table	-	6	flatten	(crafted at foot)	-	-	Brown glossy	-	craftsman	-
11	wood	cabinet	-	9		crafted	-	-	Brown glossy	-	craftsman	-
12	wood		Wall cornice	5.56	flatten	-	-	-	Brown glossy	-	-	factory
13	wood		Window cornice	2	flatten	-	-	-	Brown glossy	-	-	factory
14	wood		wardrobe	9.3	flatten	(crafted at edge)	-	-	Brown glossy	-	-	factory
15	wood		Luggage rack	3	flatten	(crafted at edge)	-	-	Brown glossy	-	-	factory
16	fabric	Artwork backdrop		2.88	-	Embossed	-	Songket floral	Red, purple, gold glossy	-	craftsman	-
17	fabric	bedcover	-	5.34	flatten	-	-	-	-	White matte	-	factory
18	fabric	Bed runner	-	0.96	flatten	-	-	-	-	Gold glossy	-	factory
19	fabric	4 Pillow cases	-	1.92	flatten	-	-	-	-	White matte	-	factory
20	fabric	4 Pillow cases accent	-	0.64	flatten	-	-	-	-	Gold glossy	-	factory
21	fabric	1 Pillow case accent	-	0.16		Embossed	-	floral	-	Red, Cream glossy	-	factory
22	fabric	Under bed	-	1.74	flatten	-	-	-	-	Cream matte	-	factory
23	fabric	Bench cushion	-	0.52	flatten	-	-	-	-	Cream matte	-	factory
24	fabric	Armchair cushion	-	0.52	flatten	-	-	-	-	Cream matte	-	factory
25	fabric	Working chair cushion	-	0.52	flatten	-	-	-	-	Cream matte	-	factory
26	Fabric	2 seat sofa	-	4.61	flatten	-	-	-	-	Grey matte	-	factory
27	fabric	-	Curtain blackout	30	flatten	-	-	-	-	White matte	-	factory
28	fabric	-	Curtain transparent	30	flatten	-	-	-	-	White matte	-	factory
29	fabric		carpet	70	-	embossed	-	-	-	Grey matte	-	factory
30	fabric	2 cups table lamp	-	1.44	flatten	-	-	-	-	Cream matte	-	factory
31	fabric	2 cup floor lamp	-	1.44	flatten	-	-	-	-	Cream matte	-	factory
32	wallpaper		Backdrop wallpaper	9	-	embossed	Horizontal line	-	-	Lime glossy	-	factory
33	paint	-	Wall painting	76	flatten	-	-	-	-	Rose white matte	-	factory
34	marble	-	Marble flooring	5.74	flatten	-	-	-	-	Cream glossy	-	factory
35	Clay	2 table lamps	-	1.92	-	Curved	-	-	-	Brown glossy	craftsman	-
36	clay	2 "anggrek" vases	-	0.72	-	Curved	-	-	-	Brown glossy	craftsman	-

Four Point by Sheraton Bali

- *Junior suite*

Figure 7. Junior Suite, interior perspective.

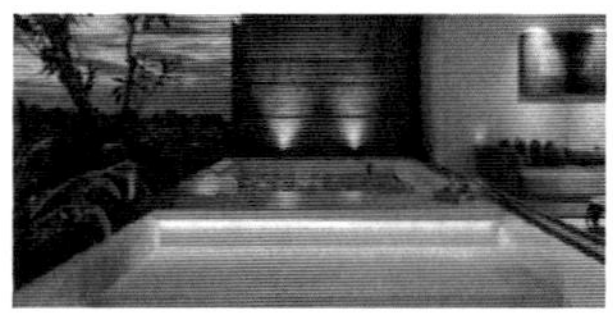

Figure 8. Junior Suite, exterior perspective.

Figure 9. Junior suite plan.

The some elements have been find in this room are:

Table 3. The elements in junior suite.

					The Elements in Junior Suite							
NO	material	type of element		measurement (surface area)	Surface		Pattern		Colour		Craftsmanship	
		Loose element	Building element		2D Flatten	3D Crafted	Geometric	Organic	Natural	Artificial	Manual	Machine
1	fabric	-	Wallpaper background	0,12	flatten	-	-	Batik Pattern, floral	-	Purple, gold, black matte	-	Factory
2	fabric	bedcover	-	5.34	flatten	-	-	-	-	White matte	-	factory
3	fabric	4 Pillow cases	-	1.92	flatten	-	-	-	-	White matte	-	factory
4	fabric	Under bed	-	1.74	flatten	-	-	-	-	White matte	-	factory
5	fabric		carpet	5	-	embossed	-	-	-	Grey, gold outline matte	-	factory
6	fabric		carpet	6,25	-	embossed	-	-	-	Grey, black grey outline matte	-	factory
7	fabric	-	Curtain blackout	15	flatten	-	-	-	-	Light grey matte	-	factory
8	fabric	-	Curtain transparent	15	flatten	-	-	-	-	White matte	-	factory

9	fabric	armchair	-	1,82	(crafted at foot)	-	-	-	Blue mate, brown glossy	-	craftsman	-
10	fabric	2 seat sofa	-	4,61	flatten	-	-	-	-	Grey matte	-	factory
11	fabric	8 pillow cases	-	1,92	-	embossed	-	Batik pattern, Floral	-	Tosca, gold, purple mate	-	Factory
12	fabric	-	Wallpaper background	0,12	flatten	-	-	Batik Pattern, floral	-	Purple, gold, black matte	-	Factory
13	fabric	bedcover	-	5.34	flatten	-	-	-	-	White matte	-	factory
14	fabric	4 Pillow cases	-	1.92	flatten	-	-	-	-	White matte	-	factory
15	fabric	Under bed	-	1.74	flatten	-	-	-	-	White matte	-	factory
16	fabric		carpet	5	-	embossed	-	-	-	Grey, gold outline matte	-	factory
17	fabric		carpet	6,25	-	embossed	-	-	-	Grey, black grey outline matte	-	factory
18	fabric	-	Curtain blackout	15	flatten	-	-	-	-	Light grey matte	-	factory
19	fabric	-	Curtain transparent	15	flatten	-	-	-	-	White matte	-	factory
20	fabric	armchair	-	1,82	crafted at foot)	-	-	-	Blue mate, brown glossy	-	craftsman	-
21	fabric	2 seat sofa	-	4,61	flatten	-	-	-	-	Grey matte	-	factory
22	fabric	8 pillow cases	-	1,92	-	embossed	-	Batik pattern, Floral	-	Tosca, gold, purple mate	-	factory
23	Wood	Coffee table	-	2,33	flatten	-	-	-	Brown glossy	-	-	factory
24	Wood	-	Partition	7,4	-	Crafted	-	Floral	Brown mate	-	craftsman	-
25	Wood	TV panel	-	1,2	Flatten	-	-	-	Brown glossy	-	craftsman	-
26	Wood	Bedside	-	0,82	flatten	-	-	-	Brown glossy	-	-	factory
27	Wood	-	wood flooring	16,28	flatten	-	-	-	-	Brown glossy	-	factory
28	Wood	Coffee table	-	2,33	flatten	-	-	-	Brown glossy	-	-	factory
29	Wood	-	Partition	7,4	-	Crafted	-	Floral	Brown mate	-	craftsman	-
30	Wood	TV panel	-	1,2	Flatten	-	-	-	Brown glossy	-	craftsman	-
31	Wood	Bedside	-	0,82	flatten	-	-	-	Brown glossy	-	-	factory
32	Wood	-	wood flooring	16,28	flatten	-	-	-	-	Brown glossy	-	Factory
33	Wood	-	Wall background	5,4		Crafted	-	-	Brown mate	-	craftsman	-
34	metal	2 table lamps	-	1,92	-	Curved	-	-	-	Brown glossy	craftsman	-
35	Glass	Flower vase	-	0,6	-	Curved	-	-	-	Trans parent glossy	-	factory
36	marble	-	Marble flooring	17,6	flatten	-	-	-	-	Cream glossy	-	Factory

- *Vasanti Suite*

Figure 10. Vasanti suite perspective

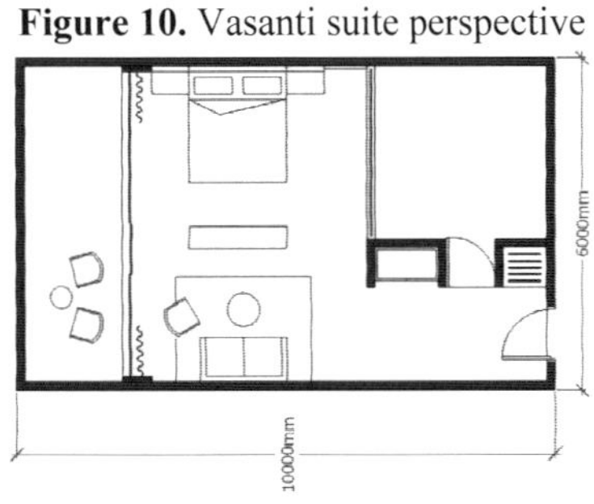

Figure 11. Vasanti suite plan.

The some elements have been find in this room are:

Table 4. The elements in vasanti suite.

The Elements in Vasanti Suite													
NO	material	type of element		measurement	Surface		Pattern		Colour		Craftsmanship	
		Loose element	Building element	(surface area)	2D Flatten	3D Crafted	Geometric	Organic	Natural	Artificial	Manual	Machine
1	fabric	-	Wallpaper background	0,12	flatten	-	-	Batik Pattern, floral	-	Purple, gold, Black, white matte	-	Factory
2	fabric	bedcover	-	5.34	flatten	-	-	-	-	White matte	-	factory
3	fabric	4 Pillow cases	-	1.92	flatten	-	-	-	-	White matte	-	factory
4	fabric	Under bed	-	1.74	flatten	-	-	-	-	White matte	-	factory
5	fabric		carpet	4,75	-	embossed	-	-	-	Grey, gold outline matte	-	factory
6	fabric	armchair	-	1,67	(crafted at foot)	-	-	-	Blue mate, brown glossy	-	craftsman	-
7	fabric	2 seat sofa	-	4,61	flatten	-	-	-	-	cream mate	-	factory
8	fabric	2 pillow cases	-	1,92	-	embossed	-	Batik pattern, Floral	-	cream mate	-	Factory
9	fabric	-	Curtain blackout	15	flatten	-	-	-	-	Light grey matte	-	factory
10	fabric	-	Curtain transparent	15	flatten	-	-	-	-	White matte	-	factory
11	Wood	Coffee table	-	1,6	flatten	-	-	-	Brown glossy	-	-	factory
12	Wood	-	Wall accent	11,8	Flatten	-	-	-	Brown glossy	-	craftsman	-
13	Wood	Bedside	-	0,82	flatten	-	-	-	Brown glossy	-	-	factory
14	Wood	-	cabinet	9	-	Crafted	-	Floral	Brown glossy	-	Craftsman	-
15	metal	2 table lamps	-	1,92	flatten	Curved	-	-	-	Brown glossy	craftsman	-
16	Glass	Flower vase	-	0,6	flatten	Curved	-	-	-	Transparent glossy	-	factory
17	marble	-	Marble flooring	25,08	flatten	-	-	-	-	Cream glossy	-	Factory

4.Discussion and Analysis

The interior work is created by the elements of the representation space. Based on this, there are five different elements that are related to spatial experience to understand the identity of the interior design works. The first one is material. There are five different materials that appears on all 4 guest rooms. The biggest percentage of appearance with 54% is wood. Clearly, all of the room uses wood as the main material. It is utilized in both type of interior objects, loose furniture and build-in objects. It also appears as the artwork of the rooms. The second presentation is fabric. The usage of fabric is created with plenty of forms, for the main material such as sofa and the complementary like pillow case. The latter is also represented as the artwork "songket fabric". After that, there are minor appearance of materials such as wallpaper (7%), clay (4%), metal (2%). These are the materials that usually are used as the accent of interior design such as backdrop for bed, artwork, or event material lamp.

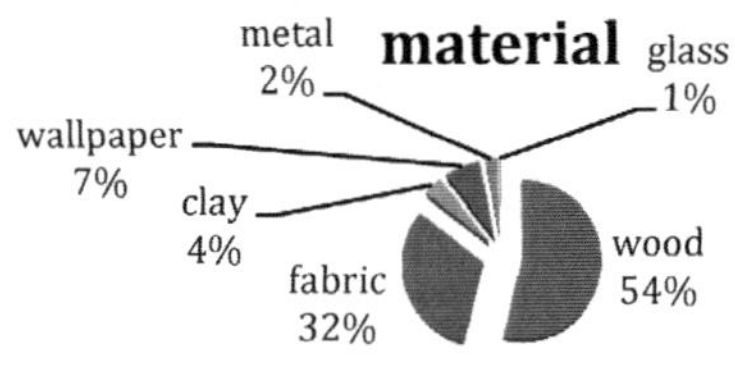

Figure 12: average material used in the study cases

After that, the second element is object. Based on the previous discussion, the most appearance of materials from the study case are woods, therefore most of objects utilize wood as the main materials. It is established that woods has strong structure but easy to form, so majority of woods are for tables with approximately 4-5 tables in one room. Then the woods are also applied in chairs with 4 items in a room. Uniquely, Dharmawangsa hotel uses "betawi" chairs, as the identity of the rooms while the Four Point hotel prefers contemporary chairs. Furthermore, all the rooms use plenty of artworks with different types. Dharmawangsa guest rooms are designed with fabric and clay artworks while Four Point hotels are assembled woods, fabrics, and glass materials.

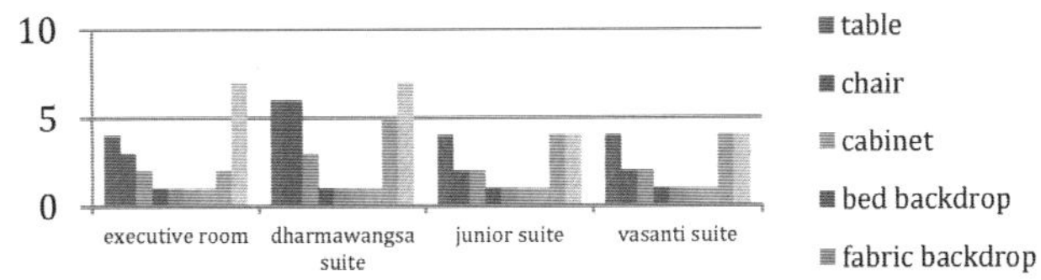

Figure 13: average objects used in the study cases

Then, the third element that is related to material is the surface. As mention before, woods are dominant in all rooms. Therefore the surface of the wood is mostly created as the surface of the rooms. Especially in Four Point hotels, the woods are also used as the flooring finish, so it creates the warmth ambience. Then most of the objects, including the ones that uses woods, had the flat surface. However, there were few objects had a combined surface between flat and crafted on a part of objects such as legs and edge. Additionally, when the object had a full crafted surface, it is dominant in the space. It happens with the cabinet at Dharmawangsa suite and wood partition at Vasanti Suite. Therefore, the usage of crafted surface should be in a right object to give a balance quality of space between objects. Moreover, the curved surface is represented as the artworks. Majority of artworks had a curved shape because it performes as accents of quality of space.

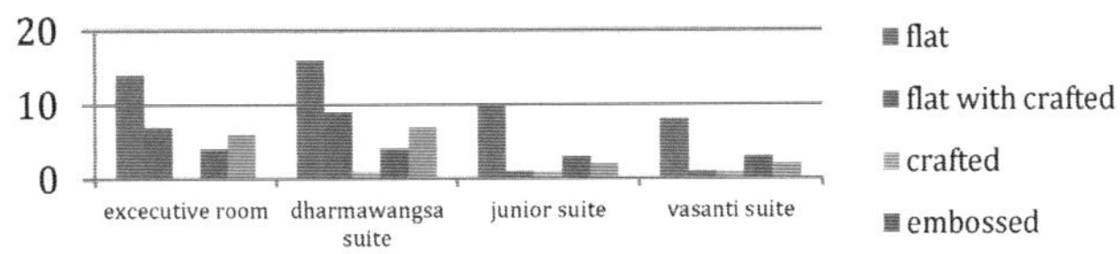

Figure 14: average surface used in the study cases

On the other hand, the fourth element which is color, depends on the finishing of the surface. Majority of surface has brown glossy as its color. This color, that represents the color of woods and clays, is applied in several objects, like chairs, tables, cabinets and most of the artworks. Then, there are the complementary colors such as gold, lime, cream, white and grey, the earth color, which balances the ambience of the color. After that, the minor colors such as red, green, blue, and purple, are utilized as the accent of the color. It usually is created as artworks or accent of the objects.

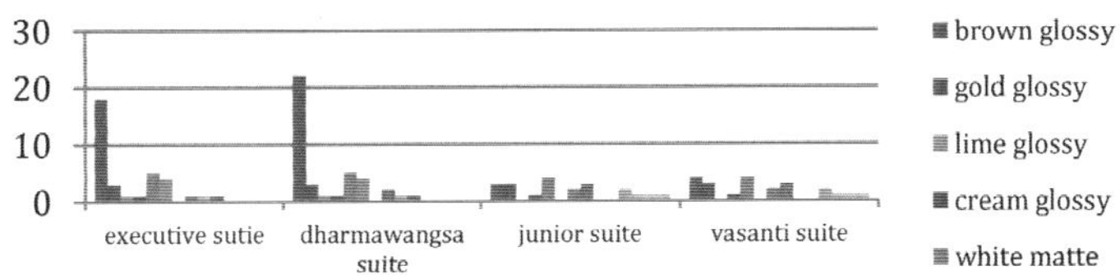

Figure 15: average color used in the study cases

Last but not a least is the pattern. This is the element that represents the diversity of Indonesia's culture. It is because the pattern comes from Indonesia's nature then it interpret as the package of pattern, floral, fauna and geometric. Unfortunately, on this study case, there are only few patterns that apply on the rooms. There is no fauna pattern on the study case. On the other hand, the floral pattern is constantly used in "batik" fabrics. It is created become a bed backdrop at Four Point hotels. Meanwhile, a bed backdrop at Dharmawangsa was used "songket" fabrics as well as the artworks. It has combined pattern the floral and geometric pattern on the fabrics, or it is called ethnic pattern

5.Conclusion

The methodology of typology has been established that woods as the local material of Indonesia is represented by the identity of design interior works of Indonesia. It constantly appears in many forms of objects and artworks. By utilizing the design with the woods, the representation of space becomes easily identified as the interior works of Indonesia.

However, there are particular categories that consider the usage of this material to identify the Indonesia's design interior works. Clearly, the form of objects should represent Indonesia's design.

The "betawi chair" or the crafted cabinet that is created by Indonesian's craftsmanship is one of object that gives Indonesia's design backgrounds. Moreover, the finishing of Indonesia's wood is glossy brown. It is because using teak wood as the materials, and it becomes dark brown after finishing. The glossiness and the flat surface often appeared to create a clean look of the objects. Meanwhile, the crafted woods are usually used in a few objects of the interior rooms to create the representation of space of Indonesia's design. There are other objects that combine woods and other materials to create contemporary objects, so not all objects that use woods as the main material is identified as Indonesia's design interior works.

On the other hand, the artworks that use materials beside woods, such as "songket" and "batik" which use fabric as the main materials, are also the objects that represent of Indonesia's design by the pattern. Ethnic pattern which are floral and geometric are gradually used at Indonesia's design works. It is represented as the identity of interior design works of Indonesia.

References

[1] Hidayatun, M I, Prijotomo J, and Rachmawati M. 2013. *Material as Part of the Identity of Indonesian Architecture: Locality and Universality Value in Mangunwijaya and Eko Prawoto Works.*

[2] Kuma K. 2014. *Patterns and Layering, Japanese Spatial Culture, Nature and Architecture.* Tokyo: Gestalten.

[3] Deplazes A (Ed.). 2005. *Constructing Architecture: Materials Processes Structures.* Switzerland: Birkhauser.

[4] Ingold. 2007. *Materials against Materiality.* Cambridge: Cambridge University Press.

[5] Leydecker S (Ed.). 2013. *Designing Interior Architecture: Concept, Typology, Material, Construction.* Germany: Birkhauser.

[6] Booth S, Plunkett D. 2014. *Furniture for Interior Design.* London: Laurence King Publishing.

[7] Brooker G, Stone S. 2009. *Basics Interior Architecture: Elements/Objects.* Switzerland: AVA Publishing SA.

[8] Simitch A, Warke V. 2014. *The Language of Architecture: 26 principles every architect shpuld know.* USA: Rockport Publisher

[9] Power, Jaqualine. 2014. *Interior Space: Representation, Occupation, Well-Being and Interiority.*

[10] Forty A. 2000. *Words and Building.* New York: Thames & Hudson Inc.

[11] Moneo R. 1978. *On typology.*

[12] Roesmanto T. 2007. *Pemanfaatan Potensi Lokal dalam Arsitektur Indonesia.* Semarang: Badan Penerbit Universitas Diponegoro.

[13] Four points Bali by Seraton Bali, Seminyak. Was accessed at Mei 21, 2016 from http://www.fourpointsbaliseminyak.com/gallery.

[14] The Dharmawangsa Jakarta. Was accessed at Mei 21, 2016 from http://www.the-dharmawangsa.com.

Prefabricated house in real estate business development in Jabodetabek

Raka Gumilang Raksamala Basmara Putra*, **Dalhar Susanto**
Faculty of Engineering, Universitas Indonesia Depok 16424 , Indonesia

*Corresponding e-mail : raka.gumilang@ymail.com

Abstract. Within constructional technology prefabricated house is known as a house whose most of the components like walls, roofs, toilets, and structural cubes and columns are designated and produced in factory and assembled in site. Based on the experience in certain countries, prefabricated house has successfully entered commercial market. Prefabricated house shows a better infestation value than conventional houses. With prefabrication technology, building construction could be done faster and cost lower price, along with a good craftsmanship quality. Until now, Indonesia is still facing settlements issues and one of them is housing backlog whose amount revolves around 13 Million units. Prefabricated houses potentially solve the settlement issue in Indonesia in a much faster way. Yet the development of prefabricated house has not been accepted well in the market yet, both by the producers and the consumers. This writing talks about the result of the research about the causes of the lack of development in Indonesia when it comes to prefabricated houses though the method of comparison; comparing prefabricated house with conventional house when it comes to its 1) productivity rate, 2) consumers preference and 3) infestation performance. The data is collected through survey methods, interview and questionnaires. The result shows that even though the productivity rate when it comes to quality and time is high, prefabricated house shows low productivity when it comes to cost because it requires greater cost than it is in conventional houses. Other than that, the lack of business development when it comes to prefabricated houses is caused by the low preference of consumers toward the product- because of its inflexibility and monotonous appearance. The infestation of prefabricated house development shows a low NPV and IRR. With a more proper infestation approach such as mass constructions with effective numbers, good flexibility, and various appearances, prefabricated house could have a chance to be accepted in real estate commercial market in Indonesia especially in Jabodetabek.

Keywords: Property, Real Estate, Prefabricated House

1. Introduction

Prefabricated house is a house whose most of its elements (such as walls, roof, bathrooms, and structural elements) are designated and produced in factory and assembled in site [1]. Through the offsite prefabrication technology, the house construction tends to be faster and cheaper that it decreases the duration of construction and presses the cost. Based on Smith (2010) as a collective cost, the construction basic cost per unit whose system is modular might be more expensive compared to on-site built products. If so, this additional cost should be considered along with its shortened duration and quality increase [2].

In a lot of countries, business development of prefabricated houses show a really good infestation value, better than it is to conventional ones. Research by Sri Velemati found that savings resulted from the cost and items related to prefabricated house in America results and increase in NPV by 3,485 and IRR 1,56%. In Indonesia, preferences of prefabricated houses are still limited to several purposes such as worker mass settlements, temporary offices or houses for disaster casualties. The development of prefabricated houses nowadays has not been accepted yet by real estate commercial market in Indonesia, both by the producers and consumers.

Indonesia is currently facing proper settlements issue especially for the middle to lower class people. Up until now, Indonesia is still shadowed by housing backlog issue whose number revolves around 13 million units. Prefabricated houses development in commercial market could have been one of the solutions addressing the settlement issue in Indonesia, yet the prefabricated house development has not yet been accepted in the market, both by the producers and the consumers.

Looking for the causes of the lack of business development of a product could be treasured through three things, which are (1) productivity rate of product construction, (2) Preference rank of the consumers toward the product, and (3) the level of product infestation profit. In the world of construction, productivity level is an output performance of a constructional activity related to three main indicators which are cost, quality and time (Ir.Iman Soeharto, 1999). Consumers preference is a behavior of consumer towards a choice of product formed by evaluations of variety of choices available. Consumer's wants to purchase a product is basically influenced by a lot of factors that relate to each other that it creates certain pattern as long as the pattern of wants of consumers when it comes to purchasing a house, which are: (1) Knowledge : level of how well the consumers understand the product whether it is a thing or a service; (2) Liking : level of wants of the consumers toward the product available; (3) Preference : the choice of consumers toward a product compared other products available; and (4) Conviction/intention : the will to buy- how much the consumers are willing to purchase the products (Kotler dan Keller. 2009). Temporarily, the level of profit of property business could be known through the infestation simulation to know the value of NPV and IRR. NPV is a difference between spending and income after the discount using social opportunity cost of capital as discount factor, or in other words, a cash flow predicted in the future, while IRR is an indicator of efficiency level of an infestation.(Suratman, 2001)

2. Method

Looking for a cause of the lack in prefabricated house business development in Jabodetabek could be found through three steps such as (1) measuring productivity level of prefabricated house construction in Jabodetabek, (2) conducting a research on consumers preference toward prefabricated house in Jabodetabek, and (3) calculating the value of NPV and IRR through infestation simulation of prefabricated house development in Jabodetabek.

Measuring the productivity level of prefabricated house construction is done through assessments using table of productivity level calculation that covers three aspects (cost, quality and time) which consist of 11 indicators overall. Assessment is done by giving scores to each indicator following 1 (lowest scale) up to 5 (highest scale). The assessment of productivity level of prefabricated house construction is done towards three products of prefabricated houses that have been developed by some producers in Indonesia such as (1) Rumah Instan WIKA Beton, (2) Rumah Prefab BBI, dan(3) Rumah RISHA. The technical data of these three prefabricated houses is conducted through survey and direct interviews with the involved producers. As a comparison tool, the productivity level of conventional house construction is also measured. Both prefabricated and conventional houses are assumed to effectively have area of $36m^2$.

Preference research is conducted by spreading questionnaire consisting of eight questions related to four aspects such as knowledge, liking, preference and conviction of consumers. The questions within the questionnaire are both confidential and open in a way that every question is provided several options of answers yet also provided with empty space for respondents in case they have additional answer, reason or information. The questionnaire is spread randomly using google.doc. online. Around 42 respondents filled the questionnaires and gave them back. To complete or confirm the responses of these consumers, another interviews were also done to developers.

Measuring the profit level in business development infestation of prefabricated house in Jabodetabek is done through simulation of business infestation measurement of prefabricated house by finding its NPV and IRR. Within this simulation, primary data are used especially those that are factual, conducted from direct field survey (such as volumes, unit price, ets) so a realistic result is expected.

3. Discussion

3.1. Ases increases the productivity of prefabricated house construction.

The result of assessment shows that generally the productivity level of prefabricated house development is higher than the conventional one. Prefabricated house shows the excellence in two criteria which is time and quality, but it is weak on cost aspect (see. Table 1)

Table 1. Development of prefabrication and convention house productivity level result

Variabel			Rumah Instan - Wika Beton	BBI Prefab - PT Bakrie	Risha	Conventional House

Cost	Materials (m2)	Walls				
		Structure				
		Roof				
		Accessories	Rp. 3.300.000,- /m2	Rp. 3.000.000,-/m2	Rp. 2.400.000,-/m2	Rp. 2.700.000,-/m2
	Construction (m2)	Fabrication				
		Assembly				
	Transportation					
	Average /m2		Rp		2.900.000,00	Rp 2.700.000,00
	Value		3	3	5	4
Quality	Materials	Wall	hollowcore Panel K-350	Sandwich panel	Concrete	Hebel
		Value	5	5	5	1
		Structure	Concrete Precast K-350	Galvanized metal	Concrete Precast K-225	Concrete K-125
		Value	5	1	3	2
		Roof	Galvanized Metal Frame + Concrete roof	Galvanized Metal Frame + Harflex Sirap	Galvanized Metal Frame + Zincalume	Galvanized Metal Frame + Concrete Roof
		Value	3	2	3	3
	Design		Limited	Limited	Limited	Unlimited
	Value		1	1	1	5
	Fleksibility	Module	Column & Beam 3m, Wall 0,6x2,2m	Column & Beam 3m, Wall 1x3m	Module 1,5m	-
		Value	1	4	4	5
		Portability	Structure Component	Material Component	Material Component	No
		Value	4	5	5	1
		Construction Technolgy	Concrete System	Screw System	Knockdown with bolt	In situ
		Nilai	1	4	5	1
	Durability		Designed with SNI Earthquake of 2012, with 1000 year earthquake condition	Normal scale earthquakes.	Earthquake risk up to zone 6	-
	Value		5	3	5	1
	Workmanship		Fabrication	Fabrication	Fabrication	Manual
	Value		5	5	5	1
Time	Time Schedule	Fabrication	4 weeks	4 weeks	4 weeks	-
		Value	2	2	2	1
		Assembly	1 weeks	1 weeks	1 weeks	± 2 Month
		Value	2	2	2	1

On the cost aspect, an average of prefab house production is more expensive 6,8% or Rp.200.000.-/m2 than the conventional hous. One the cause of the high cost of prefabricated house is quality of material that is better than the conventional one. For example, wall material for all prefabricated house is concrete. BBI- prefab is using sandwich panel, Wika concrete is using hollow core concrete panel with K-350 quality, and RISHA is using a standard concrete. Meanwhile, the wall conventional house is made by brick and plastered concrete. For building structure, WIKA concrete prefabricated house is using cast assembly method; meanwhile RISHA is using concrete with bolt connection system. From three prefabricated house types, RISHA house is cheaper and even than the conventional house. It does not mean this house with low quality material but because the fabricated method of this house is developed by government with non-profit goals. Big costs of this prefabricated house are caused by another cost on fabrication stage in workshop, transportation to development location and assemble cost on site.

Most of construction quality (workmanship) for prefabricated house is better than conventional house. It is because of its material and it's produce in factory so it has an accurate, precise and neat product. The prefabricated house that is developed by house producer in Indonesia has a limitation on design, it is because the prefabricated house is focused on time efficiency and decreasing construction fee. Design aspect on three prefabricated house products have a monotone design. It affects on preference level on prefabricated house that is lower than the conventional one.

In durability aspect and workmanship for prefabricated house shows, that level of productivity is higher than the conventional house. Most of prefabricated house in Jabodetabek is planned for earthquake resistant design to zone 6. WIKA concrete instant house is designed based on earthquake SNI 2012, with 1000 years earthquake condition. Prefabricated house workmanship is more excellent than the conventional house because by this off-site fabricated produced a good product, more precise and less waste construction than on site construction system.

By the time aspect, prefab house development is faster than conventional prefab house in Jabodetabek, that it produce about 3-4weeks for assembling time in project location for type 36 house that is only needed 7 days, it is faster than conventional house that is need about 2 months. According to the producer of WIKA concrete prefab house and Bakrie, it only need about 7 days for 1 unit of house, because it is more easier to be done and several item job will be done in parallel, not sequential. Besides that if, the prefabricated house is in mass production so that time efficiency can be produce by fabrication in factory and construction on site can be done parallel.

Previous study in Gradute School of Clemson University by Lu Na in 2000 in USA reveals that the advantage of using the prefabricated component is to reduce the construction time and efficiency time, increasing building quality and quality of execution and labor productivity [3]. One thing that also found on prefabricated house in Jabodetabek is the higher productivity on time and quality than in conventional house. Kullman added that time efficiency on construction is about 50-70% if the construction is done in factory rather than on site [4].

On Mc Graw-Hill Construction's Survey, the material efficiency and time efficiency give an impact for project's budget efficiency [5]. 65% of the company who uses prefab/modularization method reported that they could lower project budget into 6% or more. This is not happened on prefabricated house product in Jabodetabek. The cost of prefabricated house production in Jabodetabek is still more expensive than the conventional house

3.2. Consumer's preference study on prefabricated house in Jabodetabek

Four aspects such as knowledge, liking, preference and conviction did consumer's preference study on prefabricated house, by the questionnaire random distribution that is done by google.com survey application. 43 respondents have already filled in the questionnaire. This study shows that the low result on consumer's preference level on prefabricated house in Jabodetabek, especially with some aspects such as knowledge, preference and conviction, but it is quite high on liking factor.

- *Knowledge*

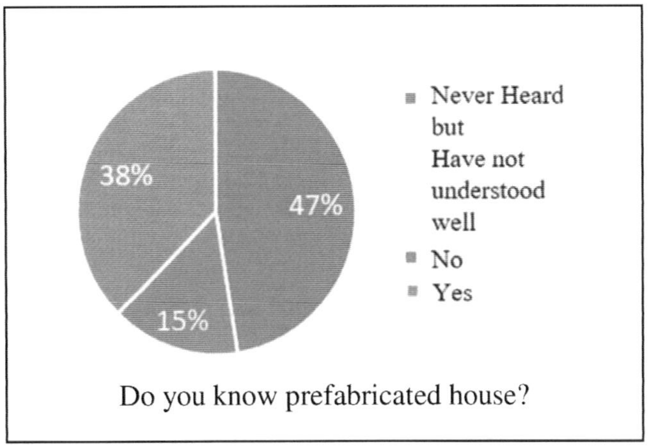

Figure 1. Result of consumer quesionnaire on knowledge of prefabricated house.

A low level of recognition of prefabricated house in jabodetabek, 47% respondents has already heard about prefab house but they are not really know it well, 38% has already known it and only 15% haven't known it at all. The society has not known well much about prefabricated house because the lack of information about prefabricated house in Indonesia.

- *Liking*

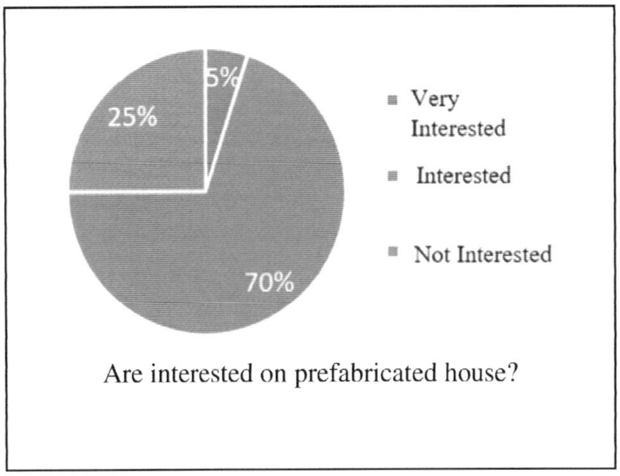

Figure 2. Result of consumer quesionnaire on liking of prefabricated house.

Even though the low result on level of recognition on prefabricated house in Jabodetabek, 70% of respondents are interested on prefabricated house concept, they're interested on a new prefab house concept. Respondent's interest on prefabricated house is on concept of technology that is expected could give an time and cost efficiency.

- *Preference*

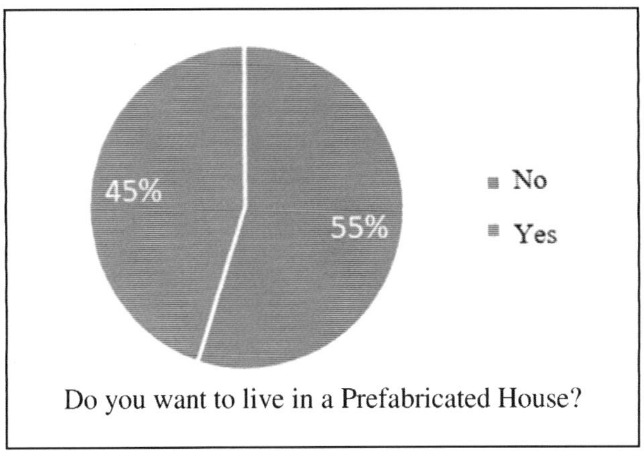

Figure 3. Result of consumer quesionnaire on preference of prefabricated house.

In addition, even though a high result on interest level on prefabricated house in Jabodetabek (70%), but we found a low result about their decision on choosing this prefabricated house. 55% of respondents said that they don't want to live in a prefab house. Some of positive respondent's reasons is about time and cost, which is the prefab house development could be done faster and cheaper than the conventional one. Meanwhile, some of negative reasons are about quality, such as monotone design, inflexible and about comfort ability.

- *Conviction*

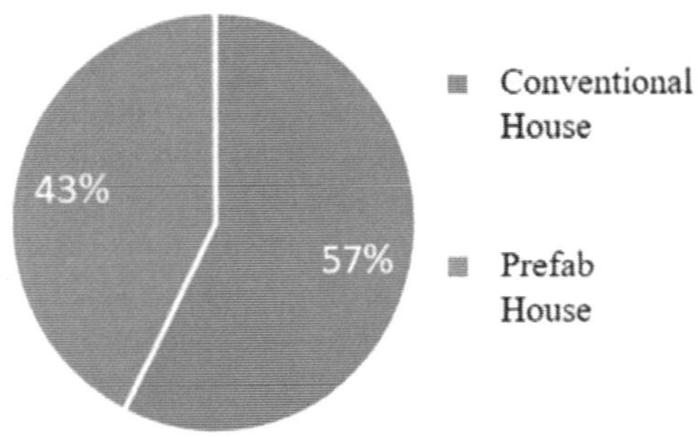

Figure 4. Result of consumer quesionnaire on conviction of prefabricated house.

The level of willingness to buy a prefabricated house is low. 57% of respondents' said that they choose conventional house even though it is 10% more expensive than prefabricated house. They are ready to pay even more expensive on design, flexibility and comfortability than prefabricated house.

The preference level on society of this prefab house is explaining on why this prefab house business hasn't been improved well yet in Jabodetabek. By preference study could be found that the society still has a negative opinion about prefabricated house, such as monotone design, inflexible and a problem of comfort ability that still blurred on their mind. This negative opinion is on middle and high-income society.

On previous study by Robert Ferber and Hugh G. Wales (1951), "The Market for Prefabricated Housing" is concluded they have a low preference on prefabricated house in America, especially for an expensive prefabricated house. This phenomenon is found on preference study that is prefabricated house product now that shows a low society's interest on this product. Meanwhile in recent situation the prefabricated house is popular in some countries, for example, Sekisui prefab house is in real estate market in Japan.

3.3 Prefabricated house business investment simulation for finding NPVand IRR value.

Based on researched data from prefabricated house's producer in Jabodetabek, and actual data from the next survey, which is made a real estate development business simulation for two types products, which is prefab and conventional house. In this simulation, potential income and sum amount of need to be paid is counted, so that will be known the profit level and is compared with investment value of this two property products (prefab house vs. conventional house). In this simulation, there are 3-investment scenario, which is (1) 100 units development in 1 year, (2) 500 units development in 2 years, and (3) 1000 units development in 3 years, with large area is about 36m^2, A capital loan of 70%, an annual interest rate of 30% and an inflation of 7%. The project is located in Jabodetabek with land price of Rp.1.000.000, - / m2. The cost of prefabricated houses development is more expensive at 6.8% compared to conventional homes. The cost of a prefabricated house development is assumed to be a 5% reduction price when it reaches 1000 units, resulting in the total investment cost incurred as follows: (see table 2)

Table. 2. Investment cost for development of cluster 100, 500, 1000 units (source : personal data)

Investation	100 unit	500 unit		1000 unit
Prefabricated House	Rp 27.183.196.800	Rp	134.991.984.000	Rp 264.532.968.000
Conventional House	Rp 26.463.196.800	Rp	131.391.984.000	Rp 262.552.968.000

In the operational cash flow, the income is fully obtained from the sale of housing unit with the selling price of Rp.300.000.000, - / unit with the assumption of purchase with DP 30%, as if the house is completed and then he customer pay another 60% and the last repayment is 10%. Meanwhile the other costs, in general operational expenditure on this project is assumed from the cost of administration, marketing, security, maintenance etc.

3.3.1 NPV (Net Present Value)

Dalam rentang waktu tersebut perkiraan okupansi penjualan diasumsikan 100% terjual, didapatkan besaran NPV sebagai berikut

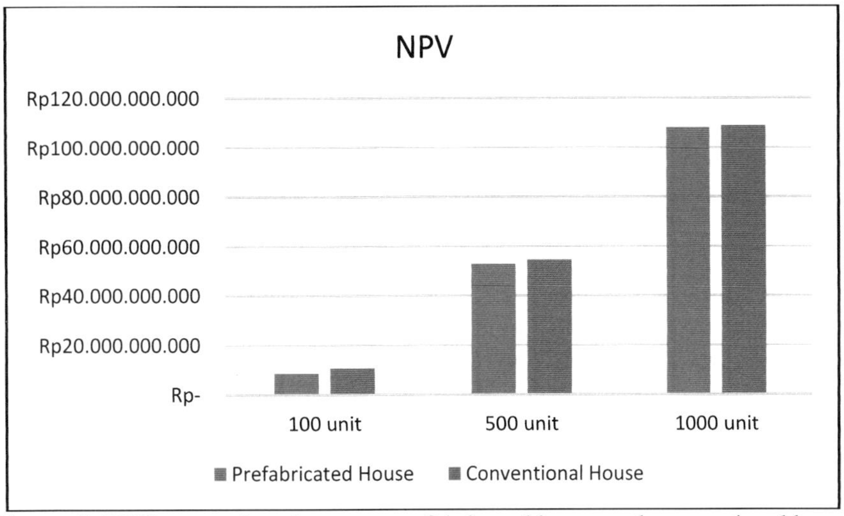

Figure 5. NPV comparison between prefabricated house and conventional house (source: personal data)

Pada setiap skenario, NPV rumah prefabrikasi lebih rendah dibandingkan dengan NPVrumah konvensional, ini disebabkan oleh *development cost* dari rumah prefabrikasi yang lebih tinggi. Namun demikian karena NPV Rumah prefabrikasi menunjukkan nilai yang positif maka dapat disimpulkan bahwa investasi rumah prefabrikasi layak untuk dikembangkan.

In each scenario, prefabricated house's NPV is lower than conventional house's NPV, it's because higher prefabricated house's development cost. However, a positive NPV value of prefabricated house is concluded feasibility of prefabricated house investment.

3.3.2 IRR (Internal Rate Return)

From the calculation data obtained comparison of IRR prefabriksi home with conventional house as follows

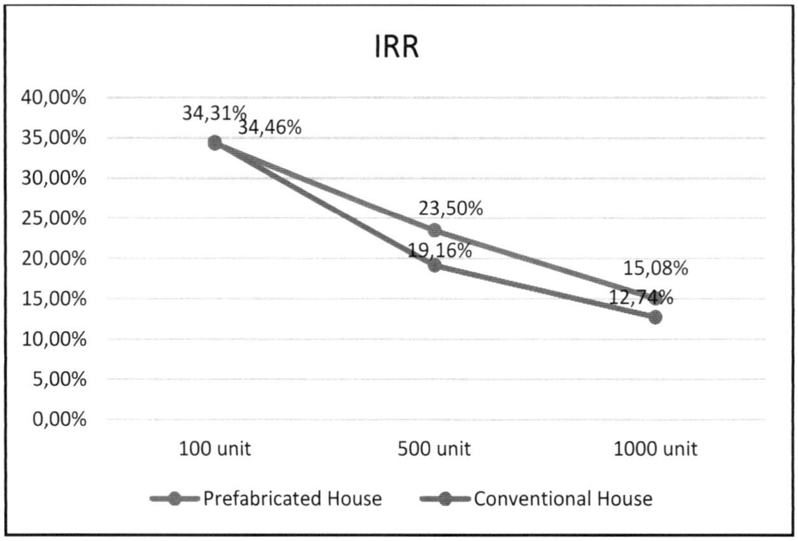

Figure 6. IRR comparison of prefabricated house and conventional house (source : personal data)

Although prefabricated house NPVs have smaller values than conventional houses but based on the cash flow simulation results, prefabricated house IRRs have a higher compared to conventional house, this is because prefabricated houses have better productivity levels compared to conventional house in terms of working time, So that its operating cash flow turns faster.

From the results of a study conducted by Sri Velemati [6] in America, the savings is generated by costs and related items resulted in an increase in NPV values of 3.485 and IRR of 1.56%. In combination, all savings can yield 5-10% of the total project cost. Meanwhile, in the simulation of prefabricated house investment in Jabodetabek did not show an increase in NPV value of prefabricated houses compared to conventional homes because it did not result in savings from total cost of projects, savings can occur if production is done very massive. However, in the simulation the increase in IRR on prefabricated home investment still occurs because the processing time is shorter so that the cash flow of the operational flow becomes faster.

4. Discussion

Productivity level of prefabricated house construction in Jabodetabek is considered higher than it is in conventional houses, especially within two factors that are quality and time. From the cost side, the cost of prefabricated house production is 6,8% higher than it is in conventional one because of its material quality that is better than it is in conventional house. When it comes to material quality, durability and workmanship prefabricated house has a higher productivity level compared to conventional houses but worse than it when it comes to flexibility. And when it comes to time, the construction of prefabricated house is faster than the construction of conventional houses.

Preference research found out that preference level of the consumers toward prefabricated house in Jabodetabek is stil low. This also explains the lack of development in prefabricated house business in Jabodetabek. Some of the reasons why the preference is low are limited design, low flexibility and production cost that is far more expensive.

Based on the cash flow simulation of prefabricated house business development, it is concluded that prefabricated house has lower NPV with a difference of Rp 1.510.666.135,- while the average of IRR from prefabricated house is higher than it is of conventional house by 2%.

The business of prefabricated house in Jabodetabek has better potentials to be one of the real estate business in Indonesia because of its construction that is far faster than the other and this can cause the business cycle faster as well, along with proper approach toward better flexibility of the product, variety of appearances, and mass production. Prefabricated house has a big potential to be one of the solutions addressing the settlements problem in Indonesia both as commercial housing products or as subsidized house products by governments.

5. References

[1] Forintek Canada Corp. & The Concordia University. (2001). Assessment of Prefabricated Building Industry: Forintek Canada Corp.
[2] Smith, R. E. (2010). Prefab architecture: A guide to modular design and construction. Hoboken, NJ: John Wiley and Sons Inc
[3] Lu, Na. 2007. Investigation of the Designers' and General Contractors' Perceptions of Offsite Construction Techniques in the United States Construction Industry, Department of Career and Technology Education, Clemson University, Clemson, GA.
[4] Modular Architecture Manual, Kullman Buildings Corporation and Garrison Architects.2008
[5] McGraw Hill Smart Market Report: "Prefabrication and Modularization Increase Productivity: Construction Report." May 11, 2011.
[6] Sri Velamati. "Feasibility, Benefits And Challenges Of Modular Construction In High Rise Development In The United States: A Developer's Perspective", Submitted to the Program in Real Estate Development in Conjunction with the Center for Real Estate in Partial Fulfillment of the Requirements for the Degree of Master of Science in Real Estate Development at the Massachusetts Institute of Technology September, 2012
[7] Mark Anderson & Peter Anderson, Prefab Prototypes: Site Specific design for offsite construction, Princeton Architectural Press, New York, 2007
[8] Soeharto, I, Manajemen proyek : Dari konseptual sampai operasional, Jakarta ; Erlangga, 1999
[9] Kotler dan Keller. 2009. Manajemen Pemasaran. Jilid I. Edisi ke 13. Jakarta: Erlangga
[10] Suratman. (2001). Studi Kelayakan Proyek : Teknik dan Prosedur Penyusunan Laporan. Yogyakarta : J&J Learning.

Sacred Space in Community settlement of Kudus Kulon, Central Java, Indonesia

Agung Budi Sardjono*, Arnis Rochma Harani
[1]Department of Architecture, Faculty of Engineering, Universitas Diponegoro, Semarang, Indonesia

*Corresponding e-mail: abskempung@yahoo.co.id

Abstract. The sacred space becomes an important part of the spatial layout of Javanese society, as well as in most houses of the archipelago. This space is related to religious activities, highly respected and usually located in the main place. Kudus Kulon community is a part of Javanese culture in the northern coastal area of Java. Known as a devout Muslim society, sacred space in Kudus Kulon community house becomes an important thing to understand the culture of living in society. The research was conducted by looking at the religious activities of the community and how the space of activity is realized. The research was conducted by qualitative research method. In the Kudus Kulon community, sacred spaces are available in the house itself, in the community grup of houses, and around Kudus Kulon area. In every house, the sacred space can be found in Gedongan and extends to Dalem. In the community group of houses, the sacred space can be found in the neighbourhood praying area, while In Kudus Kulon area, it can be found in Mosque complex and Tower tomb. Physically sacred space is realized by raising the floor height of the space around it. In architectural view, the sacred space is seen on the elevation of the roof or roof angle. Spatially the sacred space occupies a central position and visually represented by a vertical orientation.

Keywords : Sacred Space, Settlement, Kudus Kulon Community

1. Introduction

The archipelago's culture stretches, covering thousands of islands scattered between two continents and two oceans. This natural condition is one of the reasons for the diversity of local culture in Nusantara. Each culture develops and shapes its distinctive character. Intercultural local interact each other. The diversity of local culture enrich the Indonesian character that needs to be explored, known and maintained. The values of local culture as a form of adaptation of community life for a long time can be a role model and updated again to face the challenges of life in the present and future (Oliver, 2006).

Generally the form of culture can be divided into three kinds started from the most abstract of the cultural system that includes the values and views of society. The abstract values are then manifested in the social system that includes patterns of society daily life activities. From the activities of the community that took place, the space of activity as a container of the veiled event is a concrete manifestation of culture, in the form of building architecture (Koentjaraningrat, 1984; Rapoport, 1983). Pangarsa (2007) stated that although the object of worship may vary, one of the same characters in the archipelago's local cultures is the religious life itself. In Javanese culture, one of the important values of life is the balance of human life (Bawono Alit) with a larger world order (Bawono Ageng) (Tjahjono, 1989, Santosa, 1997). Thus one way of discovering the cultural character of society is to look at its religious life. The space for organizing religious activities is a sacred space for the inhabitants.

IOP Conf. Series: Earth and Environmental Science **99** (2017) 012023 doi:10.1088/1755-1315/99/1/012023

2. Kudus Community

Kudus City is the capital of Kudus District. It is located about 60 kilometers Northwest from Semarang. Kudus District the smallest area in Central Java, but has the highest Regional Original Income, mainly supported by industry and trade. Kudus culture is part of Javanese sub-culture called Coastal culture. Namely Javanese culture that developed in cities along the north coast of Central Java and East Java (Thohir, 2006). Two important things in Coastal culture character are trading and Islam.

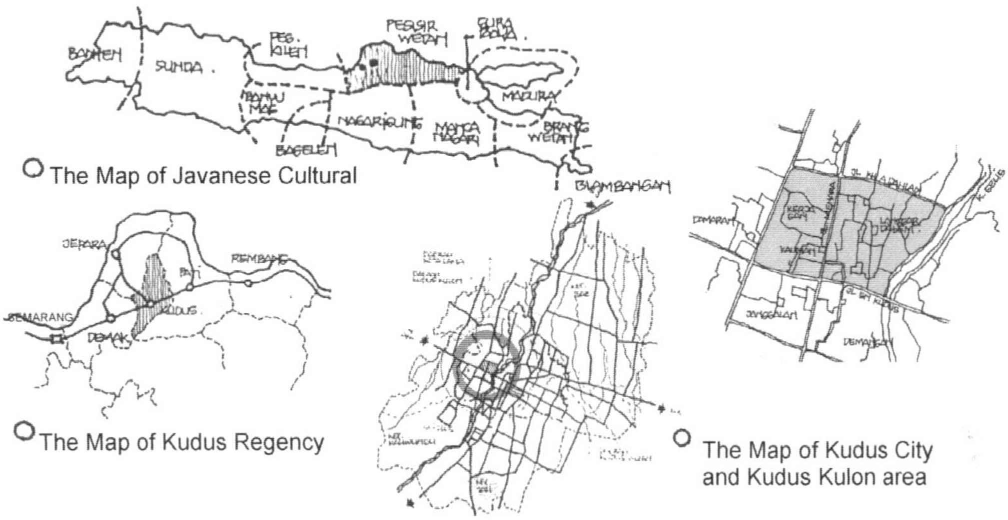

Figure 1. Map of Kudus Kulon

The history of the Kudus community cannot be separated from religious figures named Ja'far Shodiq or better known by the name of Sunan Kudus. Sunan Kudus is members of Wali Songo, nine Islam propagators in Java. Sunan Kudus withdrew from the kingdom of Demak because of political feud, then established a city that would be called the Kudus City. The early establishment of the city is marked by the building of Al Manaar mosque, located in the west of Gelis River. This environment quickly developed as the center of Islamic trade and development. In the colonial period, the government moved the city center to the east of Gelis (Castles, 1982). Currently the central development of the old city of Kudus is known as Kudus kulon or Tower area.

The Kudus Community, especially Kudus Kulon, is a society that has strong character. They are known as the Society of Merchant Traders. Their daily life is dominated by two activities that support each other, namely religious activities and trade activities. They are known as ethical and relatively puritan Islamic adherents, while in the economic aspect they are known as resilient, graceful and independent traders. They assume that prosperity must be gained by hard work in the in the field of trade and by worshiping as well (Sardjono, 1997). Strong faith and community self-reliance are believed to be imitated by Sunan Kudus during his lifetime.

The community success in trading raised many wealthy merchants in Kudus kulon area. Its economic strength is reflected individually by the look of the houses, while in community scale it can be seen by the large number of community mosques built (Bonnef, 1969).

The order of houses in Kudus kulon is quite crowded. There are still many traditional houses, beside colonial house and modern house. Most of them are high-row fenced, called kilungan, and some others are no a fence at all. The traditional houses of Kudus are oriented to the South, lined east of the East. The open space is the central courtyard of the Kilungan fenced house. At the open house, the central yard forms a short cut. At almost of home groups found Langgar or small mosque in which daily religious life held (Sardjono, 2016)

3. Religious Life

Religious activities of Kudus Kulon society dominate daily life includes daily, weekly, and annual activities. Women often held their daily worship activity in their homes. Although not a necessity, it is more commonly done in the Dalem room, the main and deepest chamber of the traditional Kudus house, while the men took a pray in congregation in Langgar (small mosque) in addition to the greater virtue and also the opportunity to meet neighbors and surrounding communities. Religious atmosphere in Langgar is felt especially during the Shubuh, Maghrib and Isha. Shalat is continued with dhikr or reading the Qur'an. At certain moments also held berjanjen (journey). People who come to langgar are not only the men surrounding but also santri (Islamic boarding school students).

Figure 2. Religious Activities in Kudus Kulon

Daily worship activities increased on Thursday night until Friday afternoon, during jumatan. Jumatan is not held in Langgar, but in big mosque. In Kudus kulon Jumatan is centered in Menara mosque. Friday is a day off for the Kudus community, so people starts to prepare jumatan since morning. Jema'ah jumatan are not only come from people around the Menara mosque but also from the other area, including from Kudus wetan (East Kudus area). Within a year, the religious activity is rapidly increased in the period closed to the end of Ramadan. In the holy month the enthusiasm of people undergoing a variety of worship is higher than usual, includes Dandangan event, a market opened since a week before Ramadan until the official announcement of Ramadhan coming. It also visited the family and ancestral tomb to pray and clean it up. Implementation of Ramadhan prayer activities in the tower mosque includes the activities of reciting after Asyar prayer while waiting Maghrib, giving takjil for breakfasting in the mosque. Continued with shalat Magrib, Isya, Tarawih, Witir and religious lectures. After Subuh, there is communal quran reading held by Kudus charismatic kiai. Religious events were also held in Langgarlanggar, although not as large and intensive as the events at the Menara mosque. Ramadhan festivities reach the peak on the day of Lebaran, as well as closing the series of activities of Ramadhan. In the morning people go to Menara mosque for shalat Iedul Fitri, then proceed with visiting neighbours and relatives, including the deceased parents and ancestors.

Another annual religious activity in Kudus Kulon community is Bukak Luwur on every 10 Dulhijjah (Muharam) held in Sunan Kudus tomb complex behind the Menara mosque. This event is basically Haul event, which is respect and pray for the soul of leader or community leader who has passed away. Luwur Sunan Kudus Proof is a series of honor programs on Sunan Kudus by replacing the mosquito net on the cupola cemetery of Sunan Kudus. The event is begun by cleaning up the Sunan Kudus relics continued with the reading of the Qur'an to finish. In the evening a recitation and recall of the Sunan Kudus history are held. On the third day of the 10th day Duhijjah the replacement ceremony of Netting is done. The ceremony distributed blessing food to the community, namely rice with buffalo meat curry. Thousands of people who come from different regions queue up the distribution of jumbo rice along the narrow alleys around the Menara mosque.

Figure 3. Religious Activities in Kudus Kulon

4. The Sacred Space

Sacred space is a place for activities related to belief or religion. It is a respected space and a high-valued space. Shalat activities at home are usually done in Gedongan, which is a closed space in the middle of Dalem (main building in Kudus traditional house). Gedongan is specifically used as a praying space and a place to store the house property. It is noticed that this space is also used as a space for the first night for newly married couples. Currently, in general, Gedongan is used by the parents to sleep and is used as the praying room as well, which is generally done by women. Physically the Gedongan is located in the middle of Dalem, rather shifted to the end. Gedongan room floor slightly elevated to Dalem. This space is a privileged and closed space.

Dalem itself is part of Kudus traditional buildings somewhat closed to other space, namely Pawon on the side and Jogosatru in front of it. Dalem is a passive family activity room, for sleeping or resting and sometimes used for praying or selametan. Pawon is the living room of the family while Jogosatru is a space to receive guests. The Dalem floor is elevated far from the floors of Pawon and Jogosatru so as to reach it is used a ladder called Bancik.

Within the house cluster, joint religious activities are held in Langgar so that the langgar is the center of the activities of the surrounding community groups, including pesantren and madrasah huts that often use Langgar for praying and Ngaji (reading quran) activities. Besides, some Langgar also held social and educational activities such as PAUD, skill counseling and health activities. Worship activities in Langgar is usually done on the main building inside with expansion when the congregation is a lot,while other activities that are not directly related to worship are held in the porch as well as in the court of the mosque. The main room floor is usually higher than the porch and separated by a wall. There is no wall between the porch and the open courtyard or only walled with a large bang. The porch floor itself is higher than the mosque's courtyard. At times between shalat and ngaji, the main room door is usually closed. Some Langgar have a grave around. These tombs are the tombs of predecessors who are respected by the people. This tomb is often visited, both by direct descendants and surrounding communities. They recite and pretend the spirit of the predecessor while asking for the blessing of salvation. This grave is often visited at the beginning of the Ramadan (nyadran) and after shalat Iedul Fitri. Likewise on certain days or months of a particular month

In Kudus area, the cultivation of large religious activities is centralized in the Menara mosque. For example, Jum'atan and shalat Idul Fitri and Idul Qurban. The mosque is also the center of the region's highly respected activities of Kudus kulon. Activities that are held are more diverse and more intensive. Praying activities are done in the main room and can extend to the porch. This space is usually closed and opened just before praying. The teachings and tadarus are usually held in the foyer of the mosque and can extend to the mosque's courtyard. The floor of the mosque's main hall is slightly higher than the porch, but the porch floor is high enough against the courtyard so there is a stair to reach it. Behind the mosque there is a complex of Sunan Kudus tomb. The complex is separated from the mosque and its surroundings with a high enough massive fence. The tomb of Sunan Kudus itself is facedown, has a porch and is separated by another tomb with a fence. Every day hundreds of pilgrims come to visit the tomb of Sunan Kudus, they dhikr in the porch cupola. The cungkup (tomb) itself is usually closed and only opened at certain moments, such as Bukak Luwur ceremony. Activities in the mosque and in the tomb are separate activities, although sometimes carried out continuously. Pilgrims do shalat in the Menara mosque and then dhikr and pray in the porch of the tomb. At the time of the ceremony Bukak Luwur, most of the activities took place in the tomb complex, while recitation activities are conducted in the porch of the mosque. The distribution of blessing food takes place along the Menara road and on the alleys of the settlements around the Menara mosque. The participation of Kudus Kulon community is huge. The men helped organize and secure the ceremony, and the mothers cooked the blessing food to be distributed. The event of Khol or Bukit Luwur Sunan Kudus is the largest rite and started the implementation of khol community leaders in the Sunan Kudus area. Bukak Luwur is then followed by a similar event on the tomb by community leaders with smaller the intensity and scope.

5. Discussion

The existence of sacred spaces in Kudus community is indicated by the existence of worship activities and activities that have to do with trust. This activity is a dominant daily activity of the society besides the economic activities. Sacred space in the traditional Javanese house (Krobongan) inside of Dalem describes transcendental vertical relationship (Tjahjono, 1989; Santosa, 1997). The same pattern is also shown in the traditional house of Kudus, but Gedongan is not intended for offerings to Dewi Sri as in the traditional Javanese house Krobongan.It is used for shalat by the house residents. The sacred space is manifested by the pattern of space placement in the center, the nature of the relatively closed space and the elevated floor surface of the space around it. Boundary element becomes an important element, in addition to protect the sacred space as well as to separate it with more profane activities. It is is in line with the opinion of Supriadi (2010) who stated that in the house of Java, a boundary plays a crucial role.

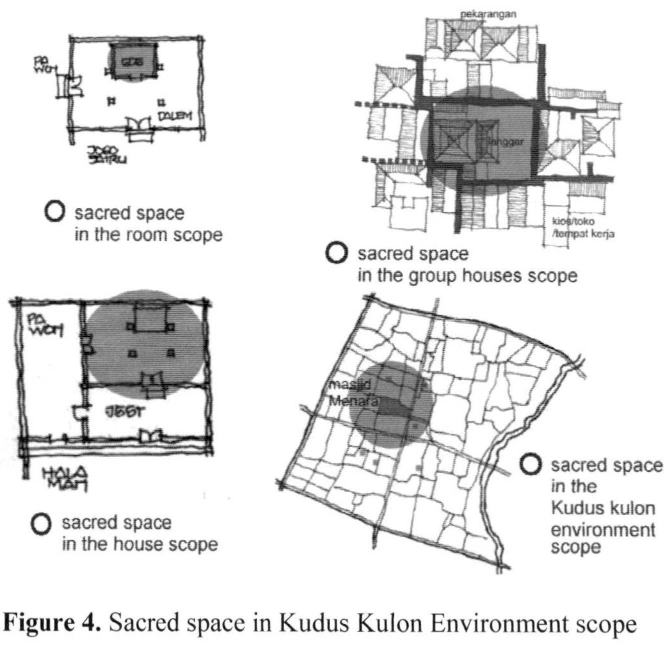

Figure 4. Sacred space in Kudus Kulon Environment scope

Even in the teachings of Islam, boundary have an important role as the separator of two different things (O'meara, 2007). Boundary separation in the form of buildings can be a thick wall and high fence (Kilungan) or separator lever on Jogosatru space that separates male and female guests. Sacred space as a place of religious worship is in the space Gedongan, Dalem, Langgar and Menara

Mosque. The sacred space also embodies the more important activities of belief and tradition in the tomb (Thohir, 2006). The sacred space in Kudus kulon also shows a tiered pattern. In the Dalem room, the sacred space is in the Gedongan room. Dalem is a sacred and the most important space in a room. In the group of houses, the sacred space is in Langgar and the grave of the precursor. Although not always in the middle, but the closure, the schedule of activities as well as the center of religious activities or beliefs show the space.

In Kudus kulon, sacred space is in the complex of Menara mosque and the tomb of Sunan Kudus. It is the center that binds and arranges the pattern of space activity around it. The existence of sacred space tiered from room, house, and environment of Kudus kulon area.

6. Conclusion

The existence of sacred space creates a strong character in Kudus Kulon area. The existence of sacred space is a form and a container of religious worship activities and traditions of belief. The influence of the teachings of the religion of Islam is enormous in realizing the sacred space, as well as the remnants of Javanese traditions and beliefs that nevertheless still exist in the life of society. The container of these activities forms the character of a sacred space that is centered, relatively closed and vertically transcendental-oriented. In relation to its environment, sacred space is manifested in stages.

7. References

[1] Bonnef, Marcel, 1983, Islam di Jawa Dilihat Dari Kudus dalam Citra Masyarakat Indonesia, Sinar Harapan, Jakarta
[2] Koentjaraningrat, 2005, Pengantar Antropologi Budaya I, Rineka Cipta, Jakarta.
[3] Oliver, Paul, 2006, Built to Meet Needs, Cultural Issues in Vernacular Architecture, Elsevier, Oxford.
[4] O'meara, Simon, 2007, Space and Muslim Urban Life, At Limits of The Labyrinth of Fez, Routlage, London and New York.
[5] Pangarsa, Galih W, 2007, Merah Putih Arsitektur Nusantara, Penerbit Andi, Yogyakarta.
[6] Rapoport, Amos, 1983, Development, Culture Change and Suportive Design, Pergamon Press, New York
[7] Santosa, Revianto B., 1997, Omah. The Construction of Meanings In Javanese Domestic Settings, Thesis, School of Architecture Mc Gill University, Montreal.
[8] Thohir, Mudjahirin, 2006, Orang Islam Jawa Pesisiran, Fasindo, Semarang
[9] Tjahjono, Gunawan, 1989, Cosmos Centre and Duality In Javanese Architectural Tradition : The Symbolik Dimention of House Shapes in Kota Gede and Surroundings, Disertasi , University of California, Barkelay.
[10] Castles, Lance, 1982, Tingkah Laku Agama Politik dan Ekonomi di Jawa : Industri Rokok Kretek Kudus, sinar harapan, Jakarta.
[11] Sardjono, Agung B., 1996, Rumah-rumah di Kota Lama Kudus, Thesis, Universitas Gadjah Mada, Yogyakarta.
[12] Supriyadi, Bambang, 2010, RuangJawa, Belajar Dari Dunia Pewayangan, Disertasi, Universitas Diponegoro Semarang.

Proposal of stack Effect technology for predicted future years

FX Teddy Badai Samodra*, Iwan Adi Indrawan
Department of Architecture, Institut Teknologi Sepuluh Nopember, Surabaya

*Corresponding E-mail : fxteddybs@gmail.com

Abstract. Recently, stack effect is a general problem solver in providing vertical ventilation for urban environmental issues. However, study on resilient technology of stack effect for future years as predicted by climate trend should be conducted. Therefore, this research proposes a design of new technology on operable and adaptable vertical ventilation to the environmental change. The research method is conducted by comprehensive simulation of Ecotect Analysis, ANSYS Fluent and Matlab. Urban environment of Surabaya, as the research location, is the representative of tropical region. The results showed that the stack effect height and area could be modified instantly adjusting the environmental condition time by time in the future years. With 1.8 m of stack width, the proposed technology could capture 40 m^3 of vertical air flow which is useful for physiological cooling and its dimension could be modified depending on the environmental condition. By providing resilient technology, predictable and sustainable ventilation method is offered to anticipate an unpredicted global warming and environmental change.

Keywords: air flow, building technology, stack effect, sustainable design, urban space

1. Introduction

In general, passive control for thermal environment could be distinguished into four basic strategies (Szokolay, 2008): (1) passive solar heating (with 0.5 and 0.7 efficiency), (2) mass effect (summer and winter + for summer with night ventilation), (3) air movement (physiological cooling) effect, for 1 and 1.5 m/s, and (4) evaporative cooling (direct and indirect). For Indonesia, the tropical country, the suitable method could be mass effect, air movement and evaporative cooling. Considering of high both air temperature and relative humidity in tropical climate and material cost, the ventilation development with accelerating of the air movement is the most possible way. The mass effect is very expensive and the evaporative cooling has risk in raising relative humidity. Nguyen and Reiter (2014) reported that natural ventilation and direct evaporative cooling have the similar effect. Natural ventilation is low-cost, easy to apply and provides good indoor air quality, but it relies strongly on natural wind and the building configuration as well as the building location. Prianto and Depecker (2003) explained that the indoor thermal comfort could not be reached by a higher air speed only. The integration between other environment factors and building design elements was required.

If solar irradiance is more than 700 W/m^2, the significance of the ambient air speed will drop (Tan and Wong, 2014). Furthermore, under the tropical weather conditions of high solar irradiance and low ambient air speed, cross ventilation performs better than a solar chimney which is recommended under a zero ambient air speed. In addition, Chungloo and Limmeechokchai (2007) reported that at high ambient temperatures and high solar intensity in the daytime, the solar chimney can reduce the indoor temperature by 1.0–3.5°C compared to the ambient air of 32.0–40.0°C. The velocity magnitude can be increased to 4–25% (Kasaeian *et al.*, 2014). Recently, stack effect is a general problem solver in providing vertical ventilation for urban environmental issues. Resilient technology of stack effect for future years as predicted by climate trend should be conducted. Therefore, this research proposes a design of new technology on sustainable vertical ventilation in future environmental change.

2. Methods

This research utilizes the existing tropical building model, Eco-house building in ITS, as the case study. The research method is conducted by comprehensive simulation of Ecotect Analysis, ANSYS Fluent and Matlab. Urban environment of Surabaya (7.2°S, 112°E, 3 m above mean sea level), as the research location, is the representative of tropical region. The field experiment using anemometer for obtaining wind speed was conducted for simulation input. As shown in Figure 1, the building stack is 1.8 x 1.8 m^2 from 3 storey building with solar penetration potency around the site (open area).

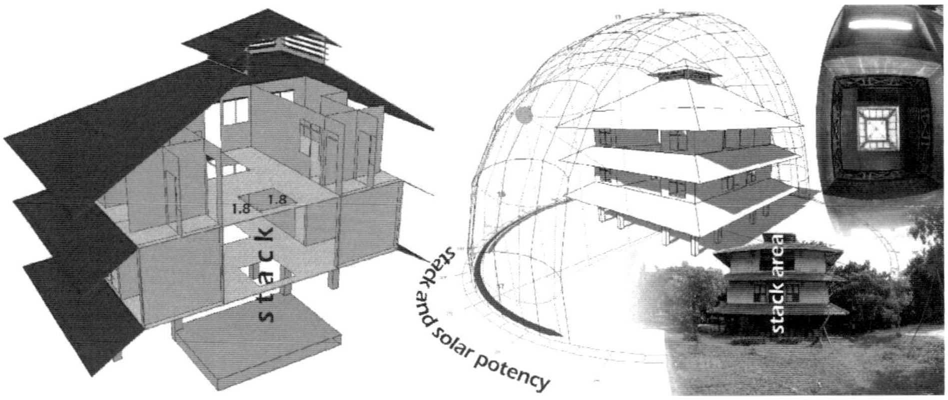

Figure 1. Building model.

3. Discussion

3.1. Environmental condition

The building model was also built in low density terrain roughness (rural) that has changed to sub-urban condition in the current time and potential of becoming urban or city in the future. The reduction of wind speed is about twice slower than before which it gives problem for present psychological cooling (Figure 2). Compared to other location, this research was conducted for critical area, lowland, which has higher wind speed decrease from past to present than highland because of faster change (The lowland condition is *city* terrain roughness while the highland is *urban* or the lowland condition is *urban* while the highland is *sub-urban).*

Figure 3 shows the climate change, increasing problem for temperature from past to present. Developed from climate data, it describes trend-line equation that could be baseline for predicting future temperature as shown in Figure 4. Based on this climate trend, the temperature tends to increase while relative humidity will decrease continually. It impacts on wind speed and its minimum requirement for physiological cooling for future years. Therefore, Figure 5 provides equation that has the capability in predicting wind speed. Developed from Aynsley et al (1977) and Aynsley and Spruill (1990), 0.3 m/s will be the minimum wind speed that will be utilized for restoring thermal comfort for year 2200 with lower wind speed for its previous years. However, it should be noted that most of all time in tropical environment needs wind speed for thermal comfort indicated by a little number (time) area in ≤ 0 (no need wind speed).

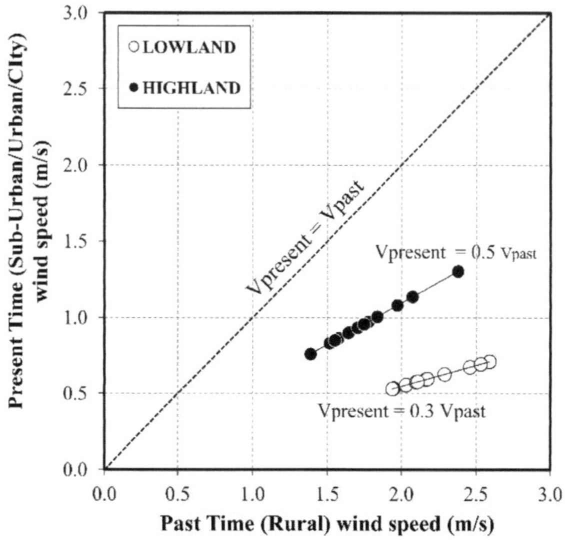

Figure 2. Terrain roughness change effect in wind speed.
(Calculation based on Climate Data from the Center of Meteorology, Climatology, and Geophysics at Juanda and Perak, Surabaya (1997-2012) for Lowland and the Center of Climatology at Karangploso, Malang (1993-2012) for Highland)

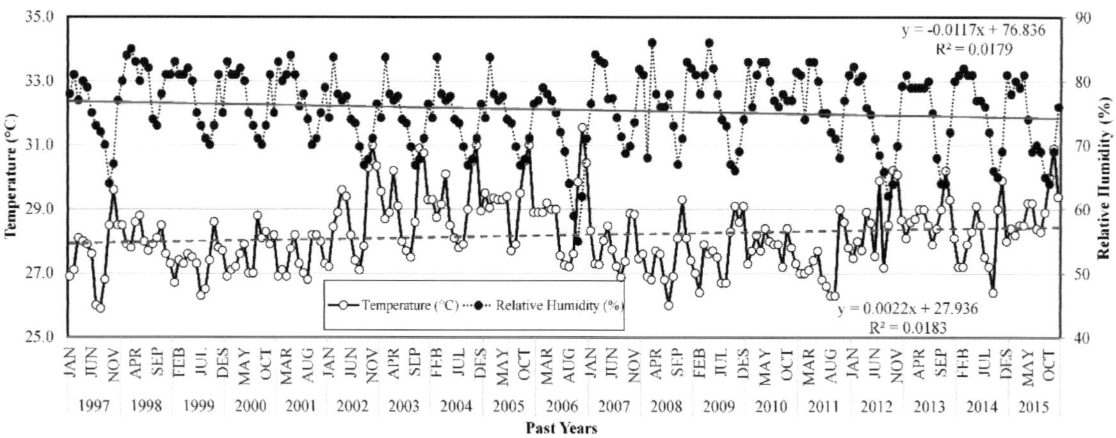

Figure 3. Climate change trend.

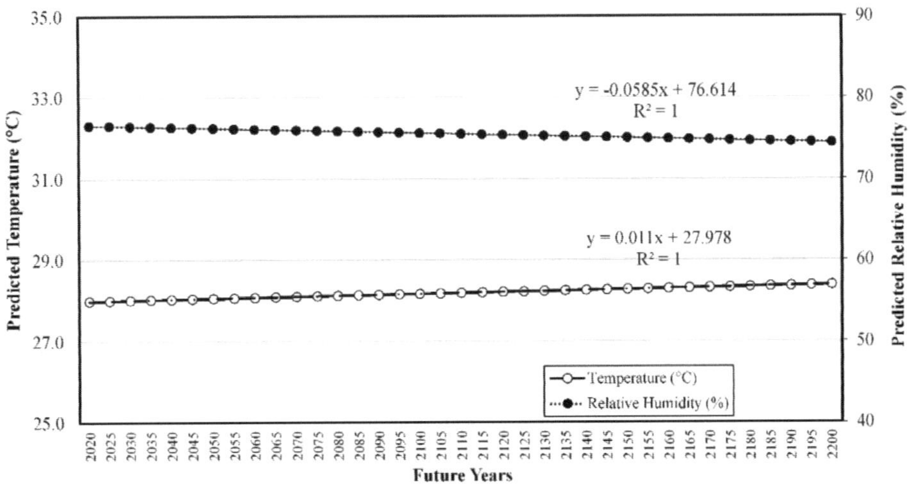

Figure 4. Predicted future temperature.

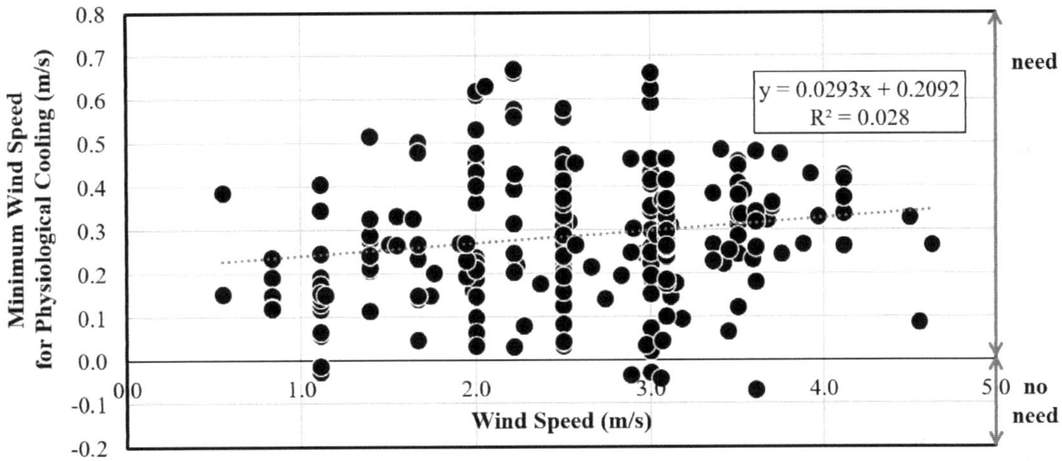

Figure 5. Wind speed and minimum wind speed for physiological cooling.

3.2. Existing building performance

For supporting stack ventilation technology, the building performance should be evaluated. Figure 6 shows the results of Ecotect Analysis simulation for annual solar radiation and its effect on Mean Radiant Temperature (MRT) which is 2/3 of environment temperature in the tropical area (Szokolay, 2008). It could be explained that the building has capability in reducing solar radiation penetration into building through designing sufficient overhang for each floor. Meanwhile, indirect solar radiation effects on the high distribution for MRT while the stack effect could be developed effectively. The building stack has highest MRT for ground floor as the consequence of opened floor. The upper floors, 2nd and 3rd have lower MRT which is good for restoring thermal comfort by vertical ventilation.

The stack environment temperature is relative low because of the roof insulation effect while its roof gets the highest temperature in the hottest time (Figure 7). There is a problem in accelerating wind. The ventilation design does not provide significant indoor thermal comfort because ventilation generates heat gains (Figure 8). The existence of horizontal ventilation results a dilemma in providing ventilation cooling. As illustrated by Figure 9, air movement through space between one room to another room reduces the role of stack effect. As the result, the existing stack ventilation only maintains the inlet air temperature (Figure 10).

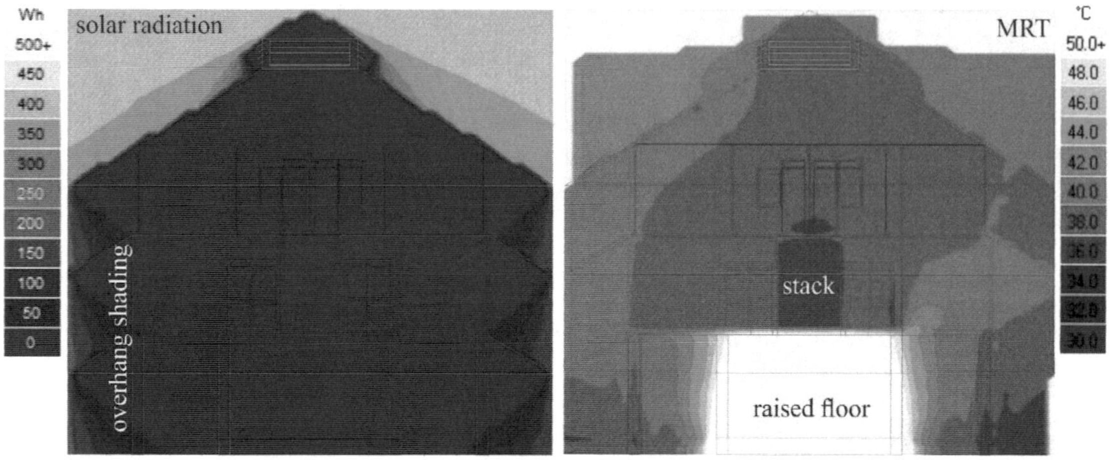

Figure 6. Solar radiation and MRT.

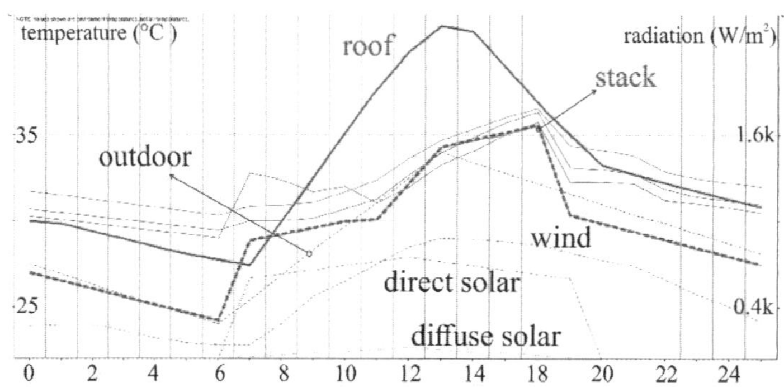

Figure 7. Stack environment temperature in hottest time.

IOP Conf. Series: Earth and Environmental Science **99** (2017) 012024 doi:10.1088/1755-1315/99/1/012024

Ventilation Gains - Qv - All Visible Thermal Zones Surabaya, Indonesia

	Jan	Feb	Mar	Apr	May	Jun	Jul	Aug	Sep	Oct	Nov	Dec
Hr	213.319	211.054	219.85	217.888	224.998	214.196	199.783	74.6426	228.734	397.365	380.547	241.449
	332.569	332.386	355.526	331.704	353.433	340.739	335.43	197.859	356.579	562.354	527.74	357.469
22	480.065	462.478	480.567	479.034	509.099	488.306	484.524	329.579	504.796	734.142	683.926	517.593
	633.061	623.01	638.472	636.428	672.914	657.562	656.064	470.133	678.166	917.801	840.568	670.417
20	789.609	776.393	799.232	794.222	852.31	831.981	828.8	655.202	870.64	1101.56	1009.28	828.579
	971.167	953.416	998.939	948.101	1026.15	1006.12	999.349	836.151	1053.9	1279.74	1189.79	996.461
18	1133.36	1101.22	1167.02	1121.86	1169.28	1160.26	1163.9	1007.91	1220.32	1449.42	1365.18	1183.61
	1316.17	1297.92	1345.56	1303.11	1352.62	1348.58	1356.95	1198.61	1403.48	1636.63	1558.7	1371.26
16	1485.33	1496.39	1538.97	1496.73	1552.49	1511.14	1518.56	1370.03	1588.19	1815.86	1750.51	1560.28
	1674.63	1662.42	1724.43	1640.52	1711.13	1677.5	1707.93	1515.41	1749.82	1994.72	1930.54	1766.47
14	1838.83	1855.64	1882.14	1821.74	1877.43	1903.3	1851.11	1668.87	1893.61	2168.25	2096.46	1900.37
	1431.39	1439.3	1442.55	1426.89	1460.42	1441.91	1437.48	1249.82	1469.15	1736.13	1677.41	1460.16
12	1005.78	1007.5	1026.81	998.699	1027.83	1002.48	990.751	824.891	1039.13	1284.45	1245.3	1035.96
	602.729	609.684	617.578	592.018	624.325	595.925	577.021	416.175	622.5	863.855	831.323	643.368
10	262.847	273.747	273.89	260.917	267.775	250.835	233.447	109.031	269.59	467.175	443.515	288.97
	0	0	0	0	0	0	0	0	14.4705	150.31	135.706	10.709
08	0	0	0	0	0	0	-0.325466	-15.3436	0	0	0	0
	-49.1704	-51.4326	-50.919	-44.027	-47.0503	-46.3745	-52.5224	-85.2626	-41.5722	-5.33999	-7.12549	-45.0086
06	-19.6607	-20.3363	-20.1688	-18.7628	-19.4704	-19.8098	-26.6497	-56.6339	-17.3334	0	0	-15.0861
	0	0	0	0	0	0	-2.07911	-27.9002	-1.81088	0	0	0
04	0	0	0	0	0	0	0	-2.54566	0	0	0	0
	0	0	0	0	0	0	0	0	0	53.9872	46.0111	0
02	0	0	0	0	0	0	0	0	22.5131	157.733	147.486	17.3273
	98.1139	97.1923	97.7291	95.1262	102.648	97.2999	75.0089	0.616798	110.356	272.004	259.047	124.642

Figure 8. Ventilation gains and losses (in Watt).

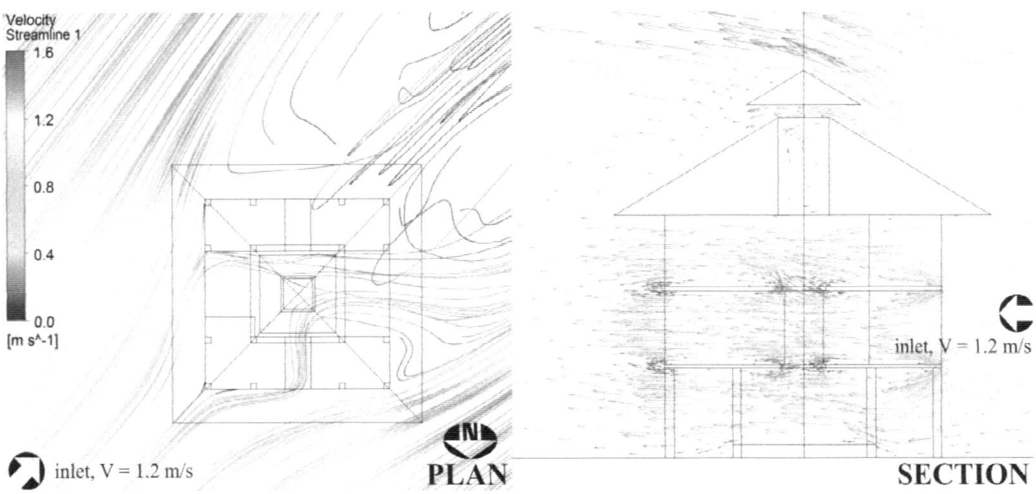

Figure 9. Simulation for existing air movement.

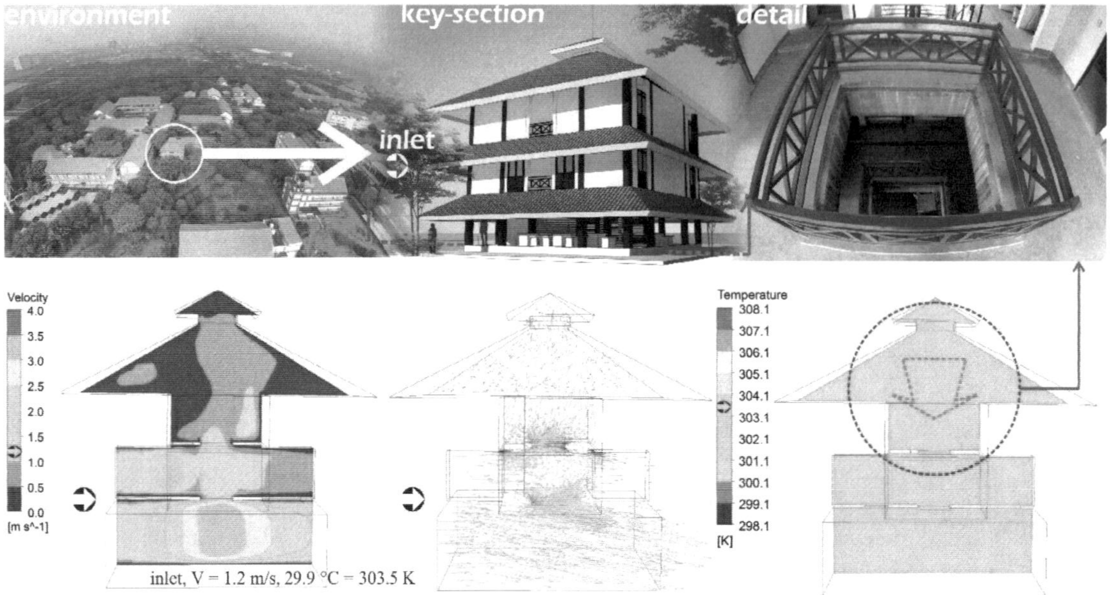

Figure 10. Existing stack ventilation.

3.3. Proposal Stack Ventilation

In general, proposal for improvement of the existing stack ventilation is explained as follows:

a) Considering the future condition which has predicted the condition in higher density by providing upper part of building stack as outlet.
b) Generating 0.3 m/s as minimum wind speed for physiological cooling as illustrated by the previous parts.
c) Reducing indoor temperature lower than inlet temperature.
d) Proposing 2 alternatives of stack technology; alternative 1: full vertical ventilation (closing direct horizontal outlet) without any treatment on ground and top of building stack; alternative 2: full vertical ventilation with ground treatment through providing cool pond and solar chimney in the top of building stack.

Figure 11 shows the implementation of proposed stack technology. Alternative 1 is more efficient than the 2nd one because it needs no treatment. It has capability for increasing wind speed. 0.5 m/s is minimum of wind speed resulted by this method. However, it has a weakness in reducing indoor temperature. The temperature is constant. Therefore, it needs an improvement and the alternative 2 is the logical method on providing lower temperature. By passive cooling for inlet with cooling pond (evaporative cooling) at the ground floor, the 20°C of air temperature could be reduced more than 5°C. The increasing flow rate by installing high material temperature as solar chimney generates wind speed. Analysis for solar chimney technology could be described in Figure 12.

The proposed stack (d = 1.8 m) receives of minimum a solar intensity more than 200 W/m^2 (see Figure 6), for each height, 1.5 m, 4.5 m, 7.5 m, and 9.0 m (top of chimney). For example, locating the glass on the top of the building for receiving solar radiation is the most effective method. It is indicated by the high volume of flow rate (40 m^3). If it is compared to d = 0.5 m and 1.0 m, for resulting higher and sufficient flow rate, wider chimney shows the higher requirement of solar intensity. The calculation is coded by Matlab programming (developed by Meshram, 2013).

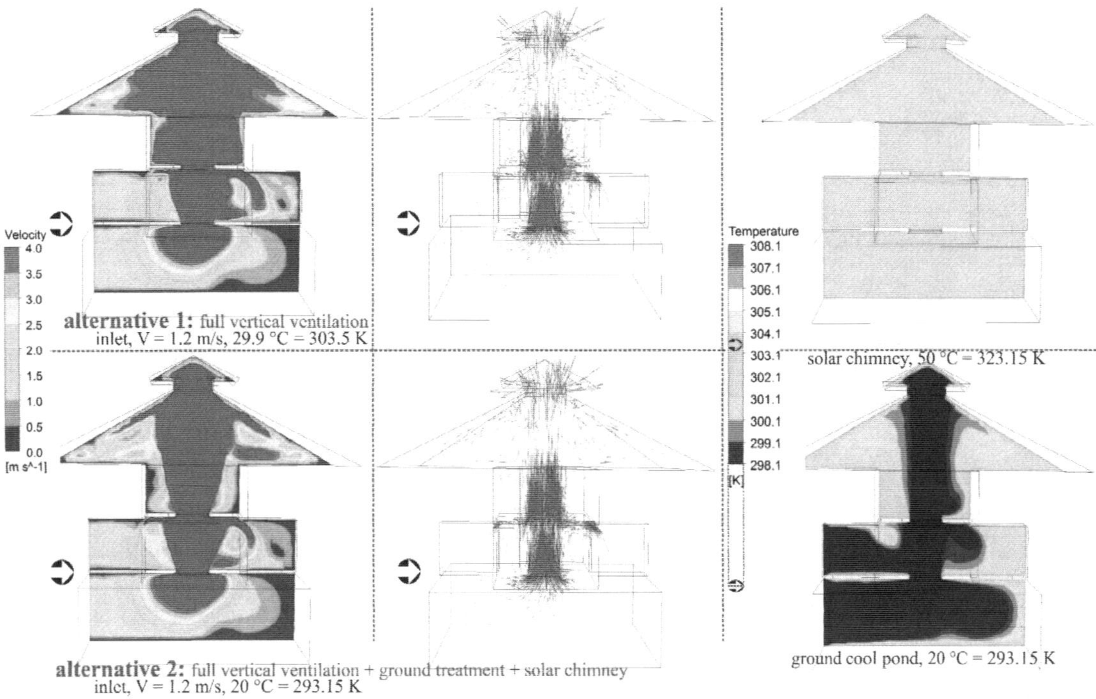

Figure 11. Alternative and proposal stack ventilation.

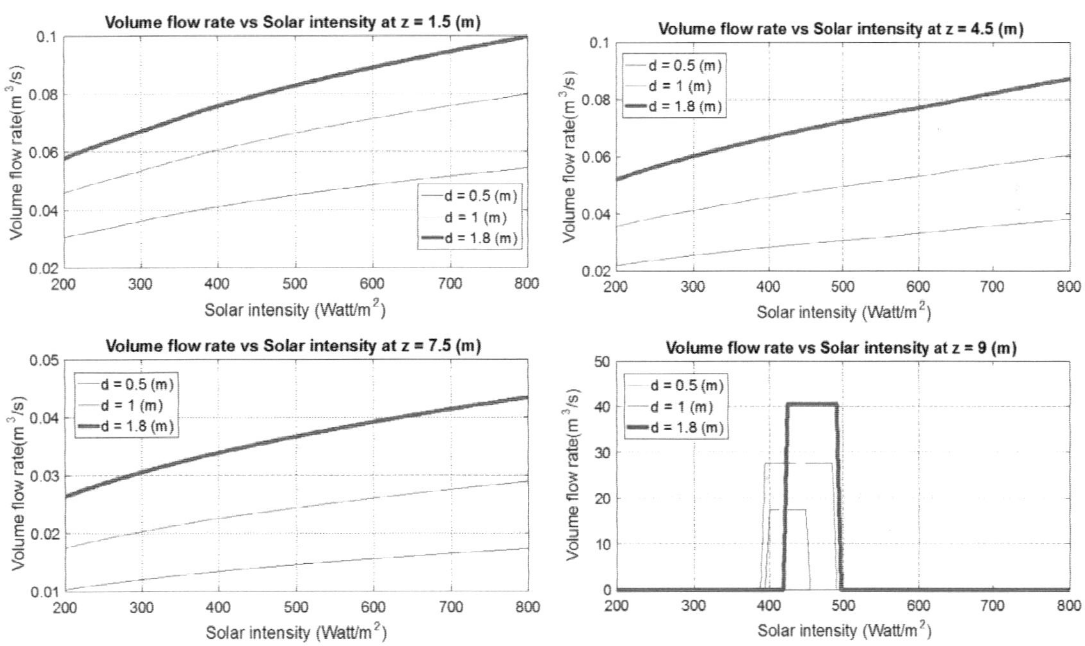

d=width of solar chimney, z=measured height in solar chimney

Figure 12. Flow rate of the solar chimney.

4. Conclusion

Different from Szokolay (2008), this paper offers evaporative cooling at ground floor combined with solar chimney for improving the exiting stack effect. As the result, the 20°C of air temperature of inlet air temperature with the same wind speed could be reduced more than 5°C. The increasing volume flow rate by installing high material temperature as solar chimney generates wind speed. The results showed that the stack effect height and area could be modified instantly adjusting the environmental condition in the future years. With 1.8 m of stack width, the proposed technology could capture 40 m^3 of vertical air flow which is useful for physiological cooling and its dimension could be modified depended on environmental condition. Proposing stack effect is one of providing resilient technology and the predictable of sustainable ventilation method is offered to anticipate an unpredicted global warming and environmental change.

5. Acknowledgments

This research is part of *Penelitian Doktor Baru* No. 881/PKS/ITS/2017 funded by *Dana Non PNBP ITS* 2017. The authors gratefully acknowledge this financial and technical supports.

6. References

[1] Ashish Meshram 2013 *Experimental Investigation on Solar Chimney for room ventilation.* http://kr.mathworks.com/
[2] Aynsley, RM, Melbourne W, Vickery, BJ 1977 *Architectural aerodynamics* Applied Science, London
[3] Aynsley, RM, and Spruill, M 1990 Thermal comfort models for outdoor thermal comfort in warm humid climates and probabilities of low wind speeds *Journal of Wind Engineering and Industrial Aerodynamics* **36** 481-488
[4] Center of Climatology at Karangploso, Malang, Indonesia 1993-2012 *Climate data (Highland)*
[5] Center of Meteorology, Climatology, and Geophysics at Juanda and Perak, Surabaya, Indonesia. 1997-2015 *Climate data (Lowland)*
[6] Chungloo, S, and Limmeechokchai, B 2007 Application of passive cooling systems in the hot and humid climate: The case study of solar chimney and wetted roof in Thailand *Building and Environment* **42** 3341-3351
[7] Kasaeian, A, Ghalamchi, M, Ghalamchi, M 2014 Simulation and optimization of geometric parameters of a solar chimney in Tehran. Energy *Conversion and Management* **83** 28-34
[8] Nguyen, AT, and Reiter, S 2014 A climate analysis tool for passive heating and cooling strategies in hot humid climate based on typical meteorological year data sets *Energy and Buildings* **68** 756-763
[9] Prianto, E., and Depecker, P. 2003 Optimization of architectural design elements in tropical humid region with thermal comfort approach *Energy and Buildings* **35** 273-280
[10] Szokolay, SV. (2008). *Introduction to architectural science.* Elsevier, Oxford, UK
[11] Tan AYK, and Wong NH 2014 Influences of ambient air speed and internal heat load on the performance of solar chimney in the tropics *Solar Energy* **102** 116-125

Universal design characteristic on themed streets

Bangun IR Harsritanto*, Indriastjario, Wijayanti
Departement of Architecture, Engineering Faculty of Universitas Diponegoro, Semarang, Indonesia

*Corresponding E-mail : bangunirh@arsitektur.undip.ac.id

Abstract. People around the world can access the streets to fulfil their daily activities regardless of their gender, age, and abilities. The streetscape is an urban public space which is built to facilitate the basic needs of people as social being. The themed street is an urban streetscape designed and built in detail with a theme or special purpose in an of urban development process. Universal design facilitates the full range of human diversity as physical appearance, perception, cognitive abilities, sizes, and shapes. By designing for the diversity, the specialized streets become more functional and user-friendly. The purpose of this study is to examine several design characteristics of themed streets in several countries from three different continents using universal design principles for giving proper directions to develop more user-friendly streets. Literature review and case study were used as research methods. The literature review was extracted and compiled from manuscripts, streetscape design books, and from universal design principles. Furthermore, the constructed theory were used to examine the case studies of themed streets. The findings indicated that themed streets' character design were strongly influenced by local cultural aspect even though the basic guidelines were universal design principles; the resumed design direction can be suggested universal along with the richness of local aspects.

Keywords: Themed streetscape, characteristics, design guideline, universal design

1. Introduction

Streets constitute a significant part of open public space and are the most important symbols of the public realm. Streets that cater to the functional, social, and leisure needs of people have been positively associated with economic growth, physical health of people, and a sense of community. Increasingly, scholars suggest a thought on streets as a social space rather than just a channel for the movement (Vikas, 2006).

The themed street is an urban streetscape which are designed and developed in detail with a special theme or purpose in urban development. However, not all people can access the specialized street easily because of their limited ability (the elderly, children and people with special needs).

The purpose of this study is to examine several themed streets' design characteristics in several countries from three different continents using universal design principles for giving proper directions to develop more user-friendly streets developments.

Literature review and case study were used as research methods. The literature review was extracted and compiled from manuscripts, books of streetscape designs, and the universal design principles, and then the constructed theory was used to examine the case studies of themed streets.

The findings indicated that themed streets'character designs strongly were influenced by local cultural aspect even though the basic guidelines were universal design principles; the resumed design direction can be suggested universal along with local aspect richness.

2. Methods
This study was performed with two research methods, such as:
1. Literature study was compiled and analysed with several related literature to construct evaluation sheet of this dissertation.
2. Case study was used to explain several phenomena happening on the case and then can be generalized into the bigger scope of research area.

3. Discussion
3.1. Universal Design Theory
The history of "universal design" word was firstly announced by Ronald L. Mace (1941-1998) as founder and former program director of The Center for Universal design at North Carolina State University. In 1997, Ron Mace, who already used a wheelchair, collaborated with a group of architects, product designers, engineers and environmental designers and developed the Seven Principles of Universal design. Universal design takes into account the full range of human diversity, including physical, perceptual and cognitive abilities as well as different sizes and shapes of human body. Designing based on this kind of diversity, we can create things that are more functional and more user-friendly for everyone.

1. Equitable Use
2. Flexibility in Use
3. Simple and Intuitive Use
4. Perceptible Information
5. Tolerance for Error
6. Low Physical Effort
7. Size and Space for Approach and Use

The seven principles of universal design are one of the grand theories in this study along with themed streets. The seven principles of universal design were chosen in this study because the seven principles are the latest detailed concept. Universal design principles were implemented on street so people could use the streets without any obstacles. Moreover, universal design was examined on every section of the themed streets to support the street function for all users

3.2. Street Theory
The Manual of Street (2007) explained more details about the streets that must have five principal functions:
1. Place;
2. Movement;
3. Access;
4. Parking;
5. Drainage, Utilities, and Street Lighting.

Themed Street is a term of the street as a part of the built environment which are designed in a theme and purpose to respond the characters of a district or an area in the frame of the urban development process. According to that terminology, the key points of a specialized streetscape are:
1. The elements of urban development process
2. Local character responses
3. Themes
4. Parts of the streetscape

In the next phase, three themed street guidelines (from Europe, Middle East and Australia) would be resumed into 8 items of specialized street which are: parking lots, public transportation modes, street furniture, street material, street utilities, signage, pedestrian space, and linkage design (see table 1). These physical items also compare their differences with the universal design principles.

Table 1. Summary of Universal Design Principles and Themed Streets

Items	Universal Design	Themed Streets
Types of theory	The product design concept	The streetscape design concept
Applications	All designed products buildings, tools, lea rning, environment, others	Themed street, upper pass, under pass, others
Samples of Application	Inclusive learning, barrier-free buildings, u niversal serial bus (USB) device, others	Monumental streets, cultural streets, shopping streets, bikeways, busways, others
Study item	Equitable Use	Parking lots
	Flexibility in Use	Bus stops/subways
	Simple and Intuitive Use	Street furnitures
	Perceptible Information	Street utilities
	Tolerance for Error	Signage
	Low Physical Effort	Street materials
	Size and Space for Approach and Use	Pedestrian spaces
		Linkage designs

3.3. Islington Street Book
The Islington guideline contributed sixteen points to this research's evaluation sheet. Furthermore, this guideline has added simplicity in design and sustainability, even though local wisdom and aesthetic aspects are also attached to the seven principles of universal design which coins as a grand theory.

3.4. Abu Dhabi Urban Street Design Manual
This Middle East specialized street manuals have contributed some elements of pedestrian priority in traffic management such as resting areas and visibility aspects with only describing guidance for passageways, and they are not area minded.

The conditions of Estidama of the Middle East urban development concept are local oriented study, which might not be adaptable to other regions. The conditions of dessert have contributed solution for street issues, especially in visual barrier and safety requirements which can be constructed as evaluation items for street examination items.

3.5. Landcome Street Design Manual
This Australia NSW Government's property developer guidelines for street, environment, and building adjusted 12 principles concerning on usability and simple perception to be recognized and accessed by people, but none of them were easily accessed by handicapped persons nor citizen with special needs. Nevertheless, the best analysis performed was to design the street heuristically

3.6. The Resumes of Street Guideline
Summarized from the 3 existing themed street guidelines (see table 2), there are substantial issues that need to be fixed and completed by this dissertation study, such as:

1. The previous studies aimed to only for tourism street facilities at Islington which produced an artificial disneyfication of specialized street without concerning on daily activities (Putra, 2013).This artificial disneyfication of tourism often forces the natural expression of the urban society to develop their neighbourhood.
2. The previous guidelines were oriented to design local district (Dubai; Lancome) which produced the unsustainable specialized street fabric from and to the site. This orientation of design built an exclusive area on the city.

Present condition analysis by universal design principles which were derived from literature studies and resumed as an evaluation sheet to compare with case studies and the site survey result

Table 2. Comparisons of Three Street Design Guidelines and the New Guideline

No	Items	Islington Street Book A borough-wide streetscape manual	Abu Dhabi Urban Street Design Manual	Street Design Guideline for Landcome	This paper
1	Purposes	To build an attractive, safe and accessible streetscape by setting out guidelines and specifications for materials, street furnitures and layout designs for the public highway and footways	To design streets that create a safe environment for all users, recognizing varying levels of driver education and cultural differences	To build environment providing routes for vehicles and public transport and accommodate utility services and drainage system	To build themed street that can be used by all user regardless sexes, ability and ages based on principles of universal design
2	Methods	Literature study	Literature study	Literature study	Literature study
3		Community-participatory researches	Participatory action researches	Action plan researches	Site surveys
4					Study case
5	Results	Tourism street books	Abu Dhabi design guidelines by Estidama's principles	Australia's streetbook guidelines	The universal design based urban streetbook
6	Cons	Concerns on local tourism	Desert climate responses	Concerns on residential areas	Concerns on urban streetscapes
7	Pros	Promote tourism, minimal material use	Strong traffic management, estidama local content	Legibility force, community-based design	Resuming and adding urban identity concept, universal based design
8	Equitability in use	0	0	0	0
	Flexibility in use			0	0
	Sizes and spaces for approach and use	0	0	0	0
	Simple, intuitive	0	0	0	0
	Perceptible information	0	0	0	0
	Low physical efforts		0		0
	Tolerance of errors		0		0

3.7. The New Guideline Criteria

Table 3. The New Guideline Criteria

No	Research objects	
	Sections	Items
1	Parking lots	Special need parking lots < 100m
2		The ramp ratio 1: 20
3		The minimum width 2m footway
4		The maximum level difference 2cm
5		Signage
6	Bus stops/subways	Local characters on decision point (information)
7		Clear view from and to the gates
8		The minimum clear height 2.1m
9		Traffic resting points
10		The minimum width 2m footway
11		Elevator or escalator
12		5-800m from public utilities
13		Ramps should have a maximum gradient of 1 in 20 or 5 per 100
14		Step marked short straight runs with a 3-12 per walk
15		Risers should have uniform height (10-15) and min 30cm deep
16		Adequate street lighting for people
17		Enclosed bus shelters
18	Linkage design	Hierarchy on street types
19		Well connected streets with good view and small street blocks
20		Avoid cross road (x) and corner greater than 90'
21		Wind breakers (landscapes)
22		Simple and directional signs on decision points
23		Local character priority
24		Less material use
25		Street furniture/rest areas on decision points
26		Adjusted street lighting, visual and audio controls
27		Future maintenance
28		Strict reinstatements
29		Public seating/rests 100-125m
30		Minimum width 2m for footway with less level
31		Traffic calming protection
32		Public art using on decision points
33	Street furniture	Street lighting for each 20m
34		Public seating for every 100-125m
35		Clear maps and local characters on decision points
36		Providing a handrail/handgrip max 0.80m
37		A simple and local character priority
38		Clear traffic and wind barriers
39		Supporting landscape vegetation
40		Public facilities lid pressure less than 2kg with lower knob
41		Braillecode/audio signal installment
42		Contrast visible
43		Permanent instalment
44		Non slippery
45	Street utilities	Simple and local character priority
46		Less material use
47		Clear traffic and wind barrier
48		The public facilities lid presser is less than 2kg with lower knob
49		Manhole cover grid less than 5cm
50		Avoid open drainage channel
51		Every 500-800m
52		Information or security booth
53	Signage	Background-sign contrast color
54		Local characters with international standards
55		Clear map with "you are here" sign
56		Location on decision point
57		Texturized sign (braille)
58		Sound signal and smell locators
59		Clear view
60		Minimum 1m from ground instalment

61		Proper lighting
62		Building information
63		Public local arts on decision points
64		Hazardous area priority location
65		Permanent instalment
66	Street materials	Less material use
67		Local materials/textures
68		Non slippery
69		Using tactile texture
70		Permanent non slip ramps instalment
71		Nontoxic materials
72		Rubberized material footway than clay or asphalt
73		Low grate paving block
74		Minimum glare or reflection
75	Pedestrian space	Minimum width 2m clear footway
76		Mix used space
77		Support landscape vegetation
78		Traffic calming protection
79		Clear traffic and wind barriers
80		Ramps way ratio 1: 20
81		Maximum level difference 2cm
82		Minimum clear height 2.1m
83		Public seating for every 100-125m
84		The manhole cover grid is less than 5cm
85		Provide handrail/handgrip max 0.80m
86		Using tactile texture
87		Well connected streets with good view and small street blocks
88		Proper lighting
89		Avoid cross roads (x) or more and corners greater than 90’

3.8. The Case Study at Jalan Asia Africa

The street cuts Bandung City into two parts, the north part and the south part. It was named after the Asia-Afrika Conference, held in Bandung back in 1950s. Jalan Asia Afrika is the part of Grote Postweg constructed in 1808. This specialized street becomes a touristic attraction right in the middle of Bandung because it is a historical building with Art Deco style and its connection to the Alun-Alun (square) of Bandung. This specialized street has a zero-kilometre monument that indicates this themed street is the centre of Bandung (see figure 1).

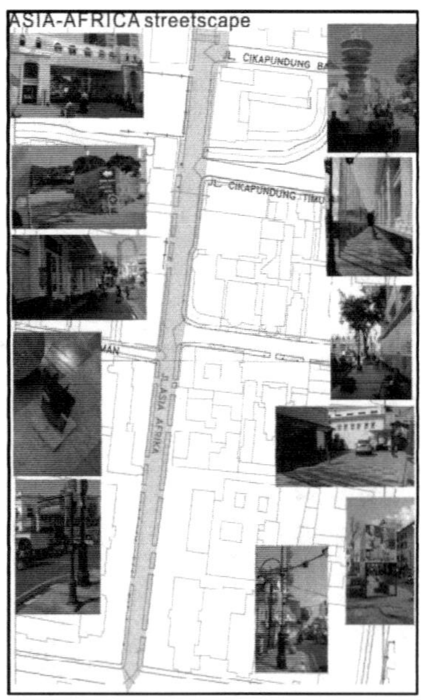

Figure 1. Condition of Jalan Asia Africa

Jalan Asia Afrika street design is straight type of street. It was dedicated as cultural street to support the big event, Asia-Africa International Conference (see figure 1). Historical buildings like Asia Afrika museum, the local newspaper office, old colonial hotels, and the historical central post office were built along this street until Bandung square and Central Mosque areas. Street materials on Jalan Asia Afrika are slightly different from the ones on Jalan Braga in which they cross on one point. The pattern of tile is checker style and the tactile paving is continuous to the pedestrian curbs. Several street furnitures such as seating, flag poles, and lightings are implanted in groups at historical site passage only which make this area become crowded with public facilities and reduce pedestrian space.

Jalan Asia Afrika has a historical meaning for Indonesia. This street was designed and built for international conference events. The evaluation has shown that Jalan Asia Afrika is good especially on linkage design (3.71) and flexibility principle (3.18). Jalan Asia Afrika street also got the lowest score in parking lot section (2.73) and low physical effort principle (3.02) (see figure 2). The Parking lots here are near the Grand Mosque and behind the street. However, the parking lot spots(2.73) did not facilitate special parking areas. The condition of the tolerated pedestrian way (3.13) such as tactile paving was not well placed even though it was already installed permanently. This condition was also worsened by the unavailability of themed street furniture installation though the surrounding buildings are the colonial types that are strong enough to be the theme of the street. Street utilities section (2.66) showed that some open drainage still can be found in the pedestrian ways that might harm passengers who accessesed the street without noticing what they will pass. In signage section, minimum signage and materials use in this area has become the old problem of the district. The pedestrian ways were not linked well (3.29) on the height and curbs. Furthermore, some building passages still obstructs the pedestrian way, and they may become barriers for the disabled people to pass this specialized street. This street is straight line type which clearly links each part of the street.

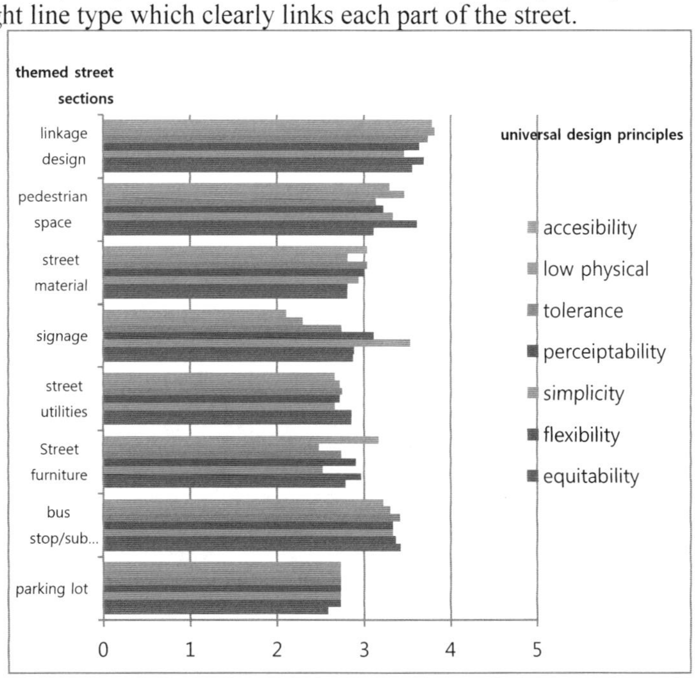

Figure 2. Evaluation scores on Jalan Asia Afrika

This examination result has showed that the street culture of universal design principles in Indonesia still needs improvement, especially for its parking lots although the linkage designs as the part of urban historical area is still acceptable and well noticed as the main character of this street.

4. Conclusion

Based on the analysis and comparisons of street guidelines on multiple countries used to examine this case study, there are some concluding points namely:

1. Indonesia needs more efforts to improve the facilities and redesign their specialized streets based on the Universal design principles
2. Cultural aspects can be more strengthened in Indonesia to support specialized street designs and developments. The cultural aspects which can be adapted are: accommodations of various cultures into one theme of pop and antique culture which are implemented material on tile, strong antique style of each building's façade on Indonesia themed streets, design of small stage at street segments to support cultural act performances, and at local traffic management to restrict vehicle access during the performances.

5. Acknowledgments

Firstly, I thank to my advisor "kakek" Prof. Cho Young-Hang. It's been an honour for me to make the first English dissertation at Architecture Department of Pukyong National University funded by NIIED KGSP and by myself; then to my family for all their love and encouragements especially for my parents, Ir. Harry Sutanto and Dra. Sri Lestari, and my sister's small family. Last but not least, I would like to thank to my friends and colleagues from various countries for giving me supports until now.

6. References

[1] Abu Dhabi Urban Planning Council 2007 Abu Dhabi Utility Corridor Design Manual ver 1, Emirates of Abu Dhabi Urban Planning Council
[2] Aslaksen 1997 Universal design: Planning and Design for All, Cornel University (http://digitalcommons.ilr.cornell.edu/gladnetcollect)
[3] Indrakusumo, Bangun 2009 Revitalisasi Taman Wilhelmina Tugu Muda Sebagai Salah Satu Node Downtown Citra Semarang. Nusantara Urban Research Institute vol 4
[4] Indrakusumo, Bangun 2016 Design Characteristics of Specialized Streets for Indonesia through Cases of South Korea Doctoral's thesis, Pukyong National University, South Korea
[5] Islington Council, 2005 IslingtonStreetbook A borough-wide streetscape manual, Islington Council London
[6] Landcome 2008 Street Project guideline, Landcome : New South Wales, Australia
[7] Mehta, Vikas 2006 Lively Streets: Exploring The Relationship Between Built Environment And Social Behavior, Doctoral's thesis, University of Maryland, College Park
[8] North Caroline State University, College of Design 1997. The Center For Universal Design, Retrieved 21 May 2010 from http://www.design.ncsu.edu/cud.
[9] Null, Roberta 2014 Universal design Principles and Models, Taylor & Francis Group
[10] Putra, Budi A 2013 Cultural Representation of Vernacular Housing in Jambi, Doctoral's thesis, Seoul National University, South Korea.
[11] Spokane, R. Arnold et all 2007 Identifying Streetscape Features Significant To Good Being, Architectural Science Review, 50:3, p234-245

Resilience of Historical Urban Multi-ethnic Settlement: Entrepreneurship and Religiosity Concept of Gresik City

Dian Ariestadi[1], Antariksa[2], Lisa Dwi Wulandari[2], Surjono[3]
[1]Doctoral Students of Civil Engineering - Faculty of Engineering, Brawijaya University & Lecturer in Civil Engineering - Universitas Negeri Malang, Malang, Indonesia
[2]Architecture Department - Faculty of Engineering, Brawijaya University, Malang, Indonesia
[3]Urban and Regional Planning - Faculty of Engineering, Brawijaya University, Malang, Indonesia

Corresponding e-mail: dian.ariestadi.ft@um.ac.id

Abstract. Important aspects in continual development include economic and social developments, as well as environment protection. Social development aspect should concern political aspiration and local socio-culture as resilience of their local wisdom features. A review on urban resilience is more focused on economic and physical concepts, without developing the social concept. Objective of the study was to find out the resilience concept of Gresik City, which was the earliest description of a big trade port city in Indonesia, for example, Jakarta, Surabaya, and Semarang. The study applied morphology approach on spatial settings at historical urban multi-ethnic settlement through physical and non-physical observations, as well as validation through historical records and archives. The descriptive analysis of morphological pattern relates to activities on social, economic, and cultural aspects in order to obtain basic concept of social life. Morphological pattern of Gresik, which is dominated by multi-ethnic settlements, such as Arabs, Chinese, ex-Dutch-colonial, and the natives of Javanese and Madurese, has attracted traders from various nations and ethnics. History of the city as the center of Islamic learning and dissemination has formed the public of Gresik to have basic religious life, which is reflected on Islamic rituals. Settlement domination, which functions as household industries, craftsmanship, and small-scale trading, shows that entrepreneurship activities as socio-economy activities have highly supported daily religious ritual activities. Entrepreneurship and religiosity concept, which is formed and developed through long history of Gresik, represent the resilience of multi-ethnic societies at cities along the North Coast of Java.

Keywords: Resilience, Entrepreneurship, Religiosity, Multi-ethnic settlement, Gresik

1. Introduction

Nowadays, urban development faces some challenges, such as, increasing population, limited environmental resources, and global climatic change. Therefore, continual urban development concept must be developed. UN's documents from World Summit 2005 stated three pillars that must be related to each other and they are considered as supporting pillars for continual development, which include economy and social developments, as well as environment protection. Social development, of course, considers socio-culture aspects of the society, whereas the development steps of continual development concept include: 1) the first phase considers economic productivity and ecological equilibrium aspects, 2) the second phase includes social welfare aspects, and 3) the third phase should consider political aspirations and socio-culture of local society. Therefore, urban development must be able to conserve its local wisdom features in an exact-new formulation for urban continuity in the future [1].

As the center of social and cultural community, cities have an important position on cultural dynamics in Indonesia. Basically, dynamic and interactive correlation between both of them cannot be separated. In fact, urban life dynamics affect the cultural dynamics, and vice versa. The historical course in Indonesia has shown that since the beginning of the establishment maritime and agricultural cities or trading towns during colonial era until the establishment of modern cities post-independence, cities in Indonesia have dynamically played important roles in social life of the Indonesian peoples, not only as the center of politics, economics, and government, but also as the place where transformation process takes place, as well as configuration of various elements of local and foreign cultures in Indonesia [2].

Cities that lie along the coast in Indonesia, particularly at the north side of Java Island, are typical of multi-cultural cities, which are established and developed along with their various ethnics. These cities formed urban morphological structure, which is heterogeneous with new culture that was established through acculturation process [3]. Cities along the North Coast of Java are categorized as initial cities in Indonesia established and developed into big and modern cities. Developments of such cities, which are recorded as big trade port cities, include: Gresik, Tuban, Lasem, Semarang, Cirebon, and Batavia.

Gresik, along with its multi-ethnic settlements that were built side by side on the same area, has formed specific character. The non-natives, such as Arabs, Chinese, and Madurese, as well as the native Javanese have different cultural backgrounds that create specific settlements. Such multi-ethnic settlements have been established and developed on a cultural and geographical environment along the North Coast of Java since colonialism era in Indonesia. Characteristics of the urban architectures and socio-culture of the societies in this area are very interesting to review as the development of a local theory concerning with concepts on historical urban multi-ethnic environment in Indonesia, particularly in Java [4–6].

Architectural and cultural reviews in Gresik City are required due to the rapid growth and development of Gresik City as a seaport and an industrial city, which will accelerate the urban physical changes including the cultural artifacts, spaces, and its architecture. Today, urban physical conditions and its society at the old city of Gresik can still be observed its historical tracks. The findings of cultural and architectural concepts, which are the local wisdom concepts of Gresik City, are used for the basic development policy of Gresik City. The founded urban concepts are also description of historical urban multi-ethnic concepts in trade port cities at the north side of Java, which have developed into big cities and metropolitan such as Jakarta, Surabaya, and Semarang.

Architectural studies show a paradigm shift. The methods and approaches used should be able to find the meanings and relationships of a phenomenon. Today, a review on new theory of architecture is interdisciplinary and uses paradigms of modernism, phenomenology, structuralism, post-modernism, and post-structuralism [7]. Development of the architectural reviews at present has put power, difference, and embodiment aspects as the important aspects of review that can be revealed by paradigm and perspective from capitalism and theory of critics, post-colonial, as well as feminist approaches [8]. Post-colonialism paradigm emphasizes on the review of power, identity, and resilience efforts to defend the existence (struggle) [9].

In historical urban of colonial era, post-colonialist paradigm did not only review the hegemonic aspect of the colonial power and its impact on the colonized (subaltern), but also reviewed its resilience concepts. Ethnic grouping and settlement separation in accordance with races, as underlined by colonial's policy, would not completely make absolute dichotomy. There are some meeting points may enable such separation to create new characteristics [10]. Thus, the architectural study with a new paradigm in Gresik City aimed to find the urban resilience concepts of Gresik city as one of the historical multi-ethnics cities in Java.

2. Methods

The research used typology-morphology analysis through architectural and spatial qualitative description on buildings and environment [11]. It is expected that the application of qualitative method could reveal socio-culture and historical background behind the physical appearances of the spaces, buildings, and environments [12]. The research was initiated by collecting data on environmental spatial morphology and buildings around the area of the research. Data were taken through map images, which were transformed into outline maps in order to show borders of the environment and building components. Map images will be validated using data from direct observation in the field. The next step was the deep-qualitative description through socio-culture and historical cases in order find out themes, concepts, and local theories. Qualitative description was conducted through observations in the field, interviews, and comparison with references and theories.

3. Discussion

Results and discussions cover: the history of Gresik City, the morphological pattern, and the qualitative description of historical, social, and cultural aspects on the topics of morphological pattern analysis results.

3.1. History of Urban Multi-Ethnic Settlement Development in Gresik

According to Chinese News Report, Gresik was so-called "*Ce-cun*" or "*T'se T'sun*". In the beginning of the 15th century, the place became a rich trade port. The populations were 1000 families and more foreign traders and the natives came to this place [13]. The history of Gresik's development is divided into several periods [14]:

- Periods of 1480-1487 was the initial development of Giri Kedaton Kingdom as the center of religious, the center for development and intensification of Islamic doctrine around Java;
- Periods of 1487-1605, The area of Gresik City started to open a port and transformed itself into city, having economic and political functions and, of course, it attracted more Islamic Kingdoms in Java and outside Java;
- Periods of 1605-1748, when Giri Kedaton Kingdom was declined as a result of the conflict of interest between Mataram Kingdom and the Dutch Colonial Government;
- Periods of 1748-1916, industrial activities at *Kampung Kemasan* as the pioneer of Gresik establishment as an industrial city, Gresik City/Grisse attracted the Dutch Government to transform the area into industrial area of tannery and other public facilities and to build Customs House at the port. Grissee City started to grow due to the development of railways network; and
- Periods of 1916-2008, which was marked by the building of the port at Surabaya City in 1911, and the trade lane was gradually moved to Tanjung Perak which made Gresik Port was left by the big traders.

The history on the establishment and development of Gresik City is related to function of the city as the center of learning and spreading of Islam. Besides Gresik had developed into a port city and the center of big trade transaction that attracted more newcomers from diverse regions and ethnics. Gresik City had grown and developed into urban multi-ethnic settlements, which was dominated by Arabian population [15]. These facts have made the society have religious life as reflected through various Islamic artifacts and cultural rituals.

3.2. The Morphology of the Old City of Gresik

The area, well-known as the old city of Gresik, belongs to Gresik Sub district of Gresik Regency that lies at the Northwest of East Java Province (Surabaya) and covers the area of 1.191.25square kilometers. Geographically, Gresik Regency is located at 112°–113° East Longitude and 7°–8° South Latitude. It comprises of 18 sub districts, and Gresik sub district comprises of 5 villages and 16 administrative villages. Based on the historical data, the center of Gresik City was located at the administrative villages of Kauman, Bedilan, Pulopancikan, and Gapuro Sukolilo. The location of those four administrative villages are around the Square. These area started to develop to the north side, west side, and to south side. Location of the Old City of Gresik is at Gresik Sub district which is presented in Figure 1.

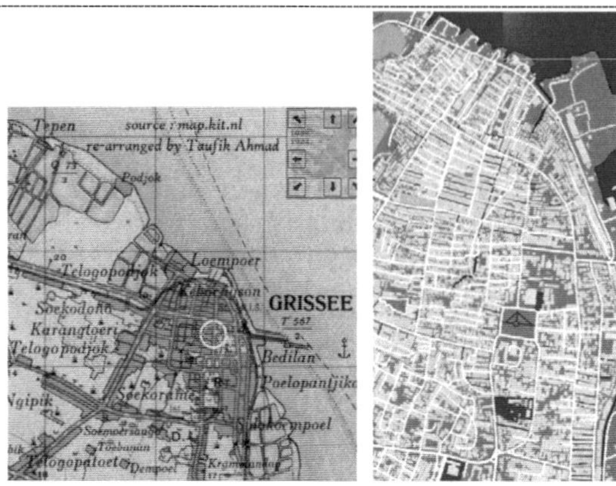

The structure of Gresik City in colonial era

The current structure of Gresik City

Figure 1. The current Gresik City structure is unchanged based on the Gresik city map of the colonial era.

Today, the Gresik's Square area and its surroundings have specific environmental layouts as heritage of the colonial period and as center of government that has hybrid pattern combining Western and Javanese traditional patterns and their physical shapes that apply colonial and local architectural styles. At the east of the square, buildings with some combined architectural styles like colonial, Arabic, Chinese, and local styles have been conserved for cultural preservation. Such morphological structures will not change at all, as recorded in historical data, which indicated that components and spatial layouts of the Old City of Gresik are as follow [16]:

- At the beginning, highways network were utilized as access to important places in that city, such as internal highway from the center of government to seaport, markets, and mosques.
- The main market was located 500 meters at the northwest of the square, near the river flow.
- The former Great Mosque of Gresik was drawn on an old map in 1775, but it was called as the temple.
- The Square as the center of the city Is against old cities in Java.
- *Dalem* means the center of government in Gresik Regency.

The most dominant functions at old city of Gresik are the groups of settlements called '*kampung*'. These groups are mixed with trade facilities and some governmental offices at the street corridors that surround them [17]. The existence of those functions have spread evenly over the whole area of the Old City of Gresik. The morphological pattern dominated by multi-ethnic settlements, such as Arabian, Chinese, Dutch, and the natives of Javanese and Madurese, has become urban historical description that attracted traders from various nations and ethnics. According to urban archeological records of Islamic Mataram [17], it was stated that the settlements at the Old City of Gresik are shown below:

- Fromthe east of the square, we can find a Chinatown at the coastal area. According to map of 1775, There was the legend of Patzinana. It was strongly presumed that Patzinana was the Chinatown.
- At the southeast of the square, we can find Arabian and *Pancikan* settlements. The Arabs are located near the grave of Maulana Malik Ibrahim.
- At the south of the square, we can find Gapura settlements. This place is used as the grave for the Regents of Gresik, who lived before 20th century, and Maulana Malik Ibrahim as one of the Nine Saints in Java.
- At the west of the square, it is a specific area that is well-known as Kauman (formerly was a specific settlement for Muslims and the pious Muslim leaders) and Tlogotrate (it was a lake where more water lilies were grown on it).

During the colonial era, morphology of the Old City of Gresik indicated that the Arabs settlements were at the north side of the city, and the Chinatown was at the south side. Meanwhile, the square was right at the center of the city, and it was surrounded by tamarind trees. At the south, west, and north sides of the Square were settlements for the Europeans. Offices and storehouses were built around the port like the Customs House. Near the south gate, we can find the Regent's house. At the other side toward to the shore, Chinatown was found with its big buildings. The morphological pattern of the Old City of Gresik is presented in Figure 2 and 3.

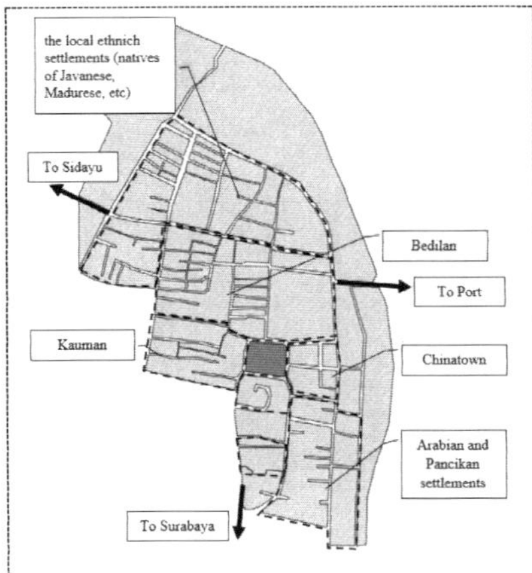

Figure 2. The urban spatial layout and functions of the Old City of Gresik.

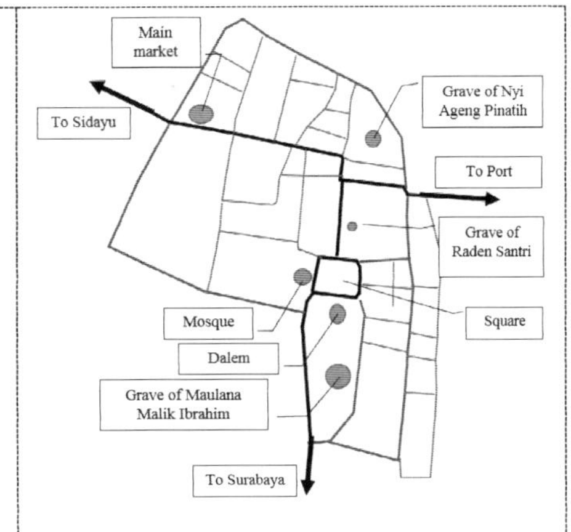

Figure 3. The block structure and street system of the Old City of Gresik.

The development of Gresik City has specific characteristic. It is generally different from other cities in Java, that began from kingdoms (Majapahit, Mataram Tuban, Surakarta, Kotagede, and others, which started from the establishment of royal palace where the king and political leader lived, and then it was followed by other buildings for the priests and workers. The Gresik City development was begun by the establishment of a village, which grew into settlements for fishermen, stopover for regional traders inter-island and cross-country, and it turned Gresik into the important trade city port at the North Coast of Java during 15th-17th centuries.

3.3. Social and Cultural Aspects of Society at the Old City of Gresik Based on Islam

Gresik has specific social and geological conditions, which makes different from other societies around it. Geological condition, lime mountain, is unsuitable for farming. The land, which surrounds this small town, is infertile and unsuitable for farming. Therefore, Gresik is surrounded by salt and milkfish embankments. Sociologically, Gresik is well-known as the place where Islamic preachers lived in the end of Majapahit era, such as Sunan Giri and Maulana Malik Ibrahim. This small town is different from other cities around it, in which population of the Arabian's descent is more prominent than the Chinese's descent.

Because the natural condition is unsuitable for agriculture and its location at the coastline, making it as the prominent port where the arab community was dominant, it makes Gresik City a unique industrial and trade city. Small-scale industries that produce bags, sandals, rimless caps, clothes, and snacks have established communities which are different from subculture of the societies around them.

Gresik has plural societies. Even though as an open-port city, which enables it to be inhabited by multi-ethnics, but they live in peace and harmony. Most of them live in groups at certain locations, inhabited by the same ethnic. Socio-culture condition of the society at the Old City of Gresik, based on demographic data, is dominated by Muslims and most of them make their livings from entrepreneurship.

The main character of Gresik peoples is the Islamic spirit, in which the Islamic teachings have deeply rooted and become the identity of Gresik people. It has enriched their views and attitudes in daily life, particularly in expressing what they feel, such as in traditional arts, in which the Islamic nuance has affected each tradition and traditional art in Gresik.

Figure 4. Socio-cultural activities of Gresik community that show influence of religious aspect and entrepreneurship.

The most prominent tradition of Gresik people is paying great respect to the ancestors, particularly the prominent Islamic figures or the Muslim leaders. Such respect is done by placing their pictures on the wall, and on public room, for example, the living room. Meanwhile, the cultural ritual activities include devotional visits to the grave of Muslim leaders and "*Khol/Haul*" to commemorate the birth of the prominent figures. Also, it is intended to reveal the history of the respected figures, to encourage the spirit of religious services, and to encourage the spirits of struggle and living in harmony.

More traditional festivals are held relating to various traditional ceremonies, for example, *Tingkeban,* proposal of marriage, *Sunatan*, and others. Among the traditional festivals, *Macapat* has still been conserved by a small number of people in Gresik. For wedding ceremony, celebration of *Maulid Nabi Khaul of Sunan Giri,* and other Islamic ceremonies are usually livened up with *Samrah* (*Yalil-valil*), *Qasidah* or *Hadrah*. All of them have Islamic nuance [18].

The development of such Islamic arts, traditional festivals, and cultures was related to the role of Sunan Giri, his descendants, and his students since mid-century of 15th. Sunan Giri was an educator who loved literary work of art. Through education as the medium of religious preach, he educated his students wisely. Sunan Giri created diverse traditional games that contained Islamic teachings (*Jelungan, Jamuran, Bendi Girit, Ilir-ilir, Jor, Gula Ganti, Cublek-cublek Suweng*), and created musical composition for gamelan ("*Asmaradhana*" and "*Pucung*") which are mostly loved by the society because they contain religious teachings.

Furthermore, other specific tradition related to socio-economic needs of the society is *Pasar Bandeng* (Market selling milkfish). The tradition at *Pasar Bandeng* (*prepekan cilik* and *prepekan gede*), which is held two days before Eid Mubarak, is the place where Gresik peoples introduce and sell their products and to strengthen the relationship between religious traditions and economy. This tradition is the preparation to welcome the day. Not only milkfishes, but also all products of Gresik people are available in this market, such as: clothes for kids, rimless caps, *terompah*, sandals, shoes, buckles, food, and special snacks from Gresik, as well as other goods, which are usually required to celebrate Eid Mubarak. There is an assumption stating that *Hari Raya* Connects to new things. For Gresik people, this activity grows, develops, and supports each other as exemplified by Sunan Giri and Nyi Ageng Pinatih, who played their roles as both religious figures and great entrepreneurs.

Before *Pasar Bandeng* is held, it is usually initiated by '*malem selawe*' (at the 25th day of Ramadhan), which is conducted at the Grave of Sunan Giri.

3.4. Entrepreneurship and Religiosity as The Resilience Concept of Society in the Old City of Gresik
The term of "resilience" is mostly used in ecology and psychology in order to describe caring capacity and ability to control oneself from pressure and stress. Societies have also strength and weakness to overcome their problems and they have to strive to continue their lives [19]. Resilience appears to be a famous word since the ecological, psychological, social and economic sciences began to use it to refer, respectively, the capacity of ecosystems, people, societies, the economy, and more recently even urban systems to overcome disturbance [20,21]. Supports, in the form of social, cultural, and environmental resources, may be developed through social network, which is reinforced as the buffer system of the community's life [22].

On historical urban multi-ethnic settlements as in Gresik, the historical aspects of urban establishment and development have formed both physical and non-physical concepts of society and city as the resilience embodiment of its society. Gresik as the center of learning and dissemination of Islam has established a religiosity concept based on Islam as the philosophy of life embodied in physical aspects, such as space, architecture, and urban environment. Meanwhile, the potency and limited resources have established an entrepreneurship concept as the basic life of socio-economy of the society, which still conforms to their religiosity principles. Such religiosity and entrepreneurship concepts have become the social resilience of historical multi-ethnics urban in Gresik.

Religiosity and entrepreneurship are the city resilience concept that has been formed in Gresik society. Religiosity and entrepreneurship principles are formed and influence the spatial and architectural shape of Gresik City. The morphological structure of the city has accommodated various religious and entrepreneur activities. The interrelation between the principles of religiousity and entrepreneurship as the social dynamic aspects and urban structures as the built environment aspect are two of the four aspects of Urban Resilience Research. Two other aspects are Metabolic Flows (Production, supply and consumption chains) and Governance Networks (Institutional structures and organisations) [22]. Thus, these aspects need to be maintained as the concept of local wisdom that has proven to form and evolve in line from the early formation of the city to this day.

4. Conclusion

The morphological pattern of Gresik City, dominated by multi-ethnic settlements, such as Arabic, Chinese, Dutch, and the natives of Javanese and Madurese, has become the urban historical description attracting traders from various nations and ethnics. The history of the city as the center of Islamic learning and dissemination has formed the people of Gresik to have basic religious life, reflected on Islamic rituals. Settlement domination, which functions as household industries, craftsmanship, and small-scale trading, shows that entrepreneurship activities as socio-economy activities have highly supported daily religious ritual activities. Entrepreneurship and religiosity concepts, which are formed and developed through long history of Gresik City, have represented the resilience of multi-ethnic societies among cities along the North Coast of Java.

5. References

[1] Childers D, Cadenasso M, Grove J, Marshall V, McGrath B and Pickett S 2015 An Ecology for Cities: A Transformational Nexus of Design and Ecology to Advance Climate Change Resilience and Urban Sustainability *Sustainability* **7** 3774–91
[2] Alfian M 2007 *Kota dan Permasalahannya* (Paper in History Discussion BPSNT. Yogyakarta)
[3] Fauzy B, Antariksa and Salura P 2012 Memahami Relasi Konsep Fungsi, Bentuk Dan Makna Arsitektur Rumah Tinggal Masyarakat Kota Pesisir Utara Di Kawasan Jawa Timur (Kasus Studi Rumah Tinggal Di Kampung Karangturi Dan Kampung Sumber Girang, Lasem) *DIMENSI J. Archit. Built Environ.* **38** 79–88
[4] Ariestadi D, Antariksa, Wulandari L D and Surjono 2017 Architecture, space and power in historical multi-ethnic city Gresik *MATEC Web of Conferences* vol 101 (EDP Sciences) p 05027
[5] Ariestadi D, Antariksa, Wulandari, Lisa D and Surjono 2016 Konsep Courtyard Pada Permukiman Multi-Etnis Historis di Kota Lama Gresik Sebagai Konsep Kearifan Lokal Berdasarkan Perspektif Post-Kolonial *Pros. Simp. Nas. RAPI XV* 310–7

[6] Ariestadi D, Antariksa and Lisa D W 2014 Teritori Ruang Hunian dan Kawasan pada Arsitektur Rumah Courtyard Di Kampung Arab Gresik SEMINAR NASIONAL ARSITEKTUR PERTAHANAN (ARSHAN) 2014 (Surabaya: Program Studi Arsitektur, UPN "Veteran" Jatim)
[7] Leach N 2005 *Rethinking architecture: a reader in cultural theory* (New York: Routledge/Taylor & Francis)
[8] Crysler C G, Cairns S and Heynen H 2012 *The SAGE handbook of architectural theory* (Los Angeles ; London ; New Delhi ; Singapore ; Washington DC: SAGE Publications)
[9] Haddad E G and Rifkind D 2014 *A Critical History of Contemporary Architecture 1960–2010* (Surrey, England: Ashgate Publishing Limited)
[10] Soekiman D 2014 *Kebudayaan Indis: Dari Zaman Kompeni sampai Revolusi* (Depok: Komunitas Bambu)
[11] Wang D, Groat and Groat L N 2013 *Architectural Research Methods* (John Wiley & Sons, Inc.)
[12] Creswell J W 2013 Qualitative Inquiry & Research Design, Choosing Among Five Approaches (3rd Ed.)- Google Scholar
[13] Mustakim 2010 *Gresik dalam Lintasan Lima Zaman* (Pustaka Eureka)
[14] Riski C, Antariksa and Surjono 2009 PELESTARIAN KAMPUNG KEMASAN KOTA LAMA GRESIK *Arsit. E-J.* **Volume 2 Nomor 2**
[15] Risbiyanto E and Antariksa S H PELESTARIAN KAMPUNG ARAB MALIK IBRAHIM DI KOTA GRESIK
[16] Adrisijanti I 2000 *Arkeologi perkotaan Mataram Islam* (Yogyakarta: Jendela)
[17] Widyastuty A A S A 2011 Identifikasi Kawasan Kota Lama Gresik *J. Tek. WAKTU* **Volume 09 Nomor 02** 6–16
[18] Widodo D I 2004 *Grissee Tempo Doeloe* (Pemerintah Kabupaten Gresik)
[19] Kirmayer L J, Sehdev M and Isaac C 2009 Community resilience: Models, metaphors and measures *Int. J. Indig. Health* **5** 62
[20] Meerow S, Newell J P and Stults M 2016 Defining urban resilience: A review *Landsc. Urban Plan.* **147** 38–49
[21] Plodinec M J 2009 Definitions of resilience: An analysis *Oak Ridge Comneg Munity Reg. Resil. Inst. CARRI*
[22] Chelleri L 2012 From the «Resilient City» to Urban Resilience. A review essay on understanding and integrating the resilience perspective for urban systems *Doc. Anàlisi Geogràfica* **58** 287–306

Spatial adaptation as the Madurese migrant resilience form at urban informal sector workers settlement : a case study of Kotalama settlement - Malang

Damayanti Asikin [1*], Antariksa[2], Lisa Dwi Wulandari[2], Wara Indira Rukmi[3]
[1]Doctoral Student of Civil Engineering Department, Faculty of Engineering, Brawijaya University, Malang 65145, Indonesia & Lecturer in Architecture Department, Faculty of Engineering, Brawijaya University, Malang 65145, Indonesia,
[2]Architecture Department, Faculty of Engineering, Brawijaya University, Malang 65145, Indonesia,
[3]Urban and Regional Planning, Faculty of Engineering, Brawijaya University, Malang 65145, Indonesia

*Corresponding e-mail : dama_asikin@ub.ac.id

Abstract. Migration is the movement of the population that will bring the change of society's behavior because of the adjustments occuring at the destination of migrant area. The availability of houses in urban area is not a proportional comparison to the increasing of limited housing space, thus it encourages urban migrants to adapt to the existing conditions. Adaptation will be closely linked to the resilience of migrants in the process of interacting with their environment. The theory of urban settlement architecture continues to grow constantly, so the used paradigm should be interdisciplinary. Thereby, the understanding of adaptation, which is used will concern to various aspects of physical and non-physical environment, and it is viewed as a process and product of human interactions with the environment holistically. Malang city is one of the migration destinations of Madurese people since 1930s, and Kotalama Malang settlement is the settlement that holds the largest Madurese migrant working in informal sector, which has been developed since 1950s. This study was conducted to determine the spatial adaptation of Madurese migrants in urban settlement area as a resilience form towards their settlement environment. The qualitative descriptive method with the discourse analysis approach of searching the data through the observation and the in-depth interview of key person were used to know the adaptation process that happened. The study result indicated that spatial adaptation as a process and product on meso and micro scale conducted by Madurese migrants was the form of resilience towards their settlement environment.

Keywords: Spatial adaptation, the resilience of Madurese migrants, urban informal sector worker settlement, Kotalama settlements, Malang

1. Introduction

Migration is a move that involves a complete change and re-alignment of the society affiliation of its individual itself. There is the change in community behavior because of the adjustments that occur at the migrant destination. Generally, it is said that migration is an important occurrence in culture. Thereby, the processes of change occuring in sociology refers to the cultural assimilation [1].

Besides food and clothing needs, housing or home is one of basic human needs. Once the basic human needs are fulfilled, the house is not only just a shelter but also a place to protect themselves from natural conditions that are not always unpredictable and as the self-actualization of its inhabitant. The existence of houses will greatly determine the quality of society and the environment in the future.

IOP Conf. Series: Earth and Environmental Science **99** (2017) 012027 doi:10.1088/1755-1315/99/1/012027

Instinctively, every person will strive to improve his own quality of life, and it will encourage people to improve the quality of their residence regarding to the value in which they believe. The efforts for improving the quality of their residences will be closely related to the process of human interactions with their environment as well as the resilience or the ability to adapt and remain strong in difficult situations.

Adaptation is a social process and a human strategy for being able to face the life, so people need to adapt to their surroundings. Adaptation has the different functions in different cultures at different times, and has the multiple definitions as the products and the processes [2]. Adaptation refers to a relatively stable change that takes place within an individual or a group in response for external demands [3]. The concept of change and adaptation is an important concept in the life, both the changes in human expectations and environmental adaptation towards the human change [4] as it affects human interaction with their environment. Adaptation in the context of human dimension of global changes usually refers to processes, actions or outcomes in the system (households, societies, groups, sectors, regions, countries) in order in order to make betterment of a system, in managing and adapting to some changing conditions, stress, danger, risk or opportunity. The human adaptation capacity both individual and group will affect the vulnerability in the adaptation process [5,6]. The theory of urban settlement architecture continues to grow perpetually, so the paradigm used should be interdisciplinary. Therefore, the understanding of adaptation which is used should pay attention to various aspects of physical and non-physical environments, and be viewed as a process and product of humans' interaction with their environment holistically.

Humans are a part of nature, but recently by the growth and development of population and the high levels of urbanization, humans are closer to the artificial environment rather than the natural environment which affects their ways of thinking, behaving and acting [7]. There is always a process of finding balance between people and their environment. As a result of this process, there is always changes of space [8] or in other words, the dynamics space.

The space dynamics is an environmental order (individual / family / group / community) that always seeks to adapt their needs to their environment potential. To accomplish that, it should endure through a process that is 'flexible' and 'dynamic', because the balance of environment is constantly shifting. The space dynamics process occurs in several stages: privatism process, clustering process, inclusion-exclusion process, categorization process, classification process, labeling process, bordering process, mediating power process, marking process / identity expression process, polarization process, quartering process, and environmental change process.

Malang is a city located in East Java Province-Indonesia, 90 km south of Surabaya. The city is the second largest city in East Java after Surabaya and one of the largest cities in Indonesia based on the population data. In addition, Malang is also the second largest city in the southern part of Java Island after Bandung. Malang is located in the plateau having relatively cool weather, and the whole area is bordered by Malang Regency. The total area of Malang is 252.10 km2 (Wikipedia, accessed 13 December, 2015).

The Kota lama settlement is The Madurese migrant residence of informal sector workers in Malang City which has been developed since 1950s, and it is one of the points of Madurese migrant agglomeration in Malang City [9]. The space dynamics that occurs in the environment and architecture itself becomes the interesting thing to be observed, considering the process of finding a balance between humans and environment that causes the changes or the space dynamics in the settlement has been going on relatively long time. By knowing the dynamics of architectural space that occurs, it will be able to provide the understanding of spatial adaptation done by Madurese migrants on the settlement architecture space formation of Madura migration in the urban environment.

2. Method

This study was conducted to determine the spatial adaptation of Madurese migrants in the urban settlement environment. The qualitative descriptive method with discourse analysis approach by searching data through the observation and the in-depth interview of key person were used to know the adaptation process that happened. The object observations on the meso scale included the environmental roads that were used to do social activities both individual and group activities. While the observation object for the micro scale was the residential Madurese migrants in the form of a single space and the house which consists of several spaces at the beginning of the inhabitated

3. Discussion

Living together with forming a community according to the origin culture is one of adaptation forms by migrants in urban settlements [10–12]. The process of settling in the mass settlements is done gradually starting from the sharing of space and doubling the function of the space which does not need the high privacy requirement to the modification of forms including the expansion of buildings, the addition of amount space (the addition of partitions) and the re-organization of space based on the private and public zones [13].

The Kota Lama settlement is an urban village where currently informal workers settling around are nearly 90% of Madurese migrants (Figure 1). This migrant settlement was established by five families who migrated from Sampang - Madura in the 1950s. The settlement location which is near to the city center , and the presence of these Madurese migrants has become a pull factor for other migrants from Madura Island. These conditions lead to privatism process namely the process of joining individuals with a certain groups and changing their personal identity into a group identity [14]. Furthermore, due to many similarities in the community groups in this region (ethnicity, occupation, and origin) there was the clustering process in which the process is undertaken by a community group that has many similarities to choose the environment as a habitable place with the special qualities in accordance with their characters, and it becomes a 'settlement sac' [15,16]. From both the adaptation processes (privatism and clustering process) toward the environment of Madurese migrants, nowadays Kotalama is known as the largest residential area of Madurese migrant in Malang City.

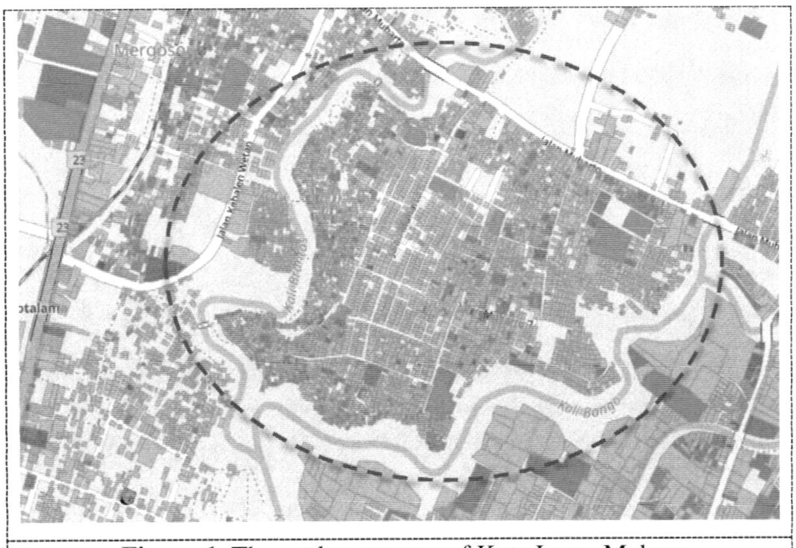

Figure 1. The settlement area of Kota Lama Malang
Source: http://peta.bpn.go.id/

The efforts done migrants to adapt or adjust to their spatial needs to the living resources in their settlement will contribute to the process of finding balance between humans and their environment. The process always causes the changes in the space or dynamics space, which are 'flexible' and 'dynamic' [8]. The spatial adaptation of Madurese migrants at the site of this study was reviewed on the meso scale and micro scale.

3.1. The meso-Scale Adaptation

The spatial adaptation of Madurese migrants in Kota lama Malang settlement began with the phase of dynamics space in the meso or environmental scale. After the stage of privatism process and clustering process in this settlement area, it continued to the next stage of space dynamics process in the meso scale, namely inclusion - exclusion process (I-EP) stage. At this stage, there was a 'boundary' that emphasized the 'social identity' and as a result of this it was divided into the society's groups 'territory' having the same characteristics or homogeneity and the groups having different characteristics or diversity. This condition caused one group to feel more powerful than the others and moreover, having the power can suppress the other groups [17]. This can be seen in the utilization of roads and corridors, in the stable settings the functional changes of space considers to the user's interest and importance. The roads become the exclusive space for the certain groups living near that road environment as a result of their social relationships, and it indirectly changes the road function as a sharing space. Thereby, the forming of space territory within the control group is based on the space and time required, and one group will feel more powerful than the others and feel that they the power to suppress the other groups [18]. In this case, the society group surrounding their road neighborhood will feel more having the sense of belonging to control the road compared to the other societies who do not live near the road.

In the activity of road utilization for various societies' interest and importance, there is the categorization process called the process of spatial classification, where the informal order does not need to be stable, it can be the term of temporary classification, and it is not bound by the formal regulation. There is also the classification process underlying the group of human to join in a certain group and chooses the order that is appropriate with their space and environment [19]. The process is seen with the existence of space classification in their environment in certain activities such as weddings and interaction in the spaces in which they agreed (the part / intersection of roads, the public facilities, and the utilization of each dwelling's terrace). The societies choose the particular order according to their space and environment and utilize the environmental roads in different ways regarding to the time of their activities. Figure 2 shows the use of road in front of the house as the place to sell, with the different sequence between morning and afternoon.

Figure 2. The utilization of roads as a business space in the morning and afternoon
Source: field survey, 2017

Similarly, if there is a wedding event as shown in Figure 3, the street around the event would be used as a place to hold the event.

Figure 3. The utilization of roads as a wedding venue
Source: field survey, 2017

The space dynamics on the meso scale that occurs in this settlement shows the adaptation of Madurese migrants in Kotalama settlements as the attempt to obtain the resilience in the neighborhoods. The process of space dynamics includes privatism process - clustering process - inclusion - exclusion process - categorization process - classification process which occurs to show the adaptation as a process, while the spaces formed in accordance with the function, activity and time based on society's needs including: roads, business areas and social activity space, are the form of adaptation as a product.

3.2. The Micro-Scale Adaptation

On the micro scale, the spatial adaptation by Madurese migrants in the study site was observed through the dynamics of residential spaces as the study object. The dwelling used as the object of study was represented by occupants whom at the first time they inhabited in the form of single space called the *mubeng kandhang* form [11] and the houses which was occupied and have several rooms at the first time. The occupancy of *mubeng kandhang* form is the result of labeling process i.e. the giving of names in the environment that they have chosen for their dwelling [19].

After the space was formed, because of it was still a single space, the inhabitants determined the type of orders that helped them to easily remember the function of each part in a single space which was formed at the first time by the stage of categorization process and classification process. Moreover, as the effort to improve their life order, the inhabitants tried to present the physical boundaries of space within their dwellings through bordering process, the creating of boundaries for one space to others for making it easier to clarify their spatial functions [17,18], forming three main functions of dwelling in the term of *emper* (living room) – *kamar* (room) – *pawon* (kitchen) is shown in figure 4.

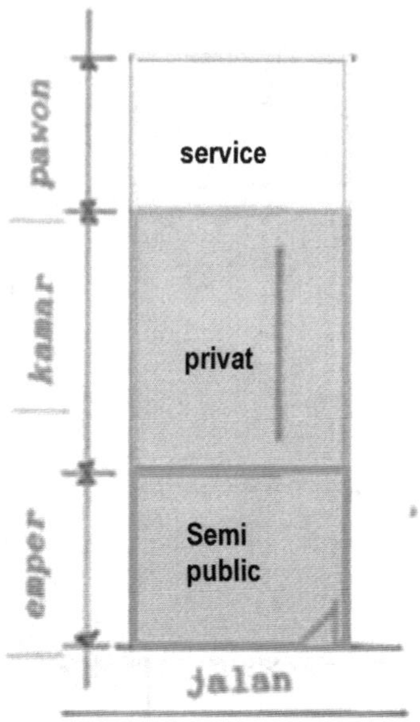

Figure 4. The spaces in the dwelling [11]

Occupants who were from the beginning occupied the house that consisted of several spaces, did the spatial adaptation with the categorization process stage, as the effort to utilize the existing space with all its limitations adjusted with the time of activities, without changing the physical occupancy itself. Because the bordering process did not longer to be done on the occupancy which has the physical bordering of space based on its function, the improvement of order and the quality of life of residents were done through the marking / identity expression process which is the personalization process and creating 'sign' to the space / place to assert that someone has mastered the space [16].

In this process, the mastery or territory towards the space becomes larger than others. This can be seen in the mastery of space outside the dwelling which is utilized based on the needs, as shown in Figure 5. The road in front of the residence becomes the territory of house owner which is used as the space of personal activity.

Figure 5. The mastery of road in front of a house as an extension of territory.
Source: field survey, 2017

Adaptation as the migration adjustment process on the micro scale is seen in the occurring spatial dynamics stages. There are the differences of spatial adaptation performed by inhabitants, due to the differences in the early conditions of their dwelling. The process of space dynamics that describes adaptation as a process is: categorization process - classification process - labeling process - bordering process - marking / identity expression process. Meanwhile the adaptation as a product seen in the form are the spaces in the dwelling in accordance with the activities that are accommodated and the space outside the dwelling territory to expand occupant mastery of the environment

4. Conclusion

The results of study showed that the resilience of Madurese migrants in the study location was achieved through the spatial adaptation in the term of space dynamics in meso and micro scales. The stages of dynamics showed that adaptation is the process associated with the resilience of Madurese migrants to live in the site of study settlements, and the adaptation as a product is demonstrated through the spatial forms for the activities as the result of dynamics space phase. Privatism process - clustering process – inclusion - exclusion process - categorization process - classification process are the meso scale dynamic space while privatism process - categorization process - process labeling - bordering process - marking/identity expression process - environmental change process are the stage of the micro scale dynamic space. The adaptation as a product is shown on the meso scale in the form of road utilization as the flexible activity space based on society's needs. While on the micro scale, the adaptation as a product is shown through the spaces within their residential units, either in the term of single space or occupancy consisting of *emper* (living room), *kamar* (room), and *pawon* (kitchen).

5. References

[1] Boyle P, Keith H. K H and Robinson V 1998 Exploring Contemporary Migration (Routledge)
[2] Hutcheon L and O'Flynn S 2013 A theory of adaptation (London ; New York: Routledge)
[3] Berry J W 2005 Acculturation: Living successfully in two cultures Int. J. Intercult. Relat. 29 697–712
[4] Moore G T, Tuttle P D and Howell S C 1984 Environmental Design Research Directions: Process and Prospects (New York: Preager)
[5] Smit B and Wandel J 2006 Adaptation, adaptive capacity and vulnerability Glob. Environ. Change 16 282–92
[6] Prabowo H and Suparman A 2005 Masalah Etnisitas Dan Tata Ruang Di Indonesia Proceeding, Seminar Nasional PESAT 2005 (Universitas Gunadarma)
[7] Bhonsle M K D and Adane V S 2013 Residential built environments: the importance of social capital. J. Humanit. Soc. Sci. IOSR-JHSS Vol. 13 13–19
[8] Leboyer C L 1998 Psychology and Environment (Sage Publications)
[9] Ubaidillah K 2014 Dinamika Perantau Madura dalam Politik Kota Malang (Suatu Kajian Antropologi Politik) (YOGYAKARTA: GADJAH MADA)
[10] Warsono 1992 Strategi Adaptif Migran Madura di Surabaya Khususnya Golongan Kenek (Universitas Indonesia)
[11] Asikin D 1996 Keragaman Spasial Rumah Tinggal di Daerah Pengaliran Sungai Brantas Kelurahan Kotalama – Kotamadya Malang (Yogyakarta: Universitas Gadjah Mada)
[12] Sasongko Y A T and Wahyuni E S 2013 Diaspora Madura: Analisis Modal Sosial Dalam Usaha Sektor Informal Oleh Migran Madura Di Kecamatan Tanah Sareal, Kota Bogor, Jawa Barat Sodality J. Sosiol. Pedesaan Vol. 01 52–63
[13] Kurniati F and Kusuma H E 2014 Adaptasi-Perilaku-dan-Modifikasi-sebagai-Proses-Menciptakan-Hunian-Ideal-Bagi-Penghuni-Perumahan-Massal.pdf Temu Ilmiah IPLBI 2014 (Program Studi Teknik Arsitektur, Fakultas Teknik, Universitas Sriwijaya: IPLBI) p E_33-E_38
[14] Ellin N 1997 Shelter from the Storm or Form Follows Fear and Vice Versa Architecture of Fear (Princeton Architectural Press) pp 13–31

[15] Rapoport A 1977 Human Aspect of Urban Form; Toward a Man – Environment Approach to Urban Form and Design vol 15 (Pergamon Press)
[16] Altman I 1980 Culture and Environment (Monterey. Ca. Brooks/Cole)
[17] Smith D M Introduction: the sharing and dividing of geographical space Shared Space, Divided Space: Essays on Conflict and Territorial Organization (London: Unwin Hyman) pp 1–21
[18] Sibley D 1995 Geographies of Exclusion: Society and Difference in the West
[19] Marcus T A and Cameron D 2001 The Words Between the Spaces: Buildings and Language (Architext) (Routledge)

Café as third place and the creation of a unique space of interaction in UI Campus

Yulia Nurliani Lukito*, Anneli Puspita Xenia

Department of Architecture, Faculty of Engineering, Universitas Indonesia, Depok

*Corresponding E-mail : yulianurliani@yahoo.com

Abstract. The emergence of modern cafés is an interesting phenomenon in the city and it is related to globalization and the creation of urban public space. This paper explores the idea of café as third place and social interaction occurs in the café with Starbucks café as the focus of discussion. We argue that modern café like Starbucks offers a comforting and neutral ground for social interaction but some other characteristic of social interaction come up as the result of modern technology. Oldenburg distinguishes third places as public places on neutral ground where people can gather and interact (Oldenburg, 1991). In contrast to first places (home) and second places (work), third places serve as places for creative and social interaction and often considered anchors of community life. The characteristic of cafés as third space relates to the social interaction occurs at the neighborhood that in terms may enhances quality of life and provide social bounding. With the development of information technology, some characteristics of third places are changed although third places still become an important medium for social interaction as well as in increasing the ability of the city to be more adaptable to changes. Conversation is still the main activity at third place but many people use their electronic devices alongside face-to-face conversations. As a method, this study uses direct observations and qualitative analyses of UI Starbucks in as case studies.

Keywords: third place, café, public space

1. Introduction

In *Great Good Place*, Ray Oldenburg (1999) emphasizes the importance of neighborhood gathering places in enhancing the lives of people. Oldenburg defines these gathering locations as third places, and further states that third places are not home nor work, but a place between home and work where people spend their time as a community. Oldenburg describes third places as "a generic designation for a great variety of public places that host the regular, voluntary, informal, and happily anticipated gatherings of individuals beyond the realms of home and work" (Oldenburg, 1999, p. 16). Beside home and work, people needs informal social spaces, or third places, that provide an opportunity to connect with the people in communities and give a balance to the increased privatization of public life.

The concept of "third place" is related to the concept of public, which is defined by Habermas as "the site of collective performance that brings together those who are different from one another precisely because they are different." (Habermas 1956). He argues that it is the architectural nature of the classic seventeenth century coffeehouses (a large room with a large table) and the nature of media found therein (gazettes read aloud and discussed) that shaped our understanding of public.

There are many coffee shops in the city and their existence is closely woven into the social lives of consumers. Coffee shops or cafés can be labeled as "third places," or public spaces that accommodate the regular, voluntary, and informal gatherings of individuals beyond the realms of home and work. Through the years, the idea of a coffee shop has changed especially in relation to the development of modernization; drinking coffee in a fancy coffee shop becomes a lifestyle. Conversation is added with or replaced by laptop, tablets, and mobile phones. As a result, the idea of third place takes different shapes and plays different roles in allowing social interaction is called into question, as well as our understanding of public. However, the main function of the coffee shops remains the same: as a space for social engagement and networking.

Oldenburg's third places have the following characteristics:

- neutral ground, in which people can enter and leave the space whenever they want and none are required to play host.
- a social leveler or a level-playing field for participants where people come and interact freely. In the third places individuals can select their associates and third places serve to expand possibilities.
- conversation is the main activity of a third place.
- good accessibility and accommodation; third places must be accessible for people and can accommodate their needs.
- the regulars; third places have some regulars that help to give the space its characteristic so that the places attract people.
- a low profile; third places have a homely feel and accept all types of individuals.
- playful mood; the tone of conversation in third places are never marked with tension but they have a playful nature.
- a home away from home or a place with psychological comfort that occupants might expect at home. Occupants of third places have the same feelings of warmth and feel a piece of themselves is rooted in the space.

There are some critiques and changes to the idea of Oldenburg's third places. As Rosenbaum (2007) asserts, for some people third places can provide a more psychological support than homes do. The concepts of "home" and "comfort" are heavily influenced by personal experience, cultural background, and social class, that means coffee shops are not necessarily homes away from home even for people who regularly go to coffee shops. Some people, such as those who do not have a designated office space or want a change of scenery, use the coffee shop as an office away from office.

Pozos-Brewer (2015) argues that there is no neutral space instead space is related to power dynamics and social hierarchies. The locations of coffee shops is not neutral anymore because the presence of coffee shop indicates a reproduction of social norms of the neighborhood. For instance, the owners of cafés have some rules and have the right to enforce those rules to guests who come to their cafés. There are some etiquettes that guide the behaviors of people who visit cafés. Other critic addresses the low profile character because in reality most people do not come to a café to meet strangers but they may plan a meeting or just want to spend some times alone. Moreover, people who come to a café create invisible boundaries so that they are not disturbed by other people. The claim that third places are for neighbors to meet each other does not necessarily apply to a city, thus, cafés are not places that necessarily lead to an interaction with complete strangers. Most third places in the city are no longer open as long as the sixteen-hours a day of the early English coffee houses. Bars open in the evening while coffee shops usually close around 9 PM. In short, cafés have their own target customers, although there are some exceptions to the norm.

In relation to the regulars or customers who come frequently to a café and help determine its social character, Pozos-Brewer (2015) argues that the baristas still count people who frequently come to get coffee to go as regulars, although those people spend only little time in the coffee shop and do not contribute as much to the social dynamic of the space. On the contrary, the fact that people come in and out as well as stay adds to the coffee shop's flexible nature. Third places no longer have low profiles but to have at least 'middle-class standards' for cleanliness and some cafés are well designed in order to brand themselves and be better able to compete. Playfulness does not dominate the atmosphere anymore but the ambiance of flexible spaces is more important because flexible spaces allow for different kind of interactions and moods. In addition, cafés today have more formal and sometimes unspoken barriers to enter. Not only that people need to pay for their drink, but usually they need to behave according to a standard such as clothing and manners. Consumerism and consumption affects our societies and creates the contours of our identity by possessing modern products as part of normal standards.

Mobile phones and laptops were rare when Oldenburg developed his definitions of third places. Today's common methods of communication includes phone call, texting, email, and some social media. The emergence of electronic devices allows for virtual connection that now dominate our communication. It is common to see people who choose to engage with social media rather than talk to people in a café. The use of communication technology like mobile phones in physical places isolates people (Anacleto & Fels 2013) and some cultures even prefer the use of new technology over physical contact with other persons (Turkle 2011). Gadgets and communication technology give virtual space for people to interact with each other, but in reality people still need physical contacts and a café offers an ability to transform online space into an offline space (Harrison and Dourish 1996; Tuan 2001). Moreover, a café can fulfill the need for and an opportunity of observing what others are doing in public spaces (Carr, Francis, Rivlin, & Stone 1992) and virtual spaces (Khalid & Dix 2009)

The paper discusses third places in a campus area and how Starbucks café – as an example of third place – becomes a space for social interaction in campus that enhance the quality of life. The study tries to connect how modern cafés appear to be a remarkably important social space just like Oldenburg's idea of third places but some other characteristic of social interaction come up as the result of modern technology. The argument in this paper is that modern café like Starbucks offers a familiar, comforting and neutral ground for social interaction – just like Oldenburg's third places, yet generic enough to follow new lifestyle that is related to modern lifestyle, especially the use of communication technology. Starbucks café located at the central area of Universitas Indonesia (UI) campus in Depok becomes the case study since the café has some characteristics of third places and their development.

2. Method

The method for this research includes literature studies on café and third place, as well as visual documentation and observation of the case study. The focus on the person-environment connection in this study requires descriptions of the place and dynamics of events happens in that place. Therefore, it is important to observe what people do and feel about space and events relate those observations with the physical conditions of the space. The goal is to find out the meaning of the space to the participants. The study took place in Starbucks café at the central area of Universitas Indonesia in Depok.

3. Discussion

3.1 Physical Conditions of UI Starbucks Café

Figure 1. Starbucks cafe at the Central library of Universitas Indonesia.
Source: www.nikeveryday.files.wordpress.com

Starbucks cafe is a franchise café located on the ground floor of the central library building, Universitas Indonesia (Figure 1). The café faces a big artificial lake that are surrounded by three important buildings: a rectorate building, a function building and a central mosque. Located at the central area of UI campus, it is relatively easy to reach the central library from all directions, especially since there is yellow loop buses that go around the campus that operate every day (Figure 2).

Figure 2. Map of UI campus and the location of central library.
Source: www.ui.ac.id

Starbucks café can be reached through the east entrance as well as through the main lobby and the central plaza (Figure 3). With approximately one-hundred-and-sixty square meters, Starbucks café occupies the biggest area among other retails in the building. Beside Starbucks café, there are some restaurants, a post office, a mini market, a merchandise shop and a bookstore.

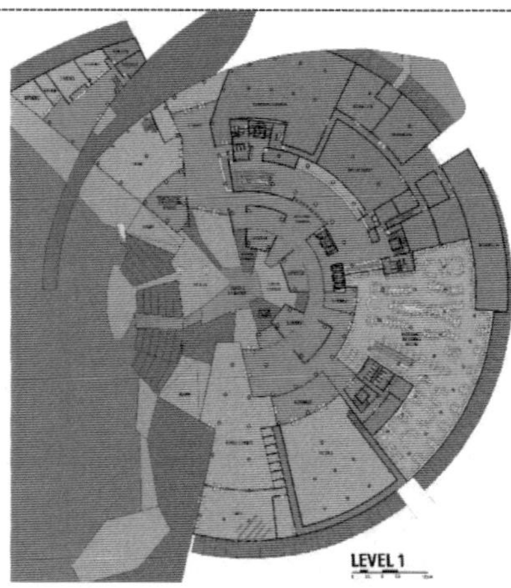

Figure 3. The plan of UI's library. Source: *www.lvshedesign.com*

UI Starbucks café has both unique exterior and interior. The exterior of the café is the skin of the building made of grey concrete walls and big glass walls so that one may enjoy the lake and the plaza from inside the café. With this kind of transparency, people at the plaza can look inside the café while visitors of Starbucks café can see clearly activities at the plaza (Figure 4 and 5). The clear glass walls allow natural lights to come to Starbucks café all day long and creates a good relation between inside and outside. The central plaza is one of favorite places for students to sit or relax at the central area.

Figure 4. The exterior of Starbucks café facing the lake and the central plaza.

Figure 5. Entrance to the Starbucks Café.

The café has a high ceiling with dominant grey color for the floor and the walls. The high ceiling makes the café looks spacious. The interior of UI Starbuck café is mainly dominated with dark wood, which gives the interior a warm atmosphere, and combined with big glass walls. The floor uses marble like dark wood color to emphasize the warm atmosphere and simple interior. The artificial lightings are set quite high and this condition causes the brightness from the lighting comfortable, in addition to natural lights that come from the glass walls. The workstation for the baristas has three types of lightings – downlight, wall lighting for working station, and some decorative lighting near the food and merchandise counter. Tables in the café are made of wood, couches are made of red leather, and the chairs look modern and simple.

3.2 Experiencing Third Spaces at UI Starbucks Café

Every time a visitor come to Starbucks café, a barista always greets him/her from a counter at the left corner of the entrance. Most of visitors who come to UI Starbucks café are students. The café opens at 8 AM and closes at 7 PM from Monday to Friday; and on Saturday the café closes at 4 PM. Starbucks café closes on Sunday because the central library is also closed. From the observation, Starbucks's visitors are still few in the morning and the afternoon is the peak time.

Figure 6. The interior of Starbucks café: Barista counter & centered tables (left) and seating area looking to the central plaza and the lake (right).

Visitors who came to Starbucks café were mostly students and they came as an individual or as a small group. Mainly, people who come there have activities such as talking, working with their laptops, or just use their mobile phones. The most favorite place to sit in the café was near the glass walls and at the corners that faced the plaza outside. People in a group sat mostly on chairs around big tables and couches around small tables for four or five people. For big tables with some chairs (see figure 6), people may share the table with other visitors up to ten people.

Figure 7. Ambience of Tables Lining Near Windows and in the Center.

Figure 8. Ambience of Tables Lining Near Entrance.

Starbucks café also offers privacy to visitors, especially through sitting arrangement. There are some tables surrounded with moveable chairs. In this way, visitors may move those chair and create boundaries necessary to create comfort zone. Between tables, there are enough space to create private space. Columns inside the café function as separators between chairs.

One of noticeable characteristics of Starbucks café is that the seating arrangement mostly facing the public space outside. With glass walls, there is no sharp boundary between the café and the public space outside (Figure 7 and 8). In this way, visitors of Starbucks café can easily observe activity in the plaza while visitors are also engaged in another activity, like talking and eating. The seating arrangement and the whole interior of Starbucks café seem to support the roles of visitors as spectators of people in the plaza while at the same time they could have their own privacy. Even though people in the plaza can look back to café's visitors, they usually do not reply to or even acknowledge their audience.

Visitors of Starbucks café have to buy a drink or a food when they want to sit inside the café, and this was a challenge to neutral characteristics of third place. From the observation, tables placed at the center of the café offer less privacy than tables at the corners and near the entrance or the windows. People can look easily at people who sat at the center of the café. The chairs arrangement around the big table allows people to sit across each other and suggests both close and open group conversation. There are also some seating arrangements that accommodate relatively more private talk such as small tables placed near glass windows and at the corners (Figure 7). The privacy was usually needed by people who come alone and those individuals mainly sit at the corners or near the glass walls. From the observation, individuals who come alone spent more time with their laptops or gadgets and they do not like other people to pay attention to what they do. When some people come as a group, they are free to choose their seats because they can create boundaries by moving some chairs and configure their sittings to allow interaction with the group members or working together with their laptops.

In contemporary third places like UI Starbucks café, there are some alteration in the characteristics of Oldenburg's third places. The presence of information and computer technology is obvious. Most visitors use one type of technology in this third place of Starbucks - mobile phones – and people are often occupied with mobile phones, instead of interacting to a person near-by. Starbucks café has a digital presence such as Wifi, and surveillance camera that allows the owner of third places to record every activity. In general, modern cafés usually offer a digital connection that at least operates locally. People very often are online connected while in third places and as an effort to attract customers third places also offer a presence of digital connection. In this way, people may have multiple focuses while talking, which is not mentioned by Oldenburg. Although people are in a physical third place that affords the ability to talk to each other, they may choose to temporarily access a virtual place or occupied with their phones and laptops or have online conversation.

4. Discussion

In the age globalization and Internet, third places remain places of contemplation and meeting for people wanting to expand their work and home activities. The connection of people to physical spaces has changed specially to accommodate modern culture such as online conversation, social media rather than cultural specificity originally outlined by Oldenburg. The characteristic of third places outlined by Oldenburg has been developed and changed especially in relation to modern culture and communication technology. Starbucks café remains as neutral where people can easily come in and out, but there are things like etiquettes and rules that guide visitors' behaviors inside the café. The physical conditions of the café such as the seating arrangement show neutrality as well as offering flexible spaces for different kind of interactions. Third places no longer have low profiles but to have at least 'middle-class standards' and the interior of the café usually enhance those standards. Modern cafés like Starbucks do not primarily encourages interactions between strangers but accommodate the need both relaxing and working and doing multiple conversation through social media. Conversation is still the main activity at third place but many people use their gadget and electronic devices alongside face-to-face conversations. Contemporary third places also give a possibility to observe others and to allow others to observe you, both physically and virtually.

5. Acknowledgments

The research for this paper is conducted partially with the support of PITTA grant from Universitas Indonesia.

6. References

[1] Oldenburg, Ray. 1989. The Great Good Place. Cambridge, MA: Da Capo Press.
[2] https://nikeveryday.files.wordpress.com/2011/09/perpustakaan-pusat-baru-fa-triatmoko.jpg?w=620
[3] http://www.lvshedesign.com/wordpress/wp-content/uploads/2012/04/201204060109.jpg
[4] http://static.panoramio.com/photos/large/68357677.jpg
[5] Pozos-Brewer, Rose. (2015. Coffee Shops: Exploring Urban Sociability and Social Class in the

Intersection of Public and Private Space. Swarthmore College. Dept. of Sociology & Anthropology.
[6] Rosenbaum, Mark S., James Ward, Beth A. Walker, and Amy L. Ostrom. 2007. A Cup of Coffee with a Dash of Love: An Investigation of Commercial Social Support and Third-Place Attachment. Journal of Service Research.
[7] Memarovic et all. 2014. Rethinking Third Places: Contemporary Design with Technology. The Journal of Community Informatics. Vol 10, No 3.
[8] Carr, S., Francis, M., Rivlin, L.G., & Stone, A.M. 1992. Public space. Cambridge: Cambridge University Press.
[9] Habermas J. 1956. The Structural Transformation of the Public Sphere: an Inquiry into a Category of Bourgeois Society. Cambridge, MA: MIT Press.
[10] Tuan, Y. 2001. Space and Place: The Perspective of Experience, 8th edition. Minneapolis: University of Minnesota Press.
[11] Du, H., Rosson, M.B., & Carroll, J.M. 2012. Café Commons: An Exploratory Study of Public Commenting in a University Café. In. CIIT'12: The 7th IASTED International Conference on Communication, Internet, and Information Technology ACTA Press (2012).
[12] Ellis, A. 1956. The Penny universities: a history of the coffee-houses. London: Secker & Warburg.

Spatial change transformation of educational areas in Bandung

Asep Yudi Permana[1*], **Karto Wijaya**[2]

[1]Universitas Pendidikan Indonesia, Indonesia
[2]Universitas Kebangsaan Bandung, Indonesia

*Corresponding e-mail : yudi.permana@upi.edu

Abstract.

Transformation is a total transfer process from a particular shape into a new figure which can be interpreted as the final stage of a changing process. Transformation, as a gradually progressive process, utilizes space and time factors becoming the aspect that greatly influences the change until reaching the ultimate stage. The changes are completed by responding to the influence of external and internal elements which will lead to a change from an early familiar form.

The purpose of this research is to build theory based on phenomenon which occurs in the field as a result of space transformation process in Bandung so that it can be seen the setting adjustment of either physical or non-physical in the area becoming the city identity. The research method is qualitative research method with descriptive survey method approach. The focus of research is aimed at discrete phenomena in the field as an effort to disclose the background essence of the resulted theory formation. The study does not use a theoretical framework, yet there is a consistency in observing a discrete phenomenon of the grand tour results. The phenomenon is raised through the snowball sampling process. The process of collecting data, analyzing, and building theories is conducted jointly and iteratively within one research time range.

The study results in the formation of space knowledge based on knitting and the value system of the users' activities and the space settings. All users' behaviors in using space are related to social, cultural, economic and creative activities in utilizing the existing space conditions. The results of this study provide an enrichment of architectural theory, planning theory and urban design theory having already existed.

Keywords: Phenomenon, Identity, Culture, Descriptive Survey Method, User Activity

1. Introduction

Nowadays, the development and growth of urban populations have impacts on various aspects of life. The issues of urban life such as environmental degradation, cultural clashes in society, living space diminishing, capitalism, and consumerism are the problems experienced by most major cities in developing countries. This development having resulted in the growth and development of the population in large cities has increased rapidly to 32% in the last 30 years [1]. The growth is supported by natural exploration boost, rapid migration and changes in rural communities.

The rapid increase of population in urban areas causes the demands of housing, which increases rapidly, especially for people with low income. The phenomenon of one-hundred-million urban residents and a half the Indonesian population having become the urban is a condition which requires to be the foundation in the development of cities in the future [2]. In developing countries, these developments are basically not matched with the procurement of housing (sluggish process) so that the number of housing shortages in urban areas is growing. It is characterized by the growth of wild huts and slums with a high population density and illegal land use.

Besides the demand of housing, people also require the presence of adequate urban infrastructures, including the provision of water, energy, telecommunication, public transportation, and the availability of other infrastructures. According to Firman and Soegijoko (2005), the other factors felt very pressing as a result of the city development are namely environmental problems, especially water, land, and air pollutions, traffic jam, and others. In addition, there are also social problems, such as poverty, slum, crime, conflict among citizens, and others[2].

The growth of urban population in developing countries has increased swiftly [1]. This phenomenon as the impact of the exchange current and conditions affects each other in various dimensions, among economic, social, political, cultural, and technological aspects. One of the contributors to the urban population growth in developing countries is the migration of students [3]. Not only at the university level but also at the senior high school level, students have separated with their parents for the preparation of achieving the goals and hoping to be able to study at their dream colleges. It is believed that by attending school in these big cities, the opportunity to enroll the country's top universities is bigger because the access to tutoring is broader, unlike in the suburban and rural schools.

The flow of student migrations results in a high demand in the fulfillment of their living space, while the supply provided by the universities (in the form of dormitories) is not balanced with the number of the demands for student residence. This resulted in the supply handed over to the community around the campus. The condition, when viewed from the positive side, causes the increase of economic growth around the university, while from the negative side it affects the development of the city unconstructively, marked with the growth of dense settlement area and the transformation of ordinary house into student lodge.

Bandung as the city of education is not just built in one night because based on the history of Bandung, the development has begun since the inception of the School for Indonesian Teacher (Hollandsch Inlandsche Kweekschool / HIK) in 1848 [4]. The milestone of high education began with the transfer of the Dutch East Indies capital from Batavia to Bandung in the early 20th century, 1920, with the establishment of Technishe Hoogeschool / THS as the forerunner of Bandung Institute of Technology. Since then the existence of higher education in Bandung is growing rapidly. It is recorded according to Bandung BPS in 2012 nearly 168 universities are in Bandung, from 168 there are 11 state universities. While the land area of Bandung is 16,731 hectares, if it is compared to the number of universities in this city, the number is too small; this is what makes one of the exceptionalities of Bandung.

2. Methods

The applied research approach is descriptive survey method. The decision of utilizing descriptive survey method is because of the intention to map the phenomenon of space usage. Architecture is a cultural phenomenon which is not only a physical reality alone, yet the architectural space consists of both static space and dynamic space. Dynamic space is viewed from the aspects of typology, scale, relationships, and identity. The main strength of this method lies in the obtained results through the recording and direct observation coming to the area and taking time to see how space is used [5][6].

The research utilizing descriptive survey method is started by conducting descriptive research from primary and secondary survey results, and performing data processing. The primary surveys are field observation, interview and questionnaire distributed to informants. The field surveys were conducted in the studied area at a single-time research (cross-sectional study) providing snapshots, namely from February to November 2016.

2.1 Research Time and Location

This research was conducted in Bandung, West Java Province covering 6 areas, namely Bojonegara, Cibeunying, Bandung Kulos, Karees, Gedebage and Ujung Berung. The research was conducted for ten months, namely from February to November 2016.

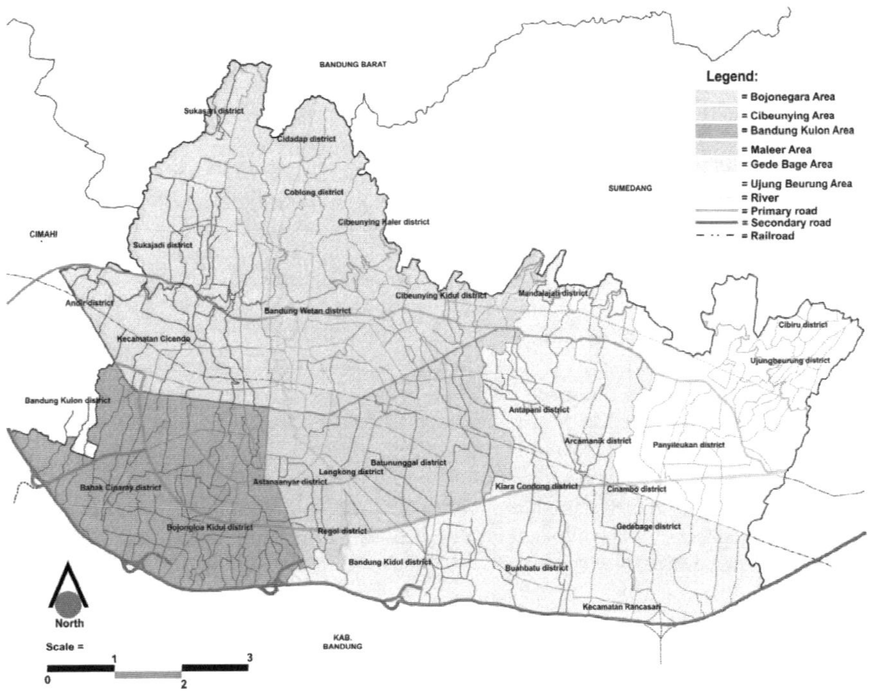

Figure 1. BANDUNG MAP
Source : Modification of Researcher from Bandung Area Spatial Planning Map 2014

2.2 Research Data Sources and Tools

The data collection tools employed for collecting research data consist of observation guides, interview guides, and documentation. Observations are used to map the spatial and space-based spaces of time-actors. Observation interviews and documentation were used to reconstruct spatial areas of research. Interviews are also used to collect primary data about the type of usage and residential function changes into student housing. In order to clarify the observation process, supporting aids are utilized to visually record the condition of the research environment by using handy cam, camera, and video player.

2.3 Research Data Analysis

The analysis phase in this research emphasizes on the process of change which occurs in spatial pattern in education area of Bandung based on several research variables consisting of land status, land and building function, and formed spatial pattern. The comparison between the spatial configuration theory and the observation of the spatial pattern is intended to reaffirm the characteristics of physicals, both form and element, in determining the identity of the area as an area of education, including explaining the various non-physical factors underlining every change.

Based on the preliminary record, ten student boarding cases were selected as a starting point as well as entry point for the next case. The selection of the ten cases was obtained based on the openness and response of the owner of the lodge encountered during the first round exploration. The sampling technique as the focus of observation is done by snow ball method, which is intended to expand the information and get new themes in it. As described by Moleong, (2006), the purpose of snow ball sampling technique is to obtain new information and themes through deepening of the research implementation, in case of repetition of information as an indication that the sample sampling has saturated then the search for samples is stopped [7].

During the induction process, in-depth interviews were conducted, which intended to obtain physical data on the pattern of space and its use, and non-physical matter includes thoughts containing the form of existing value system in accordance with the condition of the selected case. The factors observed include human both owners and students as actors in doing activities and activities as both individuals and groups, and space as a container/place to perform activities and user activities. Based on the interpretation of data that has been obtained from the mapping of the case and the results of in-depth interviews, and then it is performed the re-depiction of the lodge space settings based on the categorization found during the previous grand tour. The step is intended to facilitate the process of analysis and induction process of determining the theme and conception of existing local theories.

The understanding of the change process which occurs in an architectural space is achieved by the morphological approach. The changes that occurred in the study area is divided into 2 (two) periods of development and growth, namely: (1) The Development of Bandung in the Colonial period from 1825 to 1945 and (2) City Development in the Post-colonial Period of 1945 - 2015.

3. Discussion

A city is the identity of its citizens' community life of. According to Kostoff, (1999), growth and physical development of a city do not stand alone but always run along with non-physical changes and developments [8][9]. Therefore, a city will experience a "life cycle" by growing and developing continuously. As Rapoport, (1990) explains, the process of transformation in a city occurs as a result of changes in user activity as a consequence of changes in value, social, culture, and standpoint of its community [10]. Thus, the configuration of the physical form of a city will always change, will barely end, and will never stop as long as the city is occupied. The process of modernization in major cities in Europe and America creates a space segment having increased dramatically, but the growth of cities in Indonesia has not undergone changes drastically. This happens because it is based on the pattern and life of the Indonesian society which is different from either Europe or America.

Soetomo, (2012) states that generally cities in Indonesia are characterized by dualistic structure, where morphological structures are planned along major roads and unplanned morphological structures behind them (as urban hamlet areas, this condition is characterized by the modern life of formal sector and traditional informal sector which go together harmoniously [11][12][9]. Furthermore, Soetomo, (2012) explains that the informal sector becomes a medium of life that sustains the lives of most urban dwellers along with the formal sector [11]. The morphological dualism of the city between the modern formal and the traditional Informal is still quite prominent in Indonesian cities where the hamlets are still knitted in the center of the city, and the relationship between villages with blocks of planned settlements (formal), which are an interconnected network. According to Widjaja (2013) this environment is generally formed naturally and unintentionally, and its physical appearance was not standard[13].

Evers and& Rudiger (2002) argue that urbanization is one of the fastest processes among social changes around the world [14]. The process of the urbanization occurrence arises because, first, it refers to the inhabitant percentage in urban areas within a country; second, it refers to the growing number of urban dwellers. Firman and Soegijoko (2005) argue that demographically the source of urban population growth is due to the increase of the natural population, the migration of the population especially from rural to urban areas, and the reclassification of the status change of a particular "village" as rural locality into urban locality [2].

The city development cannot be separated from the flow of migration from village to city. This also occurs in Bandung where newcomers form a new environment. Usually they take advantage of state-owned land, riverbanks, and along the railroads, which in the future will become densely populated areas as urban hamlets. As migration flows more and more, social problems become more complicated as time goes by. The average occupants living in this area are people with low social and economic groups; they work in informal sectors, housekeepers, construction workers, pedicab drivers, merchants, and other harsh jobs.

Soetomo (2012) explains further that the definition of a city includes two great aspects where one with another cannot be separated [11]. The first aspect is the physical aspect of the city which concerns the form of space as a place formed by infrastructure and means above the natural space; and the second aspect is the human aspect as the subject of development and the user of urban space. City as a human place to live with all life activities, then the city is a part of the human settlement where human settlement consists of the content which is the human and the container which is the medium, a physical settlement, both man-made and nature as a place for human life with all his/her activities. The city is as a place formed not only by the building alone but also by the man himself. Furthermore,

Kostoff (1999) divides the city into 9 (nine) characteristics, namely: (1) Cities are places where a certain energized crowding of people takes place; (2) Cities come in cluster. A town never exists unaccompanied by other towns; (3) Cities are places that have some physical circumscription; (4) Cities are places where there is a specialized differentiation of work; (5) Cities are places favored by a source of income; (6) Cities that must rely on written record; (7) Cities are places that are intimately engaged with their country side; (8) Cities are places distinguished by some kind of monumental definition; And (9) Cities are made up of buildings and people [8].

Discussing the history of settlements of Bandung society life cannot be separated from the historical development of urban settlements in Indonesia generally, and especially the development of cities located in Java. Javanese cultural system in Indonesia influence significantly not only it is visible to the eye, such as: the systems, mechanisms, and political culture behaviors, but also it shapes various dimensions from normative to institutional, from Mataram Hindu, Majapahit to Mataram Islam and the period of the Republic of Indonesia. Culturally, the influence of urban inhabitation under Javanese tradition was implanted through the 5th century journey of Hindu-Buddhist civilization. This civilization contributes greatly to the political culture of the state.

Based on a more detailed historical record, it is mentioned that the end of Sultan Agung's empire (1613-1645) is the most obvious setback in the cities in Java. This does not mean that the cities in Java retreat in fostering a new civilization. On the contrary, since the consolidation of Dutch power from Daendels to the era of Van den Bosch in the nineteenth century, it opens the interaction between traditional and western modernity having continued during the reign of the Dutch East Indies with the establishment of *gemeente* from 1904 to 1906 in Batavia, Bandung, Semarang , Surabaya, and Makasar [15][4].

The urban living culture in Java cannot be separated from the traditional process and the understanding of urban living in the context of the state. According to Wiryomartono (1995) the contexts of urban living is that Java is not only a medium of cultural development but also the city civilization in Indonesia [16]. Discussing about civilization cannot be separated from the wider state of cultural government problems. The development of cities in Indonesia is generally developing from the port cities to the development of coastal cities, while the development of inland cities cannot be separated from the development of a city that departs from the main source where the river as a magnetic early development of inland cities. Similar condition occurs in the development of Bandung which cannot be separated from the development of the Cikapundung River which divides the city from the north to the south of the city.

Starting from the construction of Post Road of Panarukan-Anyer, Bandung development had begun which then followed by the removal of the Capital District of Bandung from Krapyak in the southern area of Bandung. The transfer of Bandung Regency government was due to the urging of Daendels, responded by the Regent of Wiranatakusumah II by choosing a place in the southern part of the road from the west of Cikapundung River (near a pair of sacred wells of Sumur Bandung). In this area, Wiranatakusumah II built his palace and the square as the center of the city. Following the traditional orientation, the Grand Mosque is placed on the south side, and the traditional market on the east side. House and City Hall as meeting places are located in the south facing the sacred mountain Tangkuban Perahu. That was the time when the Flower City (Bandung) was born [15][4].

3.1 Space Transformation

The analysis of the function transformations of building and land in general in the first period (development of Bandung City during Colonial period from 1825 to 1945) has not shown significant changes in regional structure. The changes in the functions of land and buildings cannot be separated from the policies issued by the Dutch East Indies government, as shown in Figure 2.

IOP Conf. Series: Earth and Environmental Science **99** (2017) 012029 doi:10.1088/1755-1315/99/1/012029

Year 1825 – 1882	**Year 1882 - 1905**	**Year 1905 - 1924**	**Year 1924 - 1945**
From Colonizer to Municipality	From Remote Village to City	Modern Colonial City	Japan Aggression and Independence

• During the VOC era, Bandung was made to be the coffee cultivation city. • Bandung as a new city became a temporary administration city of England. • Post Road (de Groote Postweg) was built. • The transfer of Bandung regency capital from Dayeuh Kolot to Krapyak area took place. • British invaded through Raffles military aggression. • Restoration/repair and forced planting laws were initiated. • A School of Education for Prospective Indigenous Teachers (Hollandsch Inlandsche Kweekschool/HIK) was established in 1848.	• Some Priangan areas were opened for Europeans, so Bandung became a tea plantation cultivation city. • Bandung became the center of economy, government and military by the opening of new road access.	• Bandung became an independent municipality (an autonomous government in determining government affairs, finance and administration). • The removal of forced cultivation and the influence of the Agrarian Law encouraged economic growth and demography. • Bandung built several monumental buildings, such as: Gedong Sate, Military Command Center (Kodam), Pyrotechnic Workshop • Bandung as a city of education marked by the founding of Technische Hogeschool/THS in 1920 as the forerunner of establishment of ITB.	• In 1930, it was the starting point of Bandung City planning by Thomas Karsten • Bandung became a residential area of European citizens (recorded Europeans in Bandung is the third largest among cities in Java) • Rapid growth could be seen in Bandung as a city of education (primary, secondary, and higher education) • Bandung became the capital of the Dutch East Indies in 1938 • The period of Japanese occupation • The period of preparation toward the transfer of sovereignty and the preparation of independence
Space transformation, the first period, from colonial era to independence era, 1825 – 1945			

Figure 2. Transformation of land and building functions of the colonial period 1825-1945

The analysis on the function transformation of building and land generally can be found in the second period (development of Bandung in the post-colonial era starting from 1945 to 2015), the development of Bandung escalates significantly, as shown in Figure 3.

Year 1945 – 1950	**Year 1950 - 1976**	**Year 1976 - 1980**	**Year 1980 - 2015**
Bandung after Independence	Bandung after Independence	Bandung as Education City	Bandung as Education City
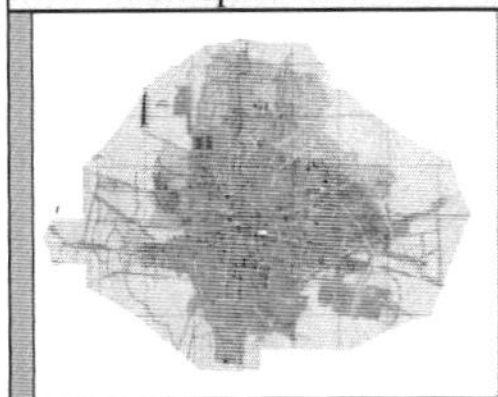		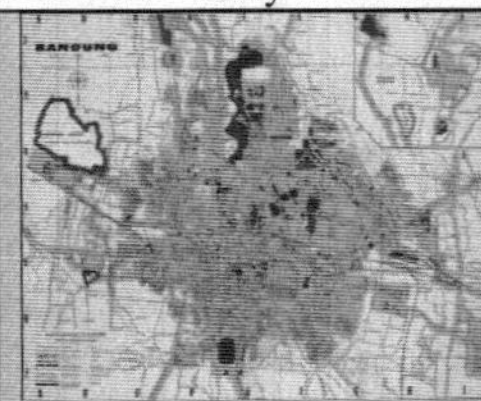	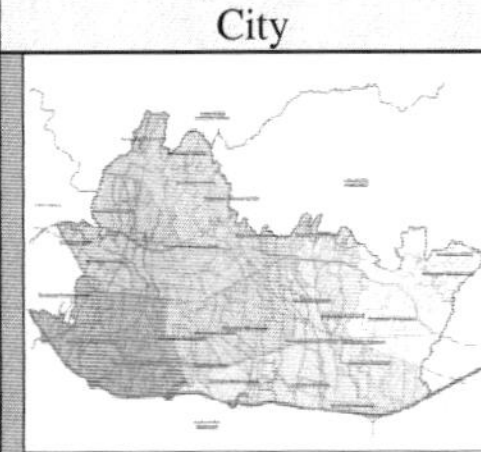
• This period was known as the physical revolution. • In this period, dualistic power occurred in which the area of Bandung in the north was controlled by the Dutch supported by allies, and the southern area of Bandung was controlled by republican freedom troops and Indonesian people. • Changes in the Indonesian state system took place after the round table conference giving birth to Pasundan state with Bandung as the capital.	• The college life in this period developed rapidly with the inception of Technical High School (STT) in 1959 as the forerunner to the establishment of ITB. • The development of the park area took place, such as Cihampelas as Gardein city and other supporting facilities of higher education.	• The growth and development of public and private universities began. • Some areas of the campus support area have changed the structure of the space (the conversion of several houses to the lodge for students). • The 1970s policy in the economic sector triggered the growth of private universities in Indonesia, which also happened in Bandung.	• The growth and development of public and private universities took place, (based on BPS Bandung there are 168 universities and 11 of them state universities). • The growth of student housing areas in the area of higher education took place as a supplier of the demands of the lodge.
City Development in Post-Colonial Era 1945 – 2015			

Figure 3. Transformation of land and building functions of the colonial period 1945-2015

3.2. Pattern Distribution of State and Private Higher Education in Bandung

Based on BPS Bandung, in 2012 it was almost no less than 168 universities both public and private in Bandung. The number is not small, and it is not owned by other big cities in Indonesia. The 168 colleges consist of: (a) 10 Institutes; (B) 22 Universities; (C) 18 Polytechnics; (D) 65 high schools; (E) 30 academies, and (f) 23 educational institutions. Of the 168 universities located in Bandung, there are 11 State Universities, namely Bandung Institute of Technology, Padjajaran University, University of Education Indonesia, State Islamic University Sunan Gunung Djati Bandung, Bandung State Polytechnic, Kemenkes Health Polytechnic Bandung, Manufacturing Polytechnic Bandung, Indonesian Art and Culture Institute of Bandung, Tourism High School of Bandung/NHI, Social Warfare High School of Bandung (STKS), and Institute of Local Government.

The growth and development of public and private universities in the city of Bandung will affect the demand of lodges for students, while on the other hand the campus does not provide adequate lodging facilities. As a result, areas of lodges were built around the campus, as shown in Figure 4.

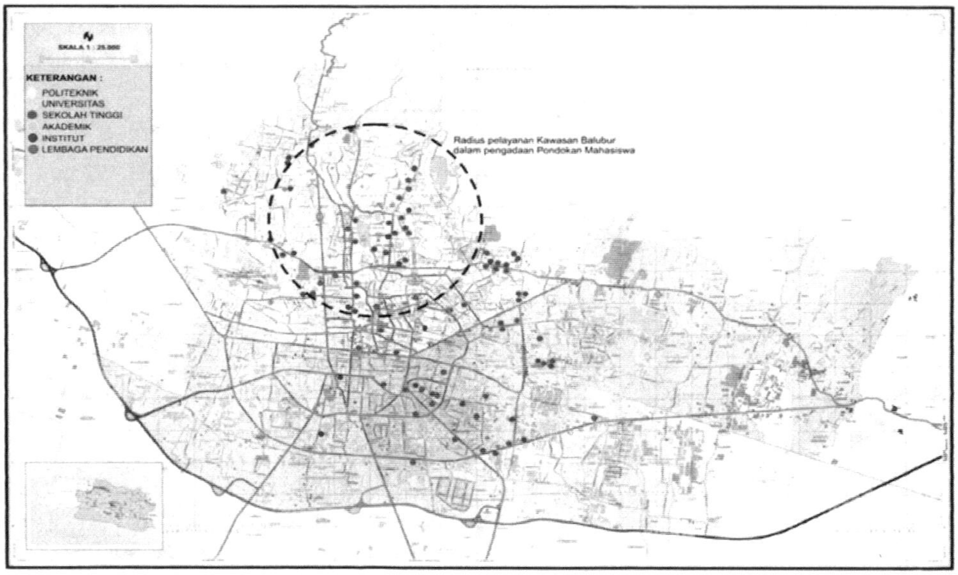

Figure 4. MAP OF HIGHER EDUCATION IN BANDUNG
Source: Researcher Modification from Spatial Planning Map of Bandung Area 2014

From figure 4 there are 8 (eight) student lodge areas as a supplier of lodge demands for students residing in the surrounding area. The 8 (eight) areas of this settlement are:

1. Bojonegara region starts from Dr. Setiabudhi Road, Sukajadi Road, Sutami Road, to Suryasumantri Road. In this area there are 5 subareas of student lodges:
   a. Setiabudhi Bawah Road, Budhi Sari, Geger Kalong Hilir, Geger Kalong Tengah, Geger Kalong Girang, Panorama, Cipaku, Ledeng, Negla, Sergeant Bajuri, Sersan Sodik. These areas serve as student lodge areas for the University of Education Campus Indonesia, NHI, UNPAS, Music High School of Bandung, AKTRIPA, Bandung State Polytechnics/POLBAN, and POST Polytechnic Indonesia.
   b. Sarijadi Complex, Geger Kalong Hilir Road, Sarikaso, Cijerokaso. These areas serve as student lodge areas for Bandung State Polytechnics /POLBAN Campus, POST Polytechnic Indonesia, Maranatha Christian University, AKTRIPA, and Bandung High School of Management.
   c. Sindang Sirna Complex, Bungur, Sukajadi. These areas serve as student lodge areas for AKTRIPA Campus and Bandung High School of Tourism, Health Polytechnic, Bandung High School of Management, Bandung Health Polytechnic, Bandung High School of Music, Indonesia High School of Technology.
   d. Setra Duta Complex, Jln. Sutami, Lemah Neundeut Complex, Suryasumantri Road, Mustank Complex / Lecturer Housing of UNPAR. These areas serve as student lodge areas for the campus of Maranatha Christian University, AKTRIPA, Polytechnic POST Indonesia, Bandung State Polytechnics (POLBAN), STIE PASIM, Academic Nurses PPNI, Aryanti Secretary and Management Academy, University of Nurtanio, Bandung Theological College, Siliwangi Aeronautical Technology Academy.
   e. Cihideung Complex, Sersan Bajuri, Negla. These areas serve as student lodge areas for UPI Campus, Advent University, and Management High School of Bandung.
2. In Cibeunying Region 1 is the oldest area to supply the demand of student lodge. In this area there are 5 areas of lodgings, namely (1) Lebak Siliwangi; (2) Dago Bengkok/Cisitu; (3) Dago Atas, Dago Pojok, Kanayakan, Tubagus Ismail; (4) Tubagus Ismail, Sekeloa, Dipati Ukur; And (5) Tamansari Bawah, Tamansari Atas - Balubur. These areas serve as student lodge areas for

ITB campus, UNPAD Dentistry, STKS, UNPAR, ABA, UNPAS, UNISBA, STDI, UNIKOM, STT INTEN, Academy of Accounting Bandung / A2B, Bandung Textile Industry Academy, Academy of Nursing TNI-AU Ciumbuleuit, Academy of Finance and Business Indonesia, Academy of Finance and Banking Merdeka, Deoostemen Institute, Harapan Bangsa Institute of Technology / ITHB, Indonesian Institute of Computer Science and Management, Indonesian Institute of International Management, Intergrace Educational Institution, Ismi Training Institute, Primagama Educational Institution, University and Bandung Business Informatics, Bandung High School of Music, and National College of Technology.

3. Cibeunying Region 2 is the northern area consisting of Ciumbuleuit, Cisatu, Bukit Jarian, Ranca Bulan, Cihampelas Atas or Lebak Siliwangi. These areas serve for UNPAR, ITB, UNPAD, UPI,
4. Cibeunying 3 starts from Pahlawan Road, Katamso Road, PHH. Mustopa Road, and Cikutra. These areas serve for ITENAS campus, Sangga Buana University, STIE, High School of Management and Padjadjaran Computer Science, High School of Management and Bandung Informatics Computer, Bandung Kridatama Polytechnic, University of Bandung Raya/UNBAR, Widyatama High School of Technology, Indonesia Mission High School of Technology.
5. Ujung Berung Area consists of Cipadung and Cibiru Complex, serving for the campus of State Islamic University (UIN) Sunan Gunung Djati Bandung, School of Economics (STIE) YAPARI.
6. Gede Bage Area consists of Jln. Kiara Condong, PSM, Jln. Soekarno Hatta, Canal Kiara Condong, Fruit Stone, Margahayu Raya Complex, Ranca Bentang. These areas serve for the campus of UNINUS, STT Mandala, Institute of Education and Training of Hotel Professions, High School of Management and Informatics Computer Indonesia, AMIK Bandung, University of ARS Indonesia, University of General Ahmad Yani (Campus 2), Education Institute Secretary AMIK Bandung, National University YPPKP, Kencana Academy of Secretary and Management, College of Textiles (formerly known as ITT).
7. Both Karees and Gedebage areas serve for Polytechnic LPKIA Bandung, Bandung Commerce and Industry Polytechnic, Tadikapuri Academy Tourism, Kebon Djati Academy of Nurses, STSI Bandung, High School of Agriculture, Bandung High School of Technology, STT TELKOM.
8. Karees 1 area consists of Martanegara Road, Lodaya Road, and Resin Road. This area serves for Binasarana Academy of Management and informatics Bandung, HASS Academy Management and Informatics, IPTB Academy of Informatics Management and Computer, National Tourism Academy Indonesia, Sandhy Putra Tourism Academy, International College, National University, Langlabuana University (UNLA), Taruna Bakti Academy of Secretary Management, Bandung Academy of Technology.
9. Karees 2 area consists of Soekarno-Hatta Road, Leuwipanjang Road, Cibaduyut Road, BKR Road. This area serves for the campus of Darul Qalam Institute of Islamic Studies, Mardira College of Information Management and Computer Indonesia, Malayahati University, Bandung Academic of Technology.

3.3 Spatial Usage Pattern as supporter facilities of Higher Education in Bandung

The more intensive observation had been conducted from the beginning to the end of 2016 covering the entire supporting areas of higher education in Bandung; this is intended as a focus of further research. Based on the initial observation as the Grand Tour on all lodgings in the residential areas, this resulted in the following lodge house categories:

1. Houses, student lodge uniting with their owners
   a. Access to the lodge is united with its owner
   b. Access to the lodge is separated from its owner
2. Home, Dorm, and Owner's Business Place
   a. Access to the lodge is united with its owner and owner's business place
   b. Access to the lodge is separated from its owner and owner's business place
3. Student Dorms whose owners are not settled in, which means he/she is not living together/where the owner does not live within the area.
4. Apartment, Flats

Table 1. Student Lodge Category in Bandung

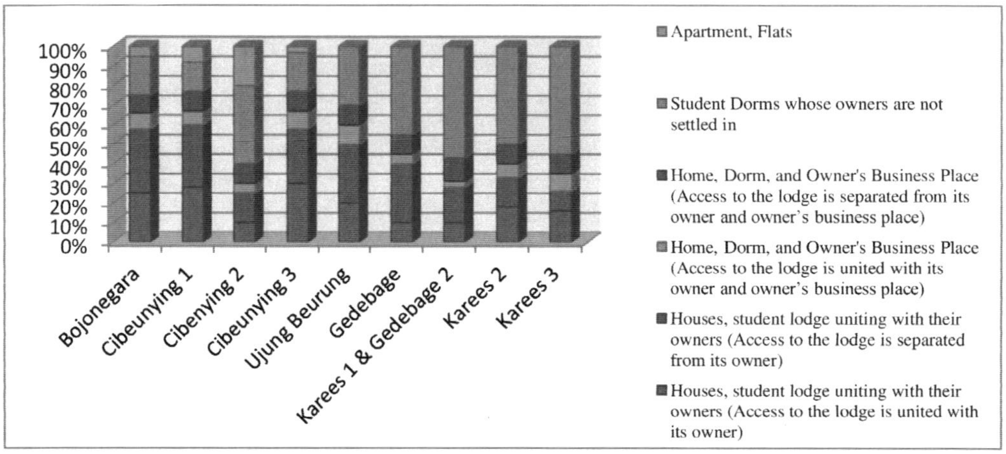

The data obtained at the initial observation was the primary data collected based on observation/direct examination and intensively in-depth interviews on the support of higher education in Bandung based on the results of Researchers Record in 2016. Based on table 1, there is a similarity of characters among Bojonegara, Cibeunying 1 and Cibeunying 3, which are dominated by student lodges uniting with their owners. This characteristic forms a densely populated area. The areas of Gedebage, Karees 1 and Gebebage 2, Karees 2 and Karees 3 are dominated by lodges whose owners do not live together with the students, where most students live by renting houses in housing. The interesting thing happened in the Cibeunying 2 area where the students live in the lodge whose owners do not live together with the students (in the form of Dormitory, Guest House, and Apartment). The area of Cibeunying 2 is dominated by dormitory, guest houses and apartments.

3.4 Student lodge as a Part of the Living Space

The transformation of space that occurs as a result of the migration of students in the city of Bandung experienced a different development among the regions. This is due to the characteristics of the development of each region. There are 2 fundamental character changes, namely:

(1) The transformation of space occurring in the area of Bojonegara, Cibeunying 1 and Cibeunying 3 taking place from a kampong to a student lodge village which still retains the village structure. Students live in one area that produces a color and pattern of village life with its own values although coming from different regions or tribes. Activity between students and society runs naturally;

(2) The transformation of space in the area of Cibeunying 2, Gedebage, Karees 1 and Gebebage 2, Karees 2 and Karees 3 occurred from a settlement area to a horizontal or vertical housing area; the community displaced from their homes (changes from the housing into horizontal and vertical housing areas (Such as apartments, flats), so that new areas form a new community. Collaboration between students and society does not happen. Environment characteristics have individualistic characteristics that are less familiar with each other.

3.5 The Pattern of Student Life in Community Environment

A space as a means for various human activities in the fulfillment of their demand produce behavior patterns (behavior setting), so that it is formed an interaction and a social communication between citizens. This occurs in the Bojonegara, Cibeunying 1 and Cibeunying 3 areas. This condition is supported by the characteristics of a unified lodge and an inherent part of the local inhabitant settlement. In this area, the existence of students as part of the community gives new color in the life of society, but physically the structure of the village survive and exist. The physical condition of student lodge almost 80% consists of limited and narrow space.

While in Cibeunying 2, Gedebage, Karees 1 and Gebebage 2, Karees 2 and Karees 3, communication between students and the local community is less, this is because students live in housing, guest house or even apartment tending to interact with the community even less.

4. Conclusion

Spatial change transformation that occurred in the city of Bandung is influenced by the characteristics and needs of users, there are 2 forms of changes that occur, namely:

(1) Bojonegara, Cibeunying 1 and Cibeunying 3 areas have resulted in a change from the form of the village having a development of a village, starting from the birth of the smallest unit of Babakan which later developes into the urban kampong. The process of the occurrence of this urban village occurs based on the history of the village. The development of urban kampong can be seen from how the process of forming, viewed from the quality of the physical environment of the settlement, which can also be viewed from the general condition of the inhabitant community. The process of village transformation into a urban kampong that occurs in this region basically in line with theories having been expressed by many experts. Judging from the process of its formation, the transformation process taking place in this region supports the opinion of (Evers and Rudiger, 2002) that the urban village can be interpreted as an indigenous and a traditional village that will develop and merge into a part of the city, but it still retains the characteristics of a village. Urban kampong is a residential area that generally comes from rural areas swallowed up by the rapid development of the city, so it transforms into inner-city settlements. While the form of urban village by referring to the general condition of its inhabitants, the urban kampong is a traditional residential environment that is specific to Indonesia, distinguished by the characteristics of life established in close family ties. The urban kampong is a lower-class society, originally formed through ethnic segregation systems, generally lacking adequate infrastructure, utilities and social facilities in both quantity and quality. The urban lifestyle evolves in line with the complex integration of formal, informal, and sub-sector economic activities. The urban kampong has little or no power in the city administration and must adapt its lifestyle to survive.

(2) Cibeunying 2, Gedebage, Karees 1 and Gebebage 2, Karees 2 and Karees 3 areas have resulted in the form of residential areas in the horizontal (residence) and vertical (apartment, flats). Their characteristics are inversely proportional to the characteristics of urban kampongs that occur in the Bojonegara, Cibeunying 1 and Cibeunying areas 3. Its community life in this region tends to be individual, communication between students and local communities is less; this is because students live in housing, guest houses or even apartments tending to interact with the community even less.

5. Acknowledgments

Appreciations are addressed to: (1) Kemenristek RI which has provided fund support through Fundamental Research scheme in 2016; (2) The Head of the University of Education of Indonesia who has supported and granted the permit for using of Interior Design Laboratory facilities during the research; And (3) The people in the education area are: Balubur-Tamansari, Geger Kalong, Ledeng, Sersan Bajuri, Panorama, Ciumbuleuit, Sarijadi, Dipati Ukur, Cipadung, Antapani areas.

6. References

[1] A. Gilbert and J. Gugler, *Cities, Poverty, and Development Urbanization in The Third World.* New York: Oxford University Press, 1983.

[2] T. Firman and B. T. S. Soegijoko, *Bunga Rampai Pembangunan Kota Indonesia Dalam Abad 21, Buku 2" Pengalaman Pembangunan Perkotaan di Indonesia.* Jakarta: URDI-YSS dan Lembaga Penelitian Fakultas Ekonomi Universitas Indonesia, 2005.

[3] A. Y. Permana, "TRANSPORMASI GUBAHAN RUANG: Pondokan Mahasiswa di Kawasan Balubur Tamansari Kota Bandung," Universitas Diponegoro, 2014.

[4] R. P. G. A. Voskuil, *No TitleBandung Citra Sebuah Kota.* Bandung: Departemen Planologi ITB bekerja sama dengan PT. Jagaddhita, 2007.

[5] R. Kier, *Urban Space.* London: Academy Edition, 1979.

[6] S. Carr, M. Francis, L. G. Rivlin, and A. M. Stone, *Public Space*. Cambridge: Cambridge University Press, 1992.
[7] L. J. Moleong, *Metodologi Penelitian Kualitatif.* Bandung: Remaja Rosda Karya, 2006.
[8] S. Kostoff, *The City Assembled: The Elements of Urban Form Through History*. London: Thames and Hudson, 1999.
[9] B. Setioko, "Integrasi Ruang Perkotaan di Kelurahan Meteseh Kawasan Pinggiran Kota Semarang," Universitas Diponegoro, 2010.
[10] A. Rapoport, *History and precedent in environmental design*. New York: Plenum Press, 1990.
[11] S. Soetomo, *Urbanisasi dan Morfologi. Proses perkembangan peradaban dan wadah ruang fisiknya: Menuju Ruang Kehidupan yang Manusiawi*, 2nd ed. Yogyakarta: Graha Ilmu, 2012.
[12] Masykur, "Karakteristik Permukiman Dualistik dan Tingkat Keberhasilan Penghunian, studi kasus Kota Bogor Jawa Barat," Institut Pertanian Bogor, 2005.
[13] P. Widjaja, *Kampung - Kota Bandung*, Pertama. Yogyakarta: Graha Ilmu, 2013.
[14] H. D. Evers and K. Rudiger, *No TitleUrbanisme di Asia Tenggara: Makna dan Kekuasaan dalam Ruang-ruang Sosial*. Jakarta: Yayasan Obor Indonesia, 2002.
[15] H. Kunto, *Semerbak Bunga di Bandung Raya*. Bandung: Pt. Granesia, 1986.
[16] Wiryomartono, *Seni Bangunan dan Seni Bina Kota di Indonesia*. Jakarta: PT. Gramedia Pustaka Utama, 1995.

Enganging the past of the city through the conservation of heritage building

Yulia Nurliani Lukito*, Amalia Nurul Rizky

Department of Architecture, Faculty of Engineering, Universitas Indonesia, Depok 16424, Indonesia

*Corresponding e-mail: yulianurliani@yahoo.com

Abstract. Built heritage is a physical representation of culture that provides a connection with the past and important for aesthetic and symbolic values for the city. The conservation of built heritage is a necessary not only to engage with the identity of but also to sustaining development in the city. However, heritage buildings are vulnerable to development and modernization. The paper examines a colonial building in Jakarta that has been converted into different functions through various levels of physical modifications. As a case study is Cut Meutia Mosque in Menteng, designed by a Dutch architect PAJ Moojen during the Dutch late colonial era. The building was initiated in 1912 as N.V. Bouwploeg, a Dutch architecture firm that developed the nearby residential area of New Gondangdia. The New Gondangdia, including Menteng as its central area, was developed according to modern garden city principles. During its lifetime, the building was used for different purposes such as a post office and a train company office. After Ali Sadikin's term as Governor of Jakarta, the building was converted into a mosque. The architecture of the building follows the Dutch Rationalist style but adapts to local climate such as a ventilation tower in the center of the building to regulate the temperature inside. Through historical and field research, this paper discusses the benefits and possible distortions of history manifest in the transformation of colonial buildings. Moreover, learning from the conservation of building heritage and urban area in the city may support the idea of livable memory of urban area and sustainable city.

Keywords: heritage building, colonial building, livable memory, building conservation

1. Introduction

During the Dutch colonization in the Dutch East Indies or present day Indonesia, the Dutch laid several foundations on settlement and infrastructure mostly with European architecture as their models. Since 1619, Batavia (the present day Jakarta) was the stronghold of the Dutch trading company (Vereenigde Oost-Indische Companie-VOC) that controlled all spice trade in Asia. To safeguard their commercial interests, the company gradually took surrounding territories. Dutch settlement in the coastal area was simply consisting of warehouse, some basic-accommodation but later on it developed into a big city with commercial and military buildings. Even after VOC was dissolved in 1800, the Dutch colonial government still took Batavia as their central administration

The city of Batavia consisted of two centers: the lower city in the north and a new district of Weltevreden in the south. There were new European suburbs that arose around the Koningsplein with spacious, monumental cityscapes emulating current trends in Europe. The essentially linear structure of Jakarta was well established with the old core and port was in the north and the new core and European residential area was in the south and kampongs filled in around it.

Architecture that was developed during the Dutch colonial era became relatively an expression of European products with little adaptation to local architecture. This kind of buildings were actually not suitable for the tropical climate and local conditions (Sidharta, 2002). There were some problems emerged from the built environment ranging from unhealthy houses to segregation issues. Approximately after 1870, when there was a rapid development of private enterprise in the Dutch Indies, there was an expansion of commercial activities and build environment. The Dutch started to build houses and building according to tropical climate with relatively high humidity and rainfalls (Ardiyanto, 2015). Since then, architecture in the Dutch East Indies was known as having their own characteristics such as having an extended roof to deal with rain water as well as high ceilings and big openings to provide cross ventilation.

Starting from the early twentieth century, the form of traditional architecture that was able to adapt to the tropical climate had inspired the architects who came from the Netherlands at this time to search for the East Indies architecture in Indonesia. The architectural works derived from the synthesis between elements of Western and Eastern cultures (Kusno, 2000; Widyarta, 2002). This idea was supported by PAJ Moojen (1907), a Dutch architect working in the East Indies who criticized the quality of architecture during that peeriod as a bad imitation of Hellenistic architecture but without spirit, nor an enthusiasm in design. Moojen, who in 1910 became the head of Art Society in the Indies, preferred modern architecture but with some adaptation to local conditions, including local construction method, traditional buildings and local materials (Sudrajat, 1991). Moreover, Moojen did not ignore modern development in architecture such as the use of concrete and modernist approach to function. The combination between modern and local elements could create good buildings suitable for tropical climate (Lukito, 2016).

Kevin Lynch in *the Image of the City* (1960) discusses the concept of public image and connects that idea to the image of the city, its elements and urban morphology. He mentions that the characteristics of a landmarks as a singularity, an easily identifiable form, and contrasts of scale, materials, and shapes that makes different figures as well as continuity with their surroundings. The image of the city can be revealed from a connection of the past, especially through conservation of buildings that give a certain identity to the city (Feiden, 1982). Although each individual creates and feels his own image, there seems to be considerable agreement among members of the same group on the image of the city. Through physical qualities, which are related to the attributes of identity and structure in the mental image, the image of urban environment may be analyzed into three components: identity, structure and meaning (Lynch, 1960). Identity means that an individual first must be able to identify a place into something specific and see the difference with the other buildings (individuality of a landmark). One can see the structure and the relationships or patterns of an object with another objects in surrounding areas. A landmark then has a meaning or significance, both functionally and emotionally.

The paper concerns with history and architecture as well as how conservation of heritage building affect the built environment. In conserving colonial buildings, historical reviews are the key to explaining the purpose of the building, the function of society at that time and the connection with the historical site. This paper is related to a building that has been converted into different functions and changes in the meaning of the building. Dodsworth (2009) describes the importance of both physical and non-physical characteristics of the site and its surrounding environment to the existing character, identity, and value of the building. Without referring to the importance of the building in the past, there will be distortion of meaning for the heritage and the loss of identity of the building.

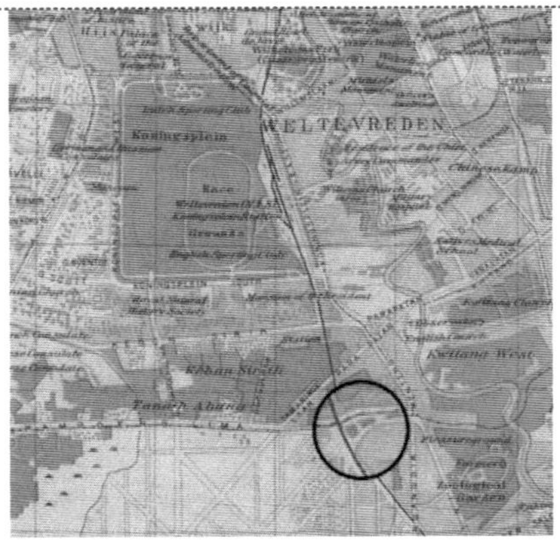

Figure 1. Maps of Batavia around 1910s and the location of Cut Meutia Mosque.

Menteng was the first modern settlement in Jakarta after a decentralized system in the early twentieth century was implemented in the Dutch East Indies. There were many traces of colonial buildings that had a symbolic value for the existence of the Dutch East Indies society in the past. Cut Meutia Mosque was one of those traces of colonial building heritage designed by a Dutch architect Pieter Adrianss Jacobus Moojen. Cut Meutia Mosque was located on Jalan Taman Cut Meutia No. 1, Kelurahan Gondangdia, District Menteng, Central Jakarta. This building functioned as a mosque in a residential area with the size of the site 1792 m^2 and the building 950 m^2. At first, this building was known as NV. De Bouwploeg, an architectural firm office built in 1912 in order to develop the nearby residential area of New Gondangdia and known as one of the gate to the New Garden City of Menteng. The strategic location of Bowploeg was not only as a gate to the garden city but also symbolically showed the direction of the development of Batavia. Before converted into a mosque, the building was used for different functions and in addition to this there were several parts of the building that had been changed. At the present time, the meaning of the building both as the gate and the landmark for the new garden city is faded.

One of the problems faced in the conservation of heritage building is the various levels of physical modification that will affects the values of the heritage building, as a physical representation of culture that provides a connection with the past and valuable for aesthetic and symbolic values for the city. Although the conservation of built heritage means not only to engage with past but also to sustain the development of the city, the heritage buildings are vulnerable to fast development and modernization. The aim of this paper is to analyze one of colonial buildings in Jakarta, Cut Meutia Mosque in Menteng, that has been converted into different functions through various levels of physical modifications. This paper discusses how conservation and transformation of colonial buildings can create some benefits as well as a possible distortion of the meaning of the building, especially as the landmark and the gate to the Menteng area. Hopefully, this paper is able to emphasize the importance of comprehensive conservation of heritage building, both physically and symbolically in order to enhance the image of the city and the idea of livable city.

2. Methods

In discussing the conservation of heritage building, we use both a literature review for heritage building in Batavia, some principles of conservation and the image of the city. The literature review helps to clarify the relationship between history and architecture, which relates to the value of buildings and the conservation. We also conducted a field research to get necessary information of the heritage building and the surrounding areas, including changes in the building and its functions, as well as its symbolical meaning as a landmarks. The method of analyzing architectural and historical data is a combination of quantitative and qualitative methods in order to examine changes, characters and symbolical meanings of the site and the building as well as the relation of the heritage building to the history and image of the city.

3. Discussion

Menteng Garden City was a new urban design developed in the 1910s located in the southern part of Merdeka Square (Koningsplein). This area became a new residential area for Dutch people. At first, the area was the first planned suburb in colonial Batavia made by Moojen, as a member of development group established the Batavia city government the *commisie van toesicht op het Geheer van het Land Menteng*. This committee was responsible for developing the larger area of Gondangdia with Menteng as its center (Silver, 2007). Moojen's plan resembled the garden city model of the English reformer Ebenezer Howard, including wide boulevards and concentric ring of streets and some squares. The plan of Menteng residential area was designed by Moojen and then changed by F.J. Kubatz during the course of the 1910s. Supported by easy access to central business district and its strategic location, this area has become one of the most expensive areas for residential real estate in Jakarta.

Figure 2. Maps of Menteng Garden City by Moojen (red) and Kubatz (yellow).

Good plan and conservation of buildings, green areas and well integrated streets make Menteng a feasible area for accommodating the collective memory of the past for the present time. Some buildings in Menteng become landmarks and have special functions for this area, such as the official residence for the Jakarta governor and ambassadors. One important part of this area is the north entrance to the area that for some decades becomes a landmark to the new development of Batavia and a connector to important buildings around *Koningsplein* or Lapangan Merdeka. Not only buildings near the north entrance play an important role as a landmark but also helps people to orient themselves to the big area and give a special impression of the expansion of Batavia into a modern city.

The main planning and construction activities of residential areas in New Gondangdia were handled directly by the architect firm *Naamloze Vennootschap de Bouwploeg,* founded by Moojen in 1912. Therefore, the name of office building for this firm was also *Bouwploeg,* which also meant a construction team. *Bouwploeg* was known not only as the design and planning office but also as the entrance to Menteng area alongside *Kunstkring* building. Moojen also designed and *Kunstkring Art Gallery* in 1914 as a landmark that welcomed visitors to the new Menteng Residential Area mainly with New Indies Style houses that was known as a modern architectural movement in the Dutch Indies. This new style combined the one-storey Indish-style villa with modern two-storey structures (Tjahjono, 1998). Kunstkring was intended to become the cultural center of Batavia during the early twentieth century. Moojen was the head of the Bataviasche Kunstkring in 1910 and he was interested in the visualc and decorative arts. In short, Moojen built two buildings that were located across each other, Bouwploeg and Kunstkring Art Gallery, as the entrance to New Gondangdia.

Moojen used Bouwploeg as the planning office for Menteng area for approximately thirteen years, and then the function of the building was changed into an office for irrigation called *Provinciale Waterstaat* for seventeen years. Bouwploeg became one of celebrated buildings in Batavia for three decades before the Dutch ended its colonial period in the East Indies. After the independence of Indonesia, the building was used as a warehouse. The building was used for different functions such as post office and train company office. From 1959 until 1960 the building was used as a mayor office of Central Jakarta and later became a drinking water department and then Department of Residential of Jakarta, and People Assembly. From 1964-1970, the building was used as the Office of Home and Religion (1964–1970). This building becomes a mosque for almost fifty years, starting from August 1987 during Ali Sadikin's term as Governor of Jakarta.

Figure 3. NV. De Bouwploeg the 1920s.

Figure 4. Cut Meutia Mosque in 2016.

3.1. Physical conditions of Cut Meutia Mosque

Cut Meutia Mosque is located on a triangular-shaped site and the building faces the north where Koningsplein is located. Cut Meutia mosque is symmetrical on each side and looks massive with a rectangular top added with a cloister vault or a segmental dome. The building has a cross plan; the middle part has higher walls up to 22 meters that functions as a ventilation tower. At the time, some Dutch architects struggled to find a way to adapt buildings to tropical climate and some used both Western and local architectural forms. The four sides of the building have gable roofs, a reminiscent of Dutch architectural form. The ventilation tower is a way to deal with local climate so that the air circulation in the building will be good. The shape of the building depicted mainly Dutch or Western architecture but it had been adapted to Indies contexts through the ventilation tower. The use of gable roof, brick and cement, and the use of columns and dome strengthened the characteristic of Dutch architecture.

The plan of Cut Meutia Mosque resembled the church plan especially with the cross plan and the dome. This building has a high roof with some large windows so that fresh air can enter. The building is considered to be a pioneer of the New Indies-style architecture that is more appropriate for tropical climates. Originally, there is a kind of grand staircase in the building. However, the lower part of the staircase is cut in order to give more space for prayer so that now there is a discontinuation of the big staircase from the second floor. At the entrance there is a *dormer* that serves as a divider between inside and outside, as well as a determinant of the building facade.

Figure 5. Existing window frame.

Figure 6. The new window frame.

There are some parts of the building changed due changes of the building function. There are some canopies added around the building. Some openings are added and the result is the loss of building pattern and material. Although the building still maintains its architectural forms, changes on the façade and the staircase cause the building to lost its originality and character. Depicted in Figure 5 are old and new window frames. The new window frame do not follow the old shape but the common window for the mosque in Jakarta. Originally, all openings in the building have the same pattern in all sides of the building. The number of openings in the building confirm that the architect tried to adapt to local climate through facilitating cross ventilation in the building. The architect also did not put too many ornaments since he preferred modern architectural principles. The building structure is a simple system of columns and beams that are shown explicitly with some geometric ornaments on the column and beams. Having changed into a mosque, the building now is decorated with calligraphy on the wall. Some ornaments remain the same such as the ventilation holes, the railing and on the interior walls.

Figure 7. Existing façade of De Bouwploeg, covering with bamboo curtain.

Figure 8. Green cover made from fiber add on façade of Cut Meutia Mosque

3.2. Symbolical Meaning of Cut Meutia Mosque

In the master plan of New Gondangdia, Moojen intended to make the building as a gate of Menteng and he considerately placed the building following the importance of the site; Bowploeg was one of the first buildings people would see when they came from the northern area. The symmetrical façade of the building was a way to welcome people from four directions, so that the building possessed a quality of a landmark. Across the building was another modern building, Kunstkring, aimed to house art pieces and exhibition that in itself bore the idea of a landmark. Both Bouwploeg and Kunstkring stood for many years to mark the area and reminded people the idea of modern part of the city.

Cut Meutia Mosque stands on a triangular-shaped land and be the one with the garden in front of the building (Figure 8). The centrality of the mosque and Kunstkring – located between two major streets

– actually can attract people. The absence of adequate sidewalks along the main streets also weakens the boundaries of the landmarks.

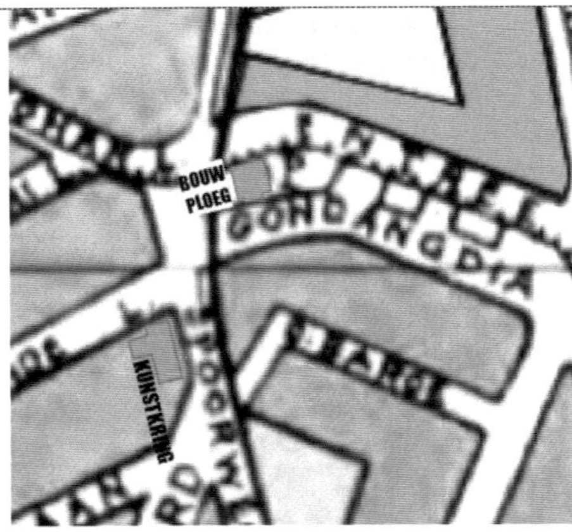

Figure 9. Bowploeg, the gate to Menteng Garden City.

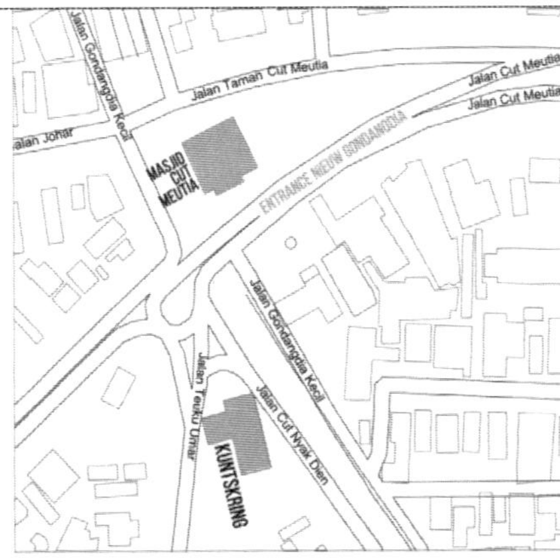

Figure 10. Cut Meutia Mosque cannot be identified anymore as a gate.

Although both buildings are still in good conditions, the idea of the buildings as the landmark and the gate to Menteng area is faded with the fast development of Menteng area. Unfortunately, there is only limited effort in conversation to relate the buildings with its site and to conserve the buildings as a landmark to Menteng area. In addition, people who enter Menteng area may be overwhelmed by crowded buildings and vehicles and cannot experience the gate to Menteng area. Since from the beginning there is only limited effort to conserve the building in relation to its symbolic meaning of the gate, there is already loss of the importance of both Bowploeg and Kunstkring as the gate to modern residential areas.

Figure 11. Surrounding Cut Meutia mosque.

A landmark creates a particular identity of the city as well as a livable memory and specific impression of the area. As Lynch mentions, people have a connection with buildings and the surrounding area that create the image of the city. Cut Meutia Mosque is a landmark – an easily identifiable form – and even a contrast to other residential buildings in Menteng. However, the continuity of the building with its surrounding cannot be seen anymore due the fast development of the surrounding areas. Bowploeg can actually give a connection to the past – to the glory of New Gondangdia. The conservation of Bowploeg can enhance the identity of the city part but unfortunately the idea of the gate to modern residential area is somehow lost. Nowadays, it is hard to understand Cut Meutia Mosque as a gate to Menteng area. The singularity of the building and the unique building shape used to be a strong landmark but now Bowploeg had lost its role as the gate to Menteng area, although the building has been well conserved.

4. Conclusion

The conservation of built heritage is necessary in engaging the physical and symbolical identity to sustain development in the city. In conserving heritage buildings, what important is not only how the buildings are able to stand for future generation but also how to connect with the past. There are physical and symbolical values of buildings and a conservation of heritage buildings should be able to include those values. In the case of Bowploeg or Cut Meutia Mosque, the building had gone through changes in functions and conservation. However, even though the building is still in a good condition, the some symbolical values may be lost, especially in relation to the importance of the building as the gate to new Menteng area during the Dutch colonial period. Through historical and field research, this paper demonstrates the benefits and possible distortions of history manifest in the conservation of colonial buildings as well as a possibility of conservation of building heritage to support the idea of livable memory of urban area and sustainable city.

5. Acknowledgments

The research for this paper is conducted partially with the support of PITTA grant from Universitas Indonesia.

6. References

[1] Boyer, M. Christine. (1994). The City of Collective Memory. Cambridge Massachusetts : the MIT Press.
[2] Feiden, Bernard M. (1982). Conservation of historic buildings. London: Butterworth Scientific.
[3] Heuken, Adolf. (2001). Menteng “Kota Taman” Pertama di Indonesia. Jakarta: Yayasan Cipta Loka Caraka.
[4] Heuken, Adolf. (2007). Historical sites of Jakarta. Jakarta: Cipta Loka Caraka.
[5] Lukito, Yulia. (2016). Exhibiting Modernity and Indonesian Vernacular Architecture: Hybrid Architecture at Pasar Gambir of Batavia, the 1931 Paris International Colonial Exhibition and Taman Mini Indonesia Indah, Wiesbaden: Springer
[6] Lynch, Kevin. (1960). The Image of the City. The MIT Press.
[7] Mrazek, Rudolf. (2002). Engineers Of Happyland. New Jersey : Princeton Paperbacks.
[8] Motamedmanesh, M. (2016). Authenticity and restoration: The benefits of historical studies on re-examining the implemented restorations in persepolis. *Arts, 5*(1), 2. Retrieved from https://search.proquest.com/docview/1771275653?accountid=17242
[9] Nas, Peter J.M. (2012). The Urban Anthropologist as Flaneur, The Symbolic Pattern of Indonesian Cities. Wacana Vol. 14 No. 2 hal: 429-454. Faculty of humanities, University of Indonesia.
[10] Nas, Peter J.M. (2009). Masa Lalu dalam Masa Kini Arsitektur di Indonesia. Jakarta: Erlangga.
[11] Pusat Dokumentasi Arsitektur. (2012). Tegang Bentang, seratus tahun perspektif arsitektural di Indonesia. Jakarta: Gramedia.
[12] Silver, Christopher. (2008). Planning the Megacity: Jakarta in the Twentieth Century, Routledge, London and New York
[13] Sudradjat, Iwan. (1991). A Study of Indonesian Architectural History,Unpublished Dissertation, University of Sydney,
[14] Tjahjono, Gunawan. (1998). *Architecture*. Indonesian Heritage. Singapore: Archipelago Press.

The conservation of Javanese-cultured city through visual expression study on the architecture of Keraton Yogyakarta

Ibrahim Tohar[1*], Gagoek Hardiman[2], Suzanna Ratih Sari[2]
[1]Student of Doctoral Program of Architecture and Urbanism Diponegoro University, Indonesia.
[2]Department of Architecture Diponegoro University, Indonesia.

*Corresponding e-mail: ibtohar@gmail.com

Abstract
Keraton Yogyakarta as a summit of Javanese culture has been renowned as a heritage building. As object of study, Keraton Yogyakarta is ornamented with a collection of architectural artifacts. The acculturation and merging of these different styles create a unique impression within the palace complex. This study aims to identify the pattern of acculturation of these two styles and to interpret their meaning and expression. A descriptive-qualitative method is employed in this research, which contains visual observation, documentation collection, interviews with informants, and relevant literature review. As results of study, the expression of Tratag Pagelaran, Tratag Sitihinggil, Bangsal Ponconiti, and Gedong Jene tends to widen, while the expression of Gedong Purwaretna tends to uprise. Every building has its own point of interest and ornamentation which its place and content are different.. In visual observations, there are two categories of buildings in Keraton Yogyakarta,which accommodate two styles, namely Javanese Traditional style and Dutch Colonial style. Buildings of Javanese traditional style, which hold a special concept of shading, were built without buttresses and embody a 'light' expression, while buildings of Dutch Colonial style, which hold a concept of protection, were built with massive enclosure and produce a "heavy" expression. Although visually split into two distinct styles, the acculturation process in Keraton Yogyakarta produced a unity in its overall expression. The expression pattern of Keraton Yogyakarta can be used as conservation guidance of Javanese-cultured city.

Keywords: visual expression, architecture, Keraton Yogyakarta.

1. Introduction
Architecture is such a very comprehensive science that there are many other sciences directly or indirectly intertwined with it. The other sciences can be in the form of related aspects in architecture, such as aesthetic, logic, ethics, social, cultural and so forth. Every aspect involved in the architectural design will certainly affect the design result. These many considered aspects have also been applied by designers in their design process. The more aspects are accommodated, the greater the potential for the emergence of complexity. Complexity in architecture might be in the form of physical complexity and meaning complexity. Physical complexity signifies that there is complexity in *architectural form*, such as the facade, space organization, mass configuration or ornamentation. The complexity will '**work**' on the physical level, which also would have caused the meaning complexity. A discussion about the relation between form and meaning has also been questioned by Psarra [1] and Ujang [2], how is the actual construction of meaning in a physical form, and how formscommunicate their meaning to viewers. According to him, architecture is not merely an expression of meaning, but also a construction of meaning by means of space composing associated with the social and cultural context.

The complexity of meaning can also occur in a simple architectural form. In this regard, according to Broadbent [3], meaning in architectural objects can be traced through the architectural form. Bonta [4] states that the meaning of an object is affected by significant feature which is an abstraction of the physical form. Meanwhile, according to Venturi [5], complexity arises because there is a simultaneous combination of various elements in an architectural design. More specifically, Venturi [5] argues, the complexity of meaning in an architectural object can be traced through visual observation of the object. An object can be said to contain complexity if in the visual observation of the object emerged a variety of visual perception simultaneously. Visually, on each composition of shape and space, there will be an expression of emotion from the designer. The term "expression" is also often used in fine art, thus there is a certain style called expressionist. Expression, a revelation of one's feelings, can be observed in a musical composition. A music composer, will pour his feelings in the music he made. Thus the result of his music composition can be said to be closely related to the atmosphere and the feeling when the music is composed. According to Scruton [8]: *"Expression is like a display of atmosphere, an abstract presentation of character. The distinction is not sharp, but I think it is real, and can be understood by comparing architecture with music".* Critically, Scruton [8] revealed a slight difference between the expression occuring in an artwork and the character of work designer. From this slight difference, it can be concluded that an expression of an artwork is an abstract of a designer character.

According to Scruton [8], the expression expressed in an art work is a representation of the artist or creator of the work. In this case, the expression of the artist is directly represented in the works so the expression arising from an art work is an expression of the artist or the abstract of the artist's character that is personal. Expression is part of the manifestation of one's inner life, associated with feelings as the basis, which later are being disclosed. The expression of feelings is an expression that could lead to different interpretations from one person to another. Expression contained in an object possesses a subjective nature. In one same object, an expression that is occured and captured by someone might be different with the other. It is due to the different feeling of each person. An object also allows the occurence of an uncertain expression that is called by Scruton [8] as ***ambiguous expression.*** *"A building might have an ambiguous character. At one moment it seems threatening and claustrophobic, at the next moment quiet and solemn ".*

The ambiguity of expression contained in an object can be caused by the dimension of time. Definition of time can be associated with atmosphere, such as atmosphere during the day would be different from atmosphere in the afternoon. Thus the expression of an object during the day would also be different from the expression in the afternoon. Time dimension can also be related to time period. Therefore, in a single object, the expression in the past would also be different from the expression in the present. To reduce the ambiguity caused by the expression of time dimension, it is necessary to set a determination or limitation of time when observing study objects. On top of that, the time limit should be specific enough, for example, observation is performed during the day when the air is clear. For architectural works, an expression that occurs in an architectural work is a representation of the object or the work itself. The expression is an abstract of the object's character that is impersonal. Related to expression, Weber [9] in his book entitled *On The Aesthetics of Architecture* mentions two theories that discuss expression, they are ***Theory of Empathy*** and ***Gestalt Theory of Expression***.

Theory of Empathy views expression from the side of the expressing subject. This theory is strongly influenced by feelings and past experiences of a person. One can tell that someone in front of him is sad because he also ever felt sadness and once ever saw others in sad state. In this case, there is an empathy from a subject to an object. While empathy of each person is very different so that according to the theory, the ***expression of an*** *object will be very subjective.*

While ***Gestalt Theory of Expression*** reveals that each object always contains various properties. All of the existing properties will form a particular expression. So according to this theory, the expression of an object is determined by properties contained by the object. The examples are *hue* which is the property of color and *size* which is the property of shape, therefore an expression that emerges from an object, among others, is determined by the hue and size of the object.

In the following study, Theory of Empathy will be related to the subject or author. Thus there is still a subjectivity that is influenced by the feelings and experience of the observer on the expression of the object of study. It is also in line with what is expressed by Scruton [8] that expression occured in a form owns its subjective nature so that it is necessary to utilize an instrument to assess. This study utilizes architectural instruments proposed by Rasmussen [10].

When observed visually, the architecture of Keraton Yogyakarta possesses a fusion of various design elements that can be observed through the building facades, building elements, outer layouts, ornamentation, and furnishing accessories. Each building in the palace owns a different form and atmosphere. Associated with the space order system, according to Rapoport [6], there are three important elements. The first is fixed features and the features included in this element are floors, walls, and ceilings. The second is semi-fixed features, such as furniture and space furnishings that can be moved. The third is non-fixed features, namely the space users performing activities in it. In each building of Keraton Yogyakarta, each of these elements will represent a partial meaning. Yet, when the three elements are present in a **setting**, it will convey new meaning which is a fusion of all the partial meanings.

There is a possibility that the fusion of the two styles is indeed a result of some of the concepts that had been deliberately combined, so that any element or part of the palace building has its own concept which was purposefully presented at the time and the place. Another possibility is that the fusion of each element and building is an expression of openness to cultural influences developing at that time, such as the culture of Hinduism, Buddhism, Islam, China and Europe. One of visually emerged influences is the fusion of Traditional Javanese style and Dutch Colonial style. According to Hadinoto, the Dutch Colonial style was mentioned as Indiche Empire Style [7]. The architecture of Keraton Yogyakarta certainly possesses its own aesthetic quality from the result of the fusion of some styles. The fusion of **Traditional Javanese** architectural style with **Dutch Colonial** architectural style will embody complex expressions. That expression complexity will convey the uniqueness of Keraton Yogyakarta architecture. To understand the complexity of expression in the architecture of the palace, it is substantial to carry out an in-depth study. There are two research questions arising in relation with the issue above; *Keraton* Yogyakarta is a collection of artifacts representing a fusion of Traditional Javanese style and Dutch Colonial style. How is the fusion pattern between the two styles in the architectural design of Keraton Yogyakarta?; How is the expression as a whole when both styles are applied in the architectural design of Keraton Yogyakarta?

2. Methods

This research utilizes qualitative descriptive method. This is in accordance with the purpose of descriptive research which is to make a systematic, factual, and accurate depiction about the facts of a particular population [11], in this case, a systematic and factual depiction about the architectural expression of the Keraton Yogyakarta. Meanwhile, according to Groat, qualitative method has an accuration to interpret the meaning of artifacts [12].

The depiction of expression is performed through visual observation that will be described. This research begins with visual observation as a whole of Keraton Yogyakarta complex. This observation focuses on the main buildings composed from North to South at Keraton Yogyakarta and the buildings that contain a fusion of various architectural styles. The identified main buildings from north to south of Keraton Yogyakarta are *Tratag Pagelaran, Tratag Sitihinggil, Bangsal Ponconiti, Bangsal Srimanganti, Bangsal Trajumas, Bangsal Kencono, Bangsal Kemagangan, Bangsal Kemandungan,* and *Sasono Hinggil Dwi Abad.* While the buildings with fusion of various architectural styles are *Gedong Jene, Gedong Puwaretna, Bangsal Manis, Bangsal Mandalasana* and *Regol Danapratapa*. However, the object of study will be limited only to the main buildings containing the dominant architectural fusion, namely *Tratag Pagelaran, Tratag Sitihinggil, Bangsal Ponconiti, Gedong Jene,* and *Gedong Purwaretna.*

Architectural instruments [10], namely: *proportion, rhythm in architecture, surface character,* and *colour in architecture* are applied to assess the five buildings. This study also conducts interviews with informants to understand the history and philosophy of Keraton Yogyakarta.

Some limitations occured in the observation of the object are: observation is only performed visually, observation only focuses on object façade, observation is performed during the day, and the observer does not change the object settings.

3. Discussion

3.1.Exploration of Object

The term Karaton, Keraton or Kraton was originated from the word ka-ratu-an that means Sultan Place [13]. Keraton Yogyakarta was established by Prince Mangkubumi or Sri Sultan Hamengku Buwono I (1755-1792) on Thursday Pahing, 7 October 1756 or 13 Suro Jimakir 1682 [14]. It was celebrated with ***condrosengkolo memet*** on Regol Kemagangan and Regol Gadungmlati. At Regol Kemagangan there are ornaments of 2 (two) red dragons with the expression of readyness to defend themselves against enemy's attacks. While on Regol Gadungmlati there are also ornaments of two tangled-tail dragons with symmetrical formations in hibiscus green-leaf colour [10].

Keraton Yogyakarta owns an area of 14,000 square meters. Inside the palace, there are many buildings and courtyards. These buildings are bangsal, tratag, gedong, and regol. The regol connects one courtyard to another (see Fig.1). Some of buildings in Keraton Yogyakarta were significantly renovated by Sri Sultan Hamengku Buwono VIII (1921-1939), namely tratag pagelaran, tratag sitihinggil and Regol Danapratapa [15] [16].

When observed visually, Keraton Yogyakarta's architecture owns a fusion of various design elements. It can be identified from the building facades, building elements, outer layouts, ornamentations and furnishing accessories. Each of the buildings in the palace also owns a different shape and atmosphere.

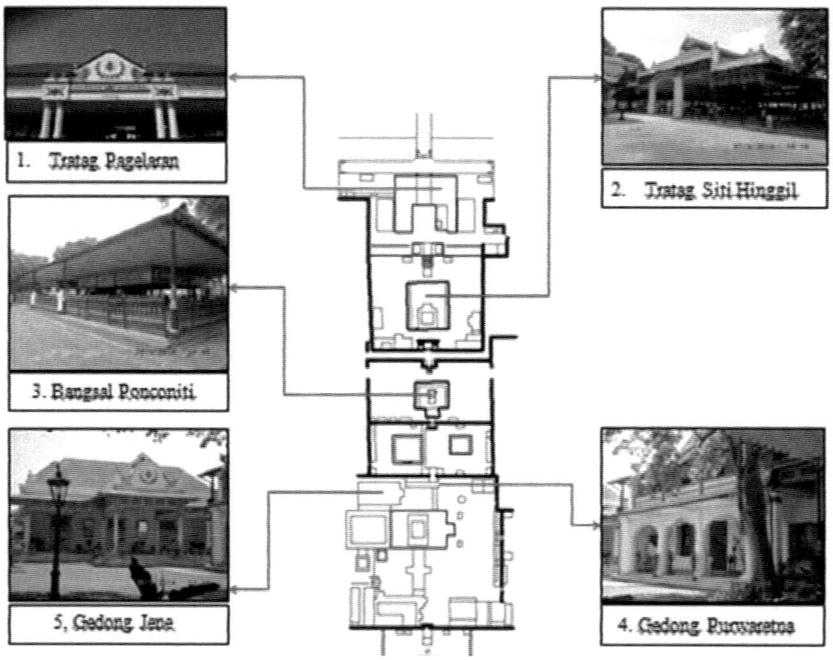

Figure.1.The Buildings as object of study at Keraton Yogyakarta

3.2. Tratag Pagelaran

The building located in the most northern area of Keraton Yogyakarta complex is Tratag Pagelaran. Tratag Pagelaran was renovated in 1934 , during the reign of Sultan Hamengku Buwana VIII [17]. The roof is of pyramid-shaped with roofing made of brown iron sheeting. There is a listplank around the roof made of white-painted bricks as high as 50 cm. The listplank is filled with alternately flower-shaped ornaments with 'column pieces' and vertical European-styled sidelines (see Fig.2.b,c,d). On the whole, the proportion tends to widen (see Fig.2.a).

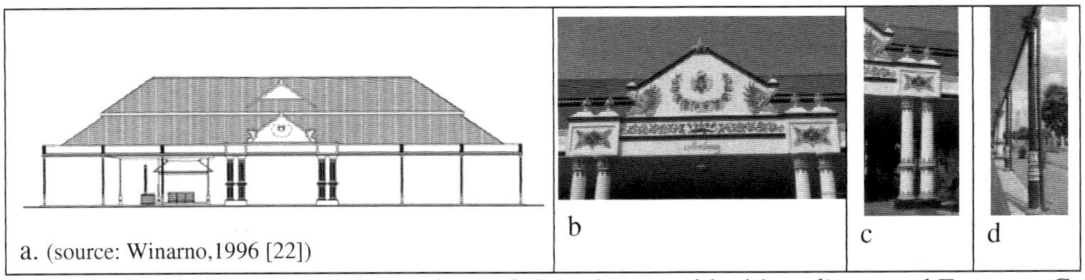

Figure 2.a,b.c, d. *Tratag Pagelaran*, the roof shape is pyramid with *pediment* and European Column

Tratag Pagelaran owns a flat ceiling, made of green-painted iron sheeting with lined patterns, with the absence of ornament and intercropping. The ceiling' height from the floor is approximately 5 meters.

In details, the large column is divided into three parts; the legs, the body and the head. On the leg part, the lower body and upper body are ornated with *tlacapan*-shaped in dark green. Tlacapan is a decoration of a row of isosceles triangle in which filled with ***lung-lungan*** ornaments, leaves or flowers which are distyled [18].

The Tlacapan upper part of column is facing the tlacapan lower part, which is connected with vertical sidelines as wide as 2 cm, while the head of the column owns a flower-shaped ornament in dark green, light green, red, and white. The form of ornaments on the column head is similar to Corinthian Capital in Greece. Doric, Ionic and Corinthian were originated from Greece Architecture style [19]. The small columns are made of iron with diameter of 30 cm painted in dark green. This column also owns vertical sidelines. The upper part and lower part of the column own flower-shaped ornaments in red and white.

At the front part of Pagelaran lies a white-painted gate ornated with flower-shaped ornaments, leaves, *kebenan*, *kemamang*, dragon, symbol of the palace, and Javanese writing letters of Pagelaran. Pagelaran gate consists of pediments supported by two pairs of large columns. The front pediment shape is isosceles triangle with leaf-shaped ornament on its peak. Both of isosceles triangles are notched by lizards' gaping mouth (see Fig.2.b). The pediment provides a visual emphasis on the object.

3.3. Tratag Sitihinggil

Tratag Sitihinggil was renovated in 1926 and functioned to receive Sultan's guests at grand ceremony. The roof is of pyramid-shaped with elevation on its ridge (*see Fig.3*). The roof cover is arranged with shingle-roof composition so as to produce a pattern of diagonal squares.

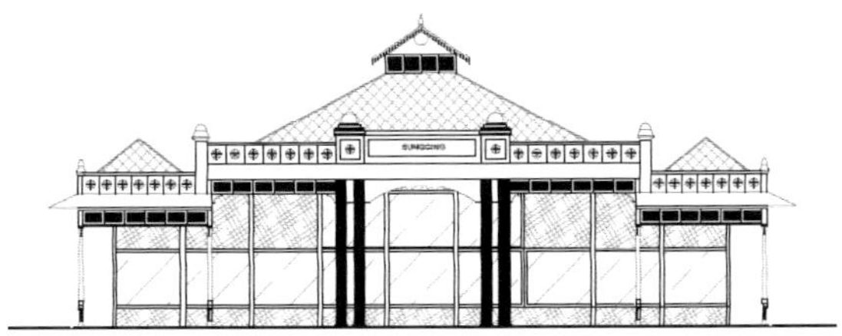

Figure 3: The roof of Tratag Sitihinggil is pyramid-shaped with elevation on its ridge.
(source: Winarno,1996 [22])

As in *Pagelaran* roof, the *Sitihinggil* roof is also surrounded by *listplank* of white-painted bricks with lotus flower-shaped ornament. At the front there are two pairs of columns with rectangle cutting. These columns possess white-painted European-styled vertical sidelines. Both pairs of columns are connected by beams with curved lower part. The beam is ornated with dragon-shaped ornaments in golden yellow and leaves in green, and tend to be ***polichromatic***, while at the interior, the form of column head is similar to Corinthian Capital in Greece with Javanese ornament *(see Fig.4.a,b,c)*

Figure 4.a,b,c: Columns & *listplank* at *Tratag Sitihinggil,* Dutch Colonial style with Javanese ornament.

This building owns a flat ceiling with a checkered pattern. The supporting columns are made of iron with vertical sidelines yielding a light impression. Because it is a 'continuation' of Pagelaran, the shape and color of the ornaments are equal to that of Pagelaran. The entire surface of the floor is flat. There is neither elevation on the floor nor"stand out" part in this section. The striking point lies in the middle with the presence of dominant *pediment.*

3.4. Bangsal Ponconiti

Bangsal Ponconiti is located in the North *Kemandungan* courtyard. It was formerly functioned as the Palace's ancestral court and the place for Gamelan relics of ***Kanjeng Kyai Gunturmadu*** and ***Nagawilaga*** in Sekaten celebration before being taken to mosque ***Kagungan Dalem*** in Kauman.
The roof shape is similar to ***Bangsal Witono***, that is ***Tajug Lambang Gantung,*** while the roof cover is shingle. This Bangsal consists of sixteen columns, four of them are as the pillars. The pillars are made of rectangular timber with **Praba** ornament in golden yellow color, while the twelve responding pillars are made of steel, with basic color of green and ornated with lotus flowers and red white-colored cabbage, yielding the soft impression (see Fig.5.a,b,c,d). The ceiling depicts a beam to 'illuminate' the Sultan when hearing a case.

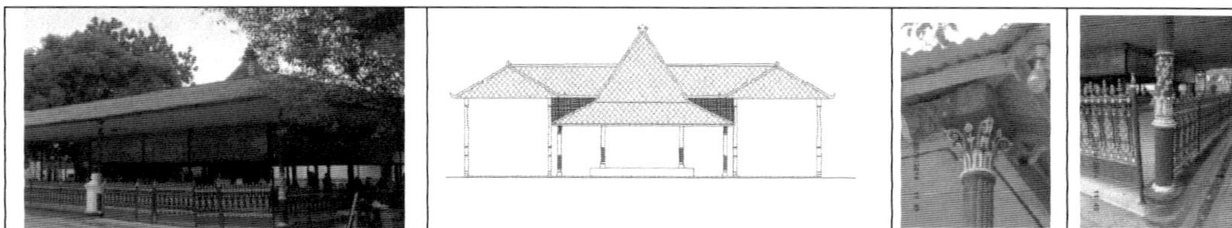

a b *c* *d*

Figure 5.a,b,c.d Columns at *Bangsal Ponconiti* with Corinthian style and Javanese ornament

Top roof of Bangsal Ponconiti has an ornament called *mustaka*. The Bangsal's interior employs two ***dada peksi*** intersecting one another. Each owns a red-based color with ornament shapes of ***saton***, ***lung-lungan***, and ***tlacapan*** in golden yellow [18].

3.5. Gedong Purwaretna

One of two-story buildings in Keraton Yogyakarta complex is Gedong Purwaretna. Gedong Purwaretna owns a pyramid-shaped roof and facing toward the South. The body part is two-story, both the first and the second floor are covered with wall with several windows and entrances. The floor of the building is rectangular with a size of 10 x 22.5 meters. The two-story building height with a pointed pyramid roof causes the building to have an ***uprise*** proportion.

At the facade on the ground floor are four cylindrical columns with European-styled ornaments, while the top including the roof shape owns a traditional Javanese style (*see Fig.6.a,b*). The number of columns on the top also consists of four pieces with the distance among columns is always the same, yet the size and style are different. It displays the acculturation of the Dutch Colonial style and the traditional Javanese style. Overall, there are not many colour used in Gedong Purwaretna. Thus, the impression arised from the building is elegant.

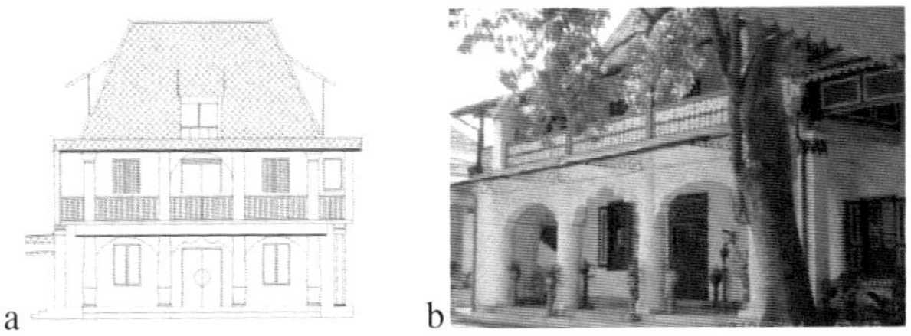

Figure 6.a,b. *Gedong Purwaretna*, *limasan* roof with *dormer* on the top floor and Dutch Colonial style on the ground floor (source: Winarno,1996 [22])

3.6. Gedong Jene

In addition to *Gedong Purwaretna*, another Dutch Colonial style building is *Gedong Jene*. It was` built at the reign of Sultan Hamengku Buwono II (1792-1810) and functioned as residence until the reign of Sultan Hamengku Buwono IX [20]. Nowdays, Gedong Jene functions as an office. *Jene* is Javanese language that means yellow. Therefore, related with its name, the doors and windows at *Gedong Jene* are coloured in yellow. Yet all its walls are painted in white and it brings the impression of monochromatic and elegant. The walls of the building were made from concrete bricks. The building owns a pyramid-shaped roof with a porch on the front (*see Fig.7.a,b*). In this porch, we can notice a kind of canopy with European *pediment* adorned by traditional Javanese ornament.

The building stretches from West to East and facing East. The body of the building is covered with white wall with several windows and entrances at the front and south side. The overall proportion tends to widen.

Figure 7.a,b,c.*Gedong Jene*, the shape of a massive building, Dutch Colonial style with Javanese ornaments, the *pediment* as point of interest at *GedongJene* (source: Winarno,1996 [22])

The floor of the building is rectangular with a size of 32.5 x 20 meters [21]. The front part of porch is open and there are six columns as the support of porch roof. The distance among the columns is always the same. These columns own various forms of traditional Javanese-styled ornaments like ***tlacapan, Praba***, calligraphy, ***putri mirong***, and ***wajikan*** [18], while the column attached to the wall and its *listplank* is Dutch Colonial style. On the whole, the mixture of ornaments yields the impression of soft decoration (*see Fig.7c*).

Table 1.The Visual Expression of Keraton Yogyakarta

object	*TRATAG PAGELARAN*	*TRATAG SITIHINGGIL*	*BANGSAL PONCONITI*	*GEDONG PURWARETNA*	*GEDONG JENE*
assessment					
PROPOR TION	widen	widen	widen	uprise	widen

RHYTM IN ARCHITEC TURE	strong emphasis in the middle	very strong emphasis in the middle	emphasis in the middle	emphasis in the middle	emphasis in the middle
SURFACE CHARAC TER	soft decorative at *pediment*	soft decorative at *pediment*	soft decorative at fence	soft decorative on the whole	soft decorative at *pediment*
COLOUR IN ARCHITEC TURE	polychromatic	polychromatic	polychro matic	monochro matic	monochro matic

Source: author analysis

4. Conclusion

As result of this study we would like to present that the proportion of *Tratag Pagelaran, Tratag Sitihinggil, Bangsal Ponconiti,* and *Gedong Jene* tends to widen, while *Gedong Purwaretna* tends to uprise. Every building has its own point of interest and ornamentation which its place and content are different. The ornaments at *Tratag Pagelaran, Tratag Sitihinggil, Bangsal Ponconiti* tend to be polychromatic, while the ornaments of *Gedong Purwaretna* and *Gedong Jene* tend to be monochromatic.

Visually, the existing styles of Keraton Yogyakarta architecture are identified in elements of *fixed* and *semi-fixed* features. From visual observations, in general, there are two categories of buildings in the Palace, which accommodate two styles, namely Traditional Javanese style and Dutch Colonial style.

At *Tratag Pagelaran*, *Tratag Sitihinggil*, and *Bangsal Ponconiti*, elements of European style are present to adorn the Javanese-styled architectural buildings that own a concept as shade, without buttresses, and 'light'.

On the other hand, at *Gedong Purwaretna* and *Gedong Jene*, buildings with Dutch Colonial style are very dominant, owning a protection function, massive enclosure, and 'heavy' expression adorned with Javanese ornamentation. Although visually the two styles possess different expressions, both styles convey the expression of unity in Keraton Yogyakarta. Various visual expressions and fusion patterns of architectural style at Keraton Yogyakarta can be utilized to exhibit local character in the conservation of Javanese-cultured city.

5. Acknowledgments
The authors would like to thank to BPPDN 2014 as the funder of the research, Dr. Sindung Haryanto, Dr. Ir. R.A Retno Hastijanti, MT., KRT Pujodiniingrat, and KRT Kusumo Negoro as the interviewees.

6. References

[1] S. Psarra, Architecture and Narrative The Formation of Space and Cultural Meaning, New York: Routledge, 2009.

[2] N Ujang, Place Meaning and Significance of the Traditional Shopping District in the City Centre of Kuala Lumpur Malaysia, International Journal of Architectural Research, Archnet-IJAR, 2014. Volume 8 – Issue 1 – March 2014 – (66-77),

[3] G. Broadbent, R. Bunt and C. Jencks, Signs, Symbols, and Architecture, New York: John Wiley, 1980.

[4] J. Bonta, Notes for Theory of Meaning in Design. Signs, Symbols and Architecture, G. Broadbent, R. Bunt and C. Jencks, Eds., New York: John Wiley & Sons Ltd., 1980.

[5] R. Venturi, Complexity and Contradiction in Architecture, New York: The Museum of Modern

Art, 1966.
[6] A. Rapoport, System of Activities and System of Setting. In Domestic Architecture and The Use of Space : An Interdisciplinary Cross-Cultural Study, S. Kent, Ed., Cambridge: Cambridge University Press, 1990.
[7] Hadinoto, Perkembangan Kota dan Arsitektur Kolonial Belanda di Surabaya (1870-1940), Yogyakarta: Penerbit Andi, 1996.
[8] R. Scruton, The Aesthetics of Architecture, Princeton University Press: New Jersey, 1979.
[9] R. Weber, On The Aesthetics of Architecture, A Psycological Approach to The Structures and The Order of percieve Architectural Space, England: Avebury Ashgate Publishing Ltd, 1995.
[10] S. E. Rasmussen, Experiencing Architecture, Cambridge: The MIT Press, 1959.
[11] S. Suryabrata, Metodologi Penelitian, Jakarta: CV Rajawali, 1997.
[12] L. Groat, D. Wang, Architectural Research Methods, New Jersey: John Wiley and Sons Ltd., 2013.
[13] F. Heryanto, Mengenal Karaton Ngayogyakarta Hadiningrat, Yogyakarta: Warna Grafika, 2007.
[14] K. P. H. Brongtodiningrat, Arti Kraton Yogyakarta, Yogyakarta: Museum Kraton Yogyakarta, 1978.
[15] J. Prijotomo, Kraton Yogyakarta: Playing The 'Architectural Game' of the colonizer dalam Dari Lamin dan Bilik Pengakuan Dosa, Surabaya: Wastu Lanas Grafika, 2004.
[16] K. J. Bangunjiwa, Keris Gagrak Kasultanan Yogyakarta, Yogyakarta: Percetakan Pohon Cahaya, 2015.
[17] R. Rintoisworo, Karaton Ngayogyakarta Hadiningrat Sebagai Pusat Budaya, Yogyakarta, 1995.
[18] S. Dakung, Arsitektur Tradisional Daerah Istimewa Yogyakarta Proyek Inventarisasi dan Dokumentasi Kebudayaan Daerah, Jakarta: Departemen Pendidikan dan Kebudayaan, 1981.
[19] C.M. Harris, Dictionary of Architecture and Construction, United States of America: McGraw-Hill, 1975.
[20] K. Sabdacarakatama, Sejarah Keraton Yogyakarta, Yogyakarta: Narasi, 2008.

[21] E. P. Hendro, Kraton Yogyakarta dalam Balutan Hindu, Semarang: Penerbit Bendera, 2001.
[22] Winarno, Suwandi, Pemugaran Bangunan Keraton Yogyakarta, Departemen Pekerjaan Umum, Daerah Istimewa Yogyakarta, 1996.

Problems of low-rise residential area of Seoul - Focused on Malmi village of Guemcheon-gu, Seoul, Korea -

Yeoryung Seo*, Saehoon Kim
Department of Landscape Architecture, Urban Design Concentration, Graduate School of Environmental Studies, Seoul National University, Seoul, Korea

*Corresponding E-mail: yeoroong@naver.com

Abstract. In Korea, there are many low-rise dwellings that are under five floors. In Seoul, for instance, about 38.2% of the total residential area is occupied by low-rise dwellings. The area has a wide variety of problems, like Malmi village. Among the many low-rise buildings, Malmi village of Guemchoen-gu, Seoul, is constructed spontaneously. Infrastructure problems such as narrow roads and the lack of parking lots, pose great challenges. When the infrastructure was initially made, the plan was based on the traditional spatial use. The infrastructure problem came out continuously due to the increases in the size of houses and density. These problems are commonly found in many low-rise dwellings in Seoul as well as in the Malmi village. Therefore, it is the time to look into the problems of the low-rise dwellings and seek solutions.

Keywords: low-rise residential area, semi-industrial district housing, multiplex housing

1. Introduction

The residential areas of Korea can be divided into low-rise residential neighborhoods and mid/high-rise apartment complexes based on the type of housing that is distributed. Low-rise residential areas are concentrated on housing such as detached and multiplex houses. According to the research of the Seoul Institue[1], the low-rise residential area is currently widely distributed among the city, accounting for 38.2% of the total residential area of Seoul. Unlike apartment complexes, the low-rise residential environment is often characterized by the lack of parks and amenities, and the residential environment is undermined. In particular, some problems emerged as a result of the surging number of households, causing inconveniences among residents.

Malmi village of Guemchoen-gu, Seoul, is located in the semi-industrial district and it's near G-valley (Seoul Digital Industrial Complex). Also, the village is mixed with diverse small size factories and houses. The village also has many problems facing the low-rise residential areas of Seoul.

Therefore, the purpose of this study is to analyze the physical problems of the low-rise residential area and the inconvenience factors experienced by the residents. The information might be used as basic data to solve the problem of residential settlements.

2. Methods

At first, the authors analyzed the contents of the village through the internet, news reports, etc., and analyzed the physical condition of the village through on-the-spot inspection. In addition, the authors looked into the previous surveys from Guemcheon-gu office, population data, and traffic accidents data, among others, to identify the problems facing the village. Afterwards, through cooperation of

Guemcheon-gu office, living lab could be proceeded for a week and in-depth investigation of the location and residents' interview was conducted.

3. Discussion

3.1. Location

Malmi village of Guemchoen-gu, Seoul, is located south-west sides of Seoul city and G-valley, which has about 150,000 employees, forms the western boundary of Guemchoen-gu[2]. There are only three bridges in Guemchoen-gu, linking the east area with the west. The transport infrastructure is congested with many cars during the peak hours. Also, one of them is connected with Malmi village.

Northern part of the village is mixed with meat market, restaurant, commercial facilities and industrial facilities. Moreover, the southern part of the village is occupied by a military camp, which has planned to move over on near future as it shows in Figure 1. Some factories around the south and nearby the village will be built with high-rise apartment complex. Therefore, the village will be surrounded by high-rise apartments, and it is feared that the gap between the rich and poor will appear.

Figure 1. Current state of Guemcheon-gu

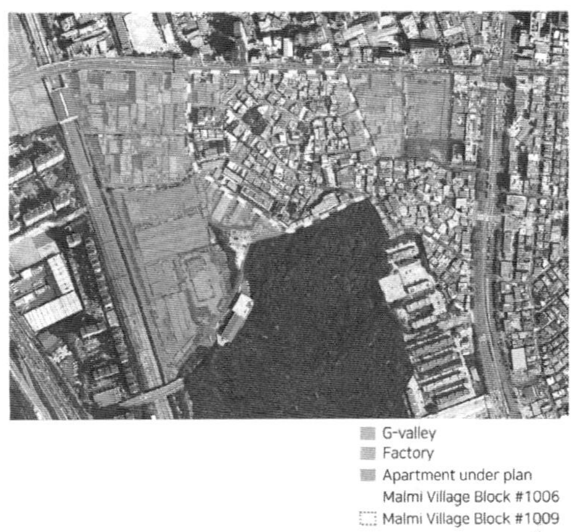

Figure 2. Current state of Malmi village

Because Malmi village is located in the semi-industrial district, a number of factories are abutting on the village as it shows in Figure 2. Also, the influence of meat wholesale markets on the north part of the village brought small meat markets into the village, causing unsightly sights and odors.

Low-rise residential area in Seoul normally has relatively low proportion of under the age 14 and high proportions of the elderly[1]. Moreover, according to the 2015 population data[3], 67% of single households live in a low-rise residential area. The situation of Malmi village is similar. There are many elderly people who lived in the village for more than decades. Recently, large number of young couples and single households who work in G-valley came to the village, too. Therefore, there are also young children in the village.

3.2. Interview

Using an in-depth interview method, the study surveyed the opinions of residents of various ages living in Malmi village. Through this, the inconveniences and conditions of the village could be analyzed. The key opinions about the village were as follows in Table 1.

Table 1. Malmi village residents interview

Interviewee	Opinions about Malmi village	Keyword
High School Student	There are plenty of old people and children in town, but there is no place for them. Because there are **no place to play**, I also move somewhere else to play with my friends.	No place to play
Housewife	Because of **nearby factories and many shoulder parking**, I am always **worried about children playing outdoors**. Plus, there are no enough space for children to play, so our family move to another area on weekends.	Factory, Shoulder Parking, Unsafe outdoor
G-Valley Worker	**I hope there are lots of parks and sports area**. Also, **there are lots of illegal parking on the street** because there are **insufficient parking spaces**. It makes narrow street much narrower. It's too inconvenient to walk.	Insufficient parking lot, sports area
Representative of Malmi village	There is **only one public space in town**. Even the public space is a playground. I hope there are diverse spaces to rest in the village.	Lack of public space
Elder	Cleanliness seems to be the most necessary issue in this village. This place is **too dirty** because of trashes from meat markets and garbage from each house. I am afraid that my grandchildren will see it.	Dirty outdoor

As shown in Table 1, the lack of a parking space was seen to be problematic by the residents. The lack of parking lots led to the illegal parking, which threatens residents' safety and even narrower roads which is already too narrow. In addition, public space and places for residents were very necessary. Trash in the village and from the meat markets were piled up on the roads, so it was important to know that cleanliness is also a key issue.

3.3. Problems

Seoul has witnessed a rapid increase in population through the process of urbanization. Due to the rapidly aging population, the housing shortage was inadequate and the government provided a large volume of new housing. To provide low-cost or affordable housing quickly, apartment buildings and multiplex housing emerged. As a result, the housing of Seoul was occupied with almost apartment house, such as apartment complexes, multiplex housing, and studio apartment buildings[4]. Malmi village is also composed with lots of multiplex houses and small size of apartments. Rapid increase of population caused the increase of vehicles and various problems occurred.

3.3.1. Road infrastructure and insufficient parking space

The village has organically-shaped roads that are naturally created. In addition, it was reorganized in the 1970s and 80s. When the infrastructure was initially made, the plan was based on the traditional spatial use. Therefore, now, with increased population and cars, roads are far narrow and parking spaces are lacking.

Figure 3. Road width and Parking space

As shown in Figure 3, in parking spaces, only 56% of the vehicles owned by the residents could be parked compared to number of households. That's how much increasing parking spaces are important in this town. Also, the roads in the village are mostly narrower than 4-6m, so only one vehicle can pass (Figure 4). As the parking space ran out, illegal parking took place frequently as it shows in Figure 5, and it caused a lot of problems. Because there are many factories near here, there are lots of trucks passing by, but illegal parked cars block the sight and occur traffic accidents. Moreover, because of narrow roads and illegal parking, emergency vehicles cannot go into the village. Plus, according to the traffic accident analysis system homepage[5], there was a constant traffic accident in this area as it shows in Figure 6.

Figure 4. Narrow Road

Figure 5. Shoulder Parking

Figure 6. Traffic Accident

3.3.2. Littering problem

One of the problems that many low-rise residential areas in Korea has is the dumping of littering. As the population grew rapidly, a lot of garbage was emitted, creating a messy village atmosphere (Figure 7). Especially in this village, meat market has been mixed with residential areas, so blood and internal organs of cows and pigs caused stinky odor and a negative image of the town as it shows in Figure 8.

Figure 7. Littering

Figure 8. Garbage from meat market

3.3.3. Lack of public spaces and facilities

There was only one park in the village, and even the park was a playground park, not suitable for use by residents of other ages, and the actual park utilization was very low. In case of apartment complexes, green spaces and various amenities are provided, but in low-residential area, only houses are dense and there are no enough facilities and public space for residents. As shown in Figure 4, there was a noticeable lack of cultural facilities within a 500-meter radius of the village.

Figure 9. Lack of amenity facilities and public space

3.3.4. Deteriorated architecture and house that are not accessible from the road

Recently, old buildings have been newly built to multiplex housings, but yet there are still many old, deteriorated houses. These old buildings are already extremely deteriorated as it shows in Figure 10. In addition, there are a great deals of architecture that is not accessible from the road, because they were illegally built like Figure 11. Therefore, they made small roads between buildings to make the entrances.

Figure 10. Deteriorated architecture

Figure 11. House that are not accessible from the road

In 2010, there was a redevelopment plan for the Malmi village, but the project was canceled due to the residents' opposition and the village is still having the same problems. Moreover, recently, the plan of building high-rise apartment complexes near the village got confirmed. Therefore, Malmi village is expected to be further isolated.

4. Conclusion

Through investigation and interviews of residents, problems of the Malmi village are, poor road infrastructure, insufficient parking spaces, littering problem, lack of public space, and deteriorated architecture as it shows in Table 2.

Table 2. Current problems of Malmi village

Problems	Detailed problems
Poor road infrastructure	Narrow width of road (almost 4-6m) / Illegal shoulder parking / Traffic accident
Insufficient parking spaces	Only 56% of households can park their vehicles / Illegal parking
Littering problem	Dumping of littering / Garbage from meat market – blood & internal organs
Lack of public space	No enough facilities and public spaces for residents
Deteriorated architecture	Deteriorated old architecture

Guemchoen-gu office is trying hard to solve these problems by proceeding shared parking system, green parking system which intends to demolish a wall separating individual parcels and to create a parking lot there, and so on. However, since the proposed measures are not a fundamental solution, Guemchoen-gu office is trying hard to improve the village by urban regeneration projects. The office is having cooperation work with Graduate School of Environmental Studies in Seoul national university and search for the better solution. There are many common problems with low-rise residential areas of Seoul, but to solve these problems, analyzing the characteristics of the area and identifying the problem should be occurred first like Guemchoen-gu office is doing right now.

5. References

[1] The Seoul Institute 2017 *A Study on the Actual Conditions and Improvement Direction of Low-rise Residential Buildings in Seoul* 54-74
[2] Guemcheon-gu Homepage (www.guemcheon.go.kr)
[3] Statistics Korea Homepage (http://kostat.go.kr)
[4] The Seoul Institute 2017 *The Future of Seoul 2, Challenging Space* 49-59
[5] Traffic accident analysis system Homepage (http://taas.koroad.or.kr)
[6] Seoul real estate information plaza Homepage (http://land.seoul.go.kr/land/)
[7] Korea Planning Association 2015 *Regeneration and Revitalization of Semi-industrial Regions* 213-230

Planning a new type of cultural community space in Goyang City, Korea

Jeonghye Kim[1*], Saehoon Kim[2]

[1]Department of Landscape Architecture, Urban Design Concentration, Graduate School of Environmental Studies, Seoul National University
[2]Associate Professor, Department of Landscape Architecture, Urban Design Concentration, Graduate School of Environmental Studies,Seoul National University

*Corresponding e-mail: rlawjdfod@snu.ac.kr

Abstract. Between March and June 2017, Seoul National University Graduate School of Environment Studies and Goyang City organized a course titled 'Environment Design and Planning'. At the beginning stage of the project, even though the Unified Korean Silicon Vally Project was initiated, our team members in the class decide to reconsider whether a large redevelopment project aimed at attracting tourists is actually a desirable goal for the residents or not. What would be truly needed for space users living in the area? How should we preserve and maintain the existing meaningful space? Meanwhile, we come up with the Gyeongui Railroad, which passes through Goyang and Ilsan. It was a railway line that shared memories of the past, the daily lives of the residents, and the joys and sorrows of them. Among the railways, 'Baek-Ma Station' was the place which was filled with the memories of the people who spent their youth in the city in the 1980s. The station is currently not in use, because this place is abandoned and empty. We wished to make this space more accessible for the residents and also to create a place that would give an opportunity to the resident's social comeback. The space could be activated again with the revival of local cultural activities. And we hope the residents will be able to create new memories of 2017 in this space.

Keywords: Culture, Community, Community exchange, Social comeback, Railroad station

1. Introduction

'Environment Design and Planning' is the class that gather 4 different majored students in Graduate School of Environment Studies and find the problem of subject space, draw up a problem resolution. Here, four different majors - Landscape Architecture, Urban Planning, Environmental Management, and Transportation - have gathered. About 10 students formed a team to work and to study together in order to find new space content for the local region. Since every single member had different academic background, each had its own perspective. It was required to coordinate the transformation of the urban space in Goyang according to the needs from the local government and the residents.

The ' Unified Korea Silicon Valley Project ' was a large-scale urban development project that developed the spaces which were not often used in Goyang City, based on four subjects : high-tech, education, housing, and culture. Here, our team focused on finding and fixing the problems of the current residents of Goyang City, rather than the original large-scale urban development project aimed at the project. This was a different approach than most of the groups that focused on other urban developments involving the city's Silicon Valley project. So we visited the housing area of Goyang, and listened to resident's stories, and focused on the ' absence of emotional community ' with the

Goyang City and the almost of modern urban area. And we devised how to form a sustainable emotional community in the city and worry about how residents can achieve positive social and economic development through it.

Today's urban residents are losing their opportunities to share the sense of place ad to form an emotional bond. Such is the case for not only the residents of Seoul, the well known capital of Korea, but also for the residents of the city of Goyang, which is a study area for Environmental Design and Planning Course at Seoul National University.

Goyang has grown as a commuter town of Seoul. As a consequence, a number of citizens of Goyang commute to Seoul or other urban cores during daytime for their social economic activity, and they do not have sufficient time to spend on their interest or sentimental communication with other people.

Albeit difficult, human interactions and emotional communications between individuals can significantly improve the quality of life and mental health of people in urban environment. Especially, depression among retired workers in their 50s, or housewives who are willing to rejoin the workforce is considered to be one of the biggest social problem. Such social problems caused by lack of communication and attention between individuals could be at least partly prevented by sufficient interaction between the residents.

For such reason, this pocused on city's long forgotten role of providing emotional and psychological community for its residents, rather than Goyang's 'Silicon Valley' project and other mass-developments. Furthermore, this research expands on the residents' biggest hope for their community: opportunity for social rehabilitation and virtuous cycle of social communities.

2.Methods

Figure 1. Gyeongui-Line 'Book Street'.

Figure 2. Gyeongui-Line 'Walk Park'.

We focused on the historical values of Baek_ma station and memories of residents with it and simultaneously tried to create spaces to meet the needs of residents. So we interviewed the residents of Baekma_Station and tried to identify the problems of the community programs that already existed, and sought to explore the characteristics of the railway station and tried to make the programs can utilize the features of the space.

Gyeongui-Line, a railway line bisecting Goyang the most patronized train in the past. For instance, Baek-Ma station was the most popular dating venue for young couples in the 80's. Today, it is the place of reminiscences for the past youths, the place where their daily memories, joys, and sorrows are left behind. Currently, the city of Seoul is designing multiple ways to utilize the old train stations, and some efforts are successfully implemented - cultural utilization efforts such as 'Book Street' and 'Forest Walk Park' are one of the successful cases. However, there are still numbers of core stations across Goyang, such as Baek-Ma, Gok-San, Pung-San, that little attention is paid to[1].

Figure 3. Figure with short caption (caption centred).

Our research group has have decided to combine the Gyeongui-Line stations, where the past memories belong, with social cultural community space, which reflects the residents' wish. Unlike the traditional method of completely demolishing the past memories and seeking to create a new space, our proposal is to preserve the histories of the long-loved space of the residents. The 'revamped' space is not just a mere commercial or rest area, but it is a place where memories stay alive, residents connect themselves with others, and communities based on those connections develop the place itself.

The importance of contemporary cities does not depend on glorious materialization of physical space per se. Cities' importance now lie on the psychological role of embracing the residents' emotional capital the resident's emotional poverty. City-wide planning and projects must, therefore, be composed after listening the resident's opinions and developed for the residents, not for the bureaucracy.

This report mainly focuses on the Baek-Ma station, one of the mostly nelected stations of Gyeongui-Line. Also, the report identified the utilization of "Gal-Gal-Gal Book Store: Teenage Book Store" within the Baek-Ma station. The report also contains an interview with residents who registered for social community programs, and records their response on what they do for their spare time, what is their hope for social communities, and what is their goal for participating in such communities.

Through the interview, we sought to formulate the social cultural community and program fully utilizing Baek-Ma station. The projected program aims to derive interaction between individual residents, creating a new experience and memory, not forcing knowledge to its attendees. After the experience, the residents shall propose a new opportunity of 'social rehabilitation,' which is the residents hope the most.

Figure 4. Interview with residentis near Baek-Ma Station.

3. Discussion

3.1. Current usage of Baek-Ma station area and surrounding environment

These days, making use of the existing infrasructure without building an overly-large structure is a major planning issue in Korea. Therefore, the already existing Baek-Ma station may serve as a valuable space that holds people's longing memories.

Baek-Ma station is surrounded by wide residential area continuing from Go-San and Pung-San station. Baek-Ma station has two levels. The first level is composed of a large square and an outlet with a size of 130m². The second floor is composed of 'Gal-Gal-Gal Book Store,' waiting room, and 100m3 wide of empty space. Huge corridors vastly lie on each levels.

The Gal-Gal-Gal Book Store and the surrounding area is used by teenagers who are willing to read books. There is also a small non-profit coffee shop inside the bookstore, which donates any profits for teenage cultural programs. However, due to its original and sole purpose as a place for teenagers, it seems quite difficult for various generations to enjoy the provided space.

The outlet store on the first floor of Baek-Ma station is leased to the store owners at a low price. According to HR personnel of Gyeongui-Line, city-sponsored projects could be done utilizing the space. For empty spaces inside the station, we discovered that the place could be lent free of cost after consulting with Korail and the city of Goyang.

3.2. Survey on current community programs with residents near Baek-Ma station

We performed an interview with residents walking in a park near Baek-Ma station and heard their opinion on existing educational programs of the civil-cultural center.

According to the interviews with the residents, we discovered that residents prefer programs in which they can interact with other residents or learn a new skill for practical uses. They believe such skills could be a milestone for their social rehabilitation. Also, the residents required counseling program for people in their 50's, to help them resolving the post-retirement depression and other mental/emotional difficulties.

Interviwee	Opinions
Elementary School Students	We don't have much choice other than the main commercial district after school or Hwakone. / We would prefer spending time at nearby Baek-Ma station if there is a program that offers culinary or art education.
Housewives	Current civic center is too small. / It would be nice if we can make a real use, such as running a coffee shop, out of what we have learned at the center.
Civic center users	The center's programs are mostly a lecture-type lessons such as musical instrument lessons, or sports lessons. We would definitely prefer to use what we learn after those lessons. Programs such as café business lessons or computer lessons could be notable ones. / Also, we would appreciate if the center could be the area where residents can interact with one another.
Elderly residents	We usually spend time walking around the park. It is sad that Baek-Ma station seems empty.

Table 1. Key contents of the residents' interview.

3.3. Projected community program based on survey results

Based on the surveyed opinions, we projected 4 main programs to fully utilize the Baek-Ma station area. Also, we sought to circulate the program sequences for registered residents to interact with each other[2][3].

Table 2. Program Table

	Mon	Tue	Wed	Thu	Fri	Sat	Sun
09:00-16:00	Retired, Housewives 1. Floral design, Dry flower, Difuser making 2. Urban agriculture: Flower pot making, Organizing empty space near Baek-Ma station, urban crop-growing 3. **Cafe-running barrista lesson**					Family, Baek-Ma station users **1. Baek-Ma square market / Open-Kitchen event** **2. Self-cooking lesson for youth** **3.** Job-experience lectures & seminars	
16:00-22:00	Students, Housewives **1. Culinary lessons with family** **2.** Art lessons with family **3.** Cafe-running barrista lesson						

These programs were allocated based on the participants characteristics and available time. Regarding the morning program during the weekdays, the programs focus on elderly, retired, and housewives. For afternoon block, the programs focus more on students after school or hawkone. The weekend block is dedicated to families and Baek-Ma station users, focusing on the square-market and free-market.

We also sought to create a system where both people and space curculate for example, products of the culinary and barrista program are sold at the square market, through event, and the products of urban agriculture program are both sold and the square market, while being used at the culinary program.

The registered residents of the programs advertise their products within the station corridors, and yet-registered people experience the products by visiting the square market. In such a manner, Baek-Ma station could be the heart of creative interaction between the residents, creating another memories as it did long time ago. This result could advertise the program itself, bringing more people to register for the program[4][6].

Table 3. Four main programs.

Culinary lessons	**Barrista lessons**
First floor, kitchen available space or outlet space	
- Recipe sharing between the residents - XX mom's recipe (interaction between people)	- Cafe business and barrista lecture - Coffee for sale while the lessons are not in session
Main office shared area	**Sqaure market**
Second floor empty space	First floor square & Second floor waiting room
- Space free for residents' use - Office/Multipurpose room for lesson preparation - Appointment-based meeting room / seminar room	- Market for urban agriculture product / floral design product - Space for advertisement, interaction between residents through events

3.4. How Baek-Ma station could lead residents to virtuous cycle of social rehabilitation opportunity

Ba

The Baek-Ma station users' first most goal was to find an opportunity to rehabilitate themselves through cultural-educational program. They also hoped that their children could dream of future, through experiencing the job-experience programs. For such reasons, we decided to let the users to sell their products while the lessons are not in session, after the cafe business lessons and culinary lessons are over. With such experience, when an individual starts a business, he/she can get significant assistance for the business. Furthermore, in long term view, the program will also include computer lectures for certifications.

4.Conclusion

The aim of our project is to reinforce city's forgotten mental/emotional role by promoting cultural community activities and utilizing valuable historical space where residents can interact with each other. Also, such community activities should be more than temporary entertainment or one-way lecture style pressured learning, but should end with social rehabilitation of the residents and promotion on local economics. The community programs and utilization of Baek-Ma station are planned based on residents' surveyed opinion and we were able to keep close communication with the workers of Goyang city for composing a tangible program. The new method is not a traditional top-down development project, but is a project that listens to the residents' idea and connects the need and requirements of the bottom and the top; consequently, the new development is closely related to the possibility of individuals' pursuit of happiness, formation of a psychologically-linked community, and space creation before novel development and transformation of the shape of the city.

We illustrated the values of Baek-Ma station, the residents' hope regarding to current situation, and proposal of programs during the Goyang final presentation. When the proposal was presented to the government officials, we received a promising support that the proposed program will substantialize at Baek-Ma station in near future. Today, we must look for emotional communication rather than one-sided development, and listen to the residents' need and work with them before we create any modification to the city. In order for this to happen, people's stance on city should change, as the residents must also be active.

5. References

[1] Northern Gyeongi, Inspecion on Major Issues in Gyeongui Line Conference plan
[2] Art town making business strategy studies, Seoul foundation for arts and culture **3-5, 91-96**
[3] Village art gallery monitoring and evaluation, Seoul research institute **54-118**

[4] Seongnam market revitalization Fountation (smr.or.kr)
[5] Seoul youth space_Zero gravity zone (http://youthzone.kr)
[6] Seongsu urban regeneration support center(sd_seongsu.blog.me)

Conservation of Semarang chinatown traditional settlement as physical characteristics of chinatown district

Bintang Noor Prabowo, Ratih Widiastuti, C. N. Bramiana*

Universitas Diponegoro, Semarang, Indonesia

*Corresponding e-mail: c.n.bramiana@arsitektur.undip.ac.id

Abstract. Historical environment must be conserved due to their contribution toward cultural, social, economic and aesthetic values. They transfer the emotion, thought and backgrounds of social live to modern societies. As one of oldest settlements, Semarang Chinatown gives great influence toward Semarang City and has endless historical values. Old Chinese settlement is very thick with Chinese culture. It becomes the primary source of physical characteristics in this historical environment. However, physical and visual problems come along with the rapid economic growth in Semarang Chinatown. The old buildings that still remains in Semarang Chinatown are in deficient condition and losing their shapes. Moreover, modernization and economic growth influence the change building facade. Several old buildings were replaced with new buildings that tend not to adopt Chinese Architecture style. Consequently, it changes the value of the buildings and the value of the historical environment.

Keywords: Old Chinese settlement, Historical environment conservation, Physical characteristics, Semarang Chinatown.

1. Introduction

Settlement as a human's built environment was formed from housing construction process as a container from the dwellers activities inside. Rapoport [1] stated that settlement area is an architectural work, which reflect its people's life, with all the meaning within. Furthermore, uniformity and diversity are the expressions of manifested figural quality forming an image as an identity, which eventually give meaning to the specific area [1].

In Semarang municipality, one of the historical areas included in List of Heritage Conservation is the Chinatown of Semarang (Decree of Semarang's Mayor No.650/157, 2005). The Chinatown with its cultural richness along with its Chinese Architectural values is the vital part in the development of Semarang as a city. However, in line with its rapid economic growth in Chinatown, there are physical and visual problems in this region. As an old area, the buildings suffer dilapidation and deformation. On the other hand, the effect of economic and modernization have also changed the existing building facades. There are many old buildings replaced by newer ones, which tend to lose its Chinese Architectural values, so this will be the problem for the area considered as a Historical area.

Based on the aforementioned background, the objective achieved in this study is to analyze the building's facades changing in the Semarang Chinatown historical area.

Figure 1. Location of Semarang Chinatown (Source: Dinas Tata Kota Semarang, 2015)

1.1. Conservation of Semarang's Historical Areas

Conservation activities is a process which includes repair, preservation, restoration, maintenance, adaptive reuse, and reconstruction [2]. In Semarang, the conservation efforts began based on the Decree of Semarang's Mayor No. 646/50, 1992 about the Historical Building Conservation in Semarang Municipality. The Decree was a form of response to the Act No. 5, 1992, regarding Conservation of Cultural Heritage in relation with 101 buildings considered as Cultural Heritage in Semarang.

The conservation efforts conducted by the Government of Semarang has been continued accoridng to The Regional Act of Semarang Municipality, No. 8, 2003, about the Building and Environment Planning Guidelines (RTBL), followed by the Regional Act No.16, 2003, regarding the Building and Environments Planning Guidelines of Semarang Old City Area, which also mentions the Chinatown area planning. This regulation also introduces the Chinatown area as the Chinese cultural center of Semarang City. This policy aims to re-enhance and develop the potency of Semarang Chinatown.

So far, the common approach being done by the Government of Semarang, in term of conservation, is the data collecting of the historical buildings in Semarang municipality. There are 144 residential buildings listed in the Heritage Conservation List, which 4 of them are located in Semarang Chinatown having particular characteristics, where the dwellers of this area still maintains their social, economic, and cultural activities. Beside the rapidly developed economic activity, the historical Chinese artifacts are still existed in this region. The glorious footprints from the past are still obvious, though modernization effects are also visible in some parts. However the characteristics of the district as a historical area can still be seen and become a distinctive factor in comparison with other districts in the city.

The Chinatown of Semarang consists of high-density array of houses. Those houses are located oppositely and there is almost no space between each house, so this condition has formed narrow alleys. There's almost no front yard found, and therefore the buildings in this area are called "sitting on the ground" the lack of space between the street and the front part of the house Chinatown of Semarang [3]. The conservation approach using historical buildings data collecting is considered insufficient to conserve this historical district [4,5].

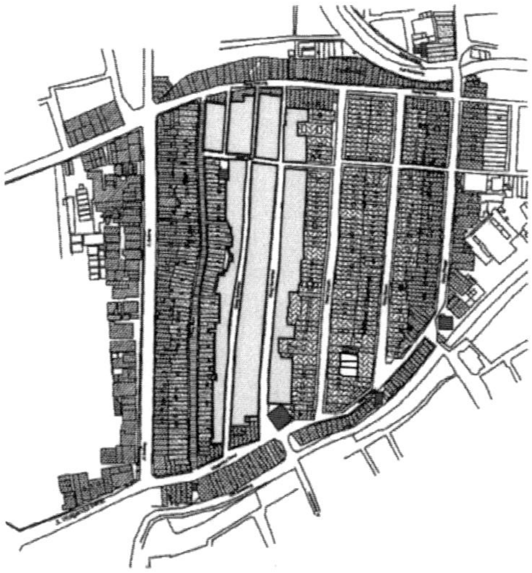

Figure 2 Traditional Residence Location in Semarang Chinatown (Source: Author analysis, 2017)

Figure 3 The Traditional Housing Facades in Semarang Chinatown

1.2. Semarang Chinatown Historical Area

The history of Semarang Chinatown was started in 1740 [6]. Semarang was a seaport visited by various ethnicities, groups, and nations. Chinese was one of them, who later on resided in Semarang. Due to the rebellion of the Chinese ethnics in Batavia (Jakarta), in 1740 the Dutch Colonial Government ordered the Chinese society to be localized in a certain region which nowadays is known as the Chinatown of Semarang. The purpose of the Dutch colonial government is to ease the supervision.

On the early stage, the buildings on these alleys belonged to the Chinese residents. Nowadays those areas are mostly developed into a commercial district. Only in Gang Tengah, Gang Cilik, and Gang Gambiran can we identify some traditional Chinese residential buildings, as seen on Figure 2 and Figure 3.

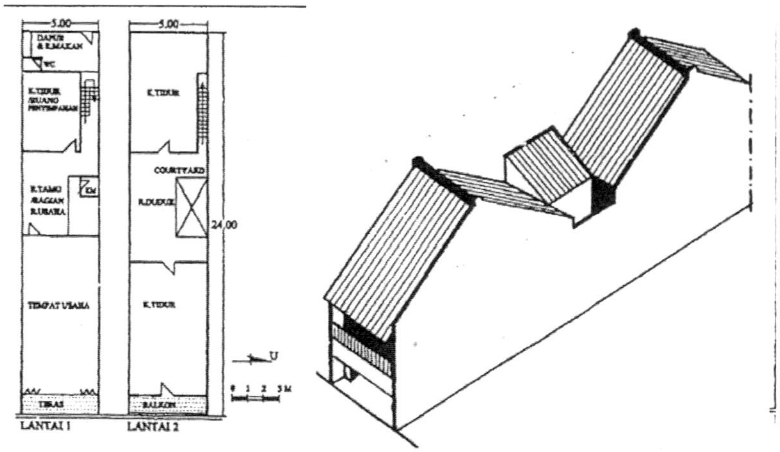

Figure 4 Illustration of a house used for business (Source: Reference [7])

According to Rosiana [7], the typology of residential buildings in the Chinatown district consists of two types:

a. The houses with multiple uses for business. They are two-storey houses where the first floor is utilized as the business place (store) and the second floor is for living place, as seen in Figure 4.
b. The purely residential houses, as seen on Figure 5 and Figure 6. The typical characteristics of the houses are: (1) Two-three storey row house in log array, and (2) Most of them were initially built in big sizes and later were turned into smaller houses

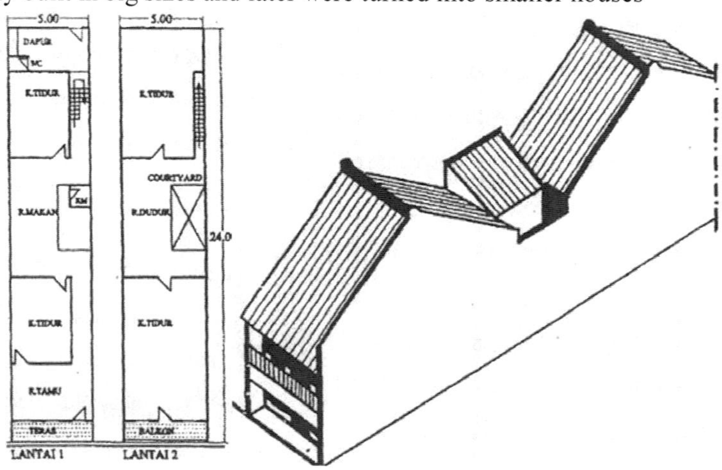

Figure 5 Illustration of pure residential house (Source: Reference [7])

Traditional Chinese houses use a folded roof, with a unique curve on their two edges. The people of Semarang called this kind of roof edges as Konde roof. As heritage buildings, the existence of these old buildings has become the physical characteristic of Semarang Chinatown district. One of the images of the district was established by the presence of the group of residential buildings with highly Chinese Architecture values. Therefore, the changing of the building façade will create a problem in the future. Although there are four buildings in the middle of the Semarang Chinatown already listed as Heritage Conservation Buildings, it still is insufficient to maintain the whole charactereristic of the Chinatown District.

IOP Conf. Series: Earth and Environmental Science **99** (2017) 012034 doi:10.1088/1755-1315/99/1/012034

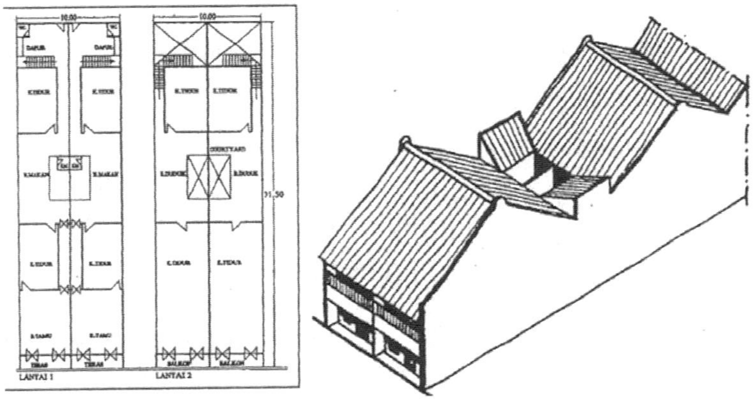

Figure 6 Illustration of Attached House Consisting of Different Units (Source: Reference [7])

2. Methods

There are 3 main research methods being used in this study, which were:

- Literature Study. Mainly concerned with literature of conservation activities taking place in Semarang Chinatown and documents of physical analysis of buildings with Chinese architecture.
- Direct observation in the Chinatown area, by identifying physical evidence of the Chinese Ethnic settlement. From the observation, we could find the specific identity and morphology of the buildings in the study area.
- Interview with the stakeholders of the Chinatown.

3. Discussion

The Chinatown as a historical place in Semarang recently face a problem concerning the district identity. The development process on the area is focusing more on the economic and efficiency factors causing a significance changing on the unique characteristics of the Chinatown. Development and modernization of the city have created similarity and uniformity of the city's image. It is a threat to the uniqueness of the Chinatown as a heritage area.

Based on the data collected from the field, the change occuring in the Chinatown area was caused by multiple factors such as:

- Deteriorating of the building elements that forced the inhabitants to change them with new building materials.
- Population growth. The division of inheritance has caused the big house to turn into smaller units, and this fact has a direct impact to the façade of the buildings.
- Economic Factor. There are residential buildings that have been fully changed into storages or business units.
- Changing of Ownership
- Lack of awareness from the society to conserve heritage buildings at their area.

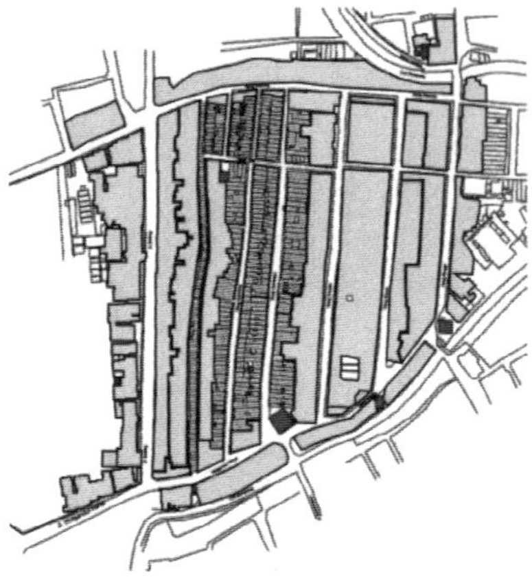

Figure 7 Location of buildings with different style

Figure 8 Deterioration of the buildings

Most of the façade changes on Chinatown were done by replacing the old-classic Chinese styles into modern and simplified ones, due to financial problems, lack of craftsmanship, and alternative materials availability.

A couple of sub-districts such as Gang Warung, Gang Pinggir, and Gang Besen have been changed with modern architecture, totally different from the typical Chinese Architecture within the area, as shown on the Figure 7, Figure 8 and Figure 9. This replacement varies from minor replacement to total conversion. There were adjustments on some cases where the owner combined the old elements with the new ones.

Figure 9. The temple located at the end of the crossroad

Figure 10 Typical *Konde* Roof

The function of the buildings has also experienced major adaptation. Most of the buildings on the main street have become commercial buildings. If the area does not have any Kong Hu Chu Temples and some building with Konde Roofs, the image of this particular area will only be a fully commercial district, same with other similar districts all over the city as shown on Figure 10. There are more building facades ignoring the unity, harmony, and architectural locality and uniqueness aspects in Semarang Chinatown, these phenomena will lead to a high variation of modern buildings that cause the decreasing of urban quality.

4. Conclusion

It is a dilemma when facing a fact that this cultural heritage area experiences physical and visual problems. The existence of classic-Chinese architecture is the historical evidence of Semarang as a city. Seemingly, economic and population growth factor are the prime causes of the characteristics decreasing at the Heritage Area. The physical identity discussed in this study referred to the metamorphosis of the façade, from the classic Chinese architecture into the modern one. The lack of guidance in maintaining the characteristics of the building facades will eventually cause the losing of Chinatown identity that was once existed. The Semarang Chinatown as a the Heritage Conservation Area is an interpretation of history, deeply filled with Chinese Architecture values that distinguishes this area from other parts of the city. Therefore, the Municipality of Semarang needs to create strategies to preserve the uniqueness of this district and to maintain the existence of Semarang Chinatown for the sake of our future generations.

5. Acknowledgments

Authors would like to express their gratitude for the assistance and the encouragement from colleagues, for the special work done by technical staff, and for the financial support from organizations should do so in an unnumbered Acknowledgments section immediately following the last numbered section of the paper.

6. References

[1] Rapoport A 1969 *House form and culture* (Prentice-Hall)
[2] Said S Y, Aksah H and Ismail E D 2013 Heritage Conservation and Regeneration of Historic Areas in Malaysia *Procedia - Soc Behav Sci* **105** 418–28
[3] BAPPEDA Kota Semarang 2006 *Senarai: Bangunan dan Kawasan Pusaka Budaya Kota Semarang* (Semarang)
[4] Idid S Z A 2005 *Urban conservation approach for a multi cultural historic cities: The urban planning and design perspective: case study, Historic City of Malacca* (Tokyo: University of Tokyo)
[5] Shamsuddin S and Sulaiman A B 2002 *The Importance of Conserving the Old Town Centres in Achieving a Sustainable Built Environment of the Future*
[6] Widodo J 1966 *The Urban History of The Southeast Asia Coastal City* (University of Tokyo)
[7] Rosiana M 2002 *Kajian Pola Morfologi Ruang Kawasan Pecinan (Studi kasus: Kawasan Pecinan Semarang)* (Universitas Diponegoro)

Towards a harmonious development between nature and culture on Walisanga religious site, Indonesia – learning from the best practices in Japan

Atiek Suprapti[1*], Edward E. Pandelaki[1], Indriastjario[1], Agung Budi Sardjono[1], Yosidha Tomohiko[2], Yagi Masao[3], Adriana P. Higashino [3]

[1]Architecture Department of Engineering Faculty of Diponegoro University Indonesia
[2] Ritsumeikan University Japan,
[3]National Institute of Technology - Akashi College

*Corresponding E-mail: atiekbudiarto@gmail.com

Abstract. Nature and culture are elements that play an important role in the development of a place. Sites of Walisanga spread all over Java since in 16 century. Muslim communities respect respect for these sites. These sites have become an attractive destination of religious tourism on the North Coast of Java. Some of these sites are very close to nature and as a part culture. Shinto shrines and Buddhist temples in Japan are the examples of the master pieces of indigenous architecture which are still maintained very well. The Japanese religious sites with the beautiful architecture and landscape in the culture format have been successfully attracting tourist from all over the world. This paper aims to determine the potency and problems associated with the development of spatial harmony between the nature and culture on the Walisanga religious sites in Indonesia by conducting a dialogue with best practices cases in Japan. This study used descriptive analysis methods. The result showed that, there are similarities between Indonesian and Japan in treating cultural heritage assets. Indonesian government gives quite significant roles to public to participate in preserving cultural heritage. Japan government has given support through the modern technology and funding, assets registration, maintenance and protection, and the assets management. Creating the harmony between nature and culture is necessary to enhance performance, facilities and infrastructure.

Keywords: nature, culture, religious sites, heritage culture.

1. Introduction

Globalization views the world is a big market, as the consequences the global expansion affects in: economic, social structure, and the built environment change [1]; [2]; [3]; [4], creating both of place and non-place [5]. Effort to strengthen the local place is required against the global expansion [6]; [7]. With that way the city identity will be stronger and the tourism benefit will be stronger too [8]. *Walisanga* in Indonesia are known as nine (9) saints who dedicated their life to spread Islam in Java in 14th century to 15th century. This era is the transition period from Hinduism to influence of Islam. This influence appears from the artifacts found showing mixed concept between Hinduism and Islam [9]. Nowadays the sites of *Walisanga* are regularly by thousands of pilgrims every day. Some of the interesting sites are in East Java Province, particularly in Surabaya and Gresik. In Surabaya, the site is located at the relics of Sunan Ampel. The Sunan Ampel Mosque was built together by Sunan Ampel and his students in 1399 in the era of Brawijaya-V King of the Majapahit Kingdom. In Gresik we can find Sunan Giri sites who which was the students of Sunan Ampel. He established the independent sultanate of Giri Kedaton and further became the centre of Islam propagation in Java and eastern Indonesia. Both of complexes are also complemented by Muslim community, who have for lived hundreds years.

In the global era where everything is considered as a market, one needs to dig deeply to improve its potency to appeal the market. And in the case of Indonesia, a country with rich historical and cultural backgrounds, one of these treasures is the urban heritage. *Walisanga* is a well known urban legend. So sacred that the sites was famously known to keep host relics of the saints that are considered sacred and visited by millions of pilgrims each year. This is the potency that needs to be reviewed and developed. The sites invite people, the people invite business opportunities, the business opportunities invite expansions, and the expansions demand developments. However, these developments are far from accommodating since most of them are done with limited resources. The purpose of this study is to determine the application of the principles of socio-cultural balance and natural balance in the practice in the Indonesian religious sites development particularly of Sunan Ampel and Sunan Giri and makes a dialogue with the best practice cases in Japan nemely which are Todaiji and Toji Temple.

The study used descriptive analysis methods supported by collecting of primary data and secondary data. Comparative study method was used to find the causes and effects of spatial design of religious sites in Indonesia through the best practice of religious sites in Japan complete with the concept of harmonious in culture and nature. The research locus in Indonesia included: (1) Sites of Sunan Ampel in Surabaya, which is the oldest of sites Walisanga, (2) the site of Sunan Giri in Gresik, the sites which was established sultanate of Giri Kedaton. Meanwhile, the religious sites in Japan included: (1) Todaiji Temple in Nara, and (2) Toji temple in Kyoto. The existence of *Walisanga* is widely known and has become the Javanese urban legend. The sites are spread all over Java, such as in East Java Province like Surabaya and Gresik, in Central Java Province like Kudus and Demak, and in West Java Province like Cirebon. There have been a myth among local inhabitant, that pilgrimage religious on Walisanga sites equals to the Haj pilgrimage activity in Mecca. While Todaiji temple is a Buddhist temple sites that was one of the 7 major temples in Nara. There is the biggest Buddha Vairocana in the world made of bronze and known as Daibutsu by the Japanese. This temple was also intended to honor the messenger of Shinto religion called the Deer which was believed to have ventured the earth. Toji temple is the centre of Buddhist learning in Shingon which preserve many excellent art. Both of temples are objects of World Heritage.

2. Result

2.1. Religious Sites of Sunan Ampel and Sunan Giri in Indonesia

Sunan Ampel site is located in Ampel Sub district in Surabaya, near Ujung Galuh port. This land was a gift from King Brawijaya V. Sunan Ampel built his first *pesantren* (Islamic dormitory school) and settled down with his students in this area. Handled by Walisanga, Islamic teachings were spread throughout the nation and to surrounding countries, such as Java, Kalimantan, Sulawesi, Ternate, Tidore, Ambon, Philippine, and Brunei [10]. Sunan Ampel Mosque was built in 1339 and was renovated and reconstructed several times: the first time was in 1926 by Regent R Aryo Niti Adiningrat (22.7m x 20.55m), then by KH Manaf in 1964 (50.7m x 50m), and lastly was extended to 11m x 120 m. The mosque was designed with a Javanese architecture: its roof was a three-tiered canopy roof (similar to Hinduism's *meru* buildings). The roof was rested on a *soko guru* (main column) located right in the middle of the building. This column was made of centuries old teak wood, believed to be brought from Bojonegoro region. The wood was also believed to have a special spiritual function. Aside of having marble flooring, the mosque also has a cylinder minaret with cone-shaped roof. The shape of the minaret is a reminiscence of a lighthouse and it is used to broadcast the *adzhan* as a call for prayers.

In this area, there are several historical artifacts from Sunan Ampel (see Figure 01), which are: (1) the cemetery which consists of the tomb of Sunan Ampel, his family and friends, religious leaders, and others, (2) the ever-flowing freshwater well among brackish water inside the mosque, (3) cauldron for the freshwater to drink and to do ablution, (4) the entrance towards the cemetery, consisting of 5 different gates symbolizing 5 pillars of Islam, (5) *mighrab* (desk for preachers) of Sunan Ampel.

Sunan Giri was the ruler of Giri Kedaton (1487 – 1517). Parts of the Sunan Giri cemetery site are: (1) Parking area in the lowest level, built in 2012, (2) staircases connecting the lower and the upper parts of the cemetery, (3) the first gate in the shape of a split *candi bentar* (see figure 02), (4) the walls surrounding the cemetery, creating boundaries for several cemetery terraces, (5) the cemetery cupola in a *joglo* roof shape, surrounded by other cemeteries, (6) Kala statue above the entrance gate and dragon head statue guarding the front of *candi bentar*, showcasing the influence of Javanese, Hinduism, and Chinese cultures, (7) the gate in the shape of the split *candi bentar* signaling the entrance to a holier space.

2.2. Religious Site Todaiji Temple & Toji Temple in Japan

The beginning of the construction of the temple was in 728 AC when Shomu Emperor implemented Kinshoen-ji (Todaiji's early name) as the resting place for Prince Motoi. The Prince passed away a year after his birth. Inside the Todaiji Temple complex are several buildings with complementary functions, but the main building is The Great Buddha Hall as the place of Buddha Varoicana.

Important buildings in this temple are: (1) The Great Buddha Hall, which is the main hall with the largest wooden structure in the world (50.480 meter long, 57.012 meter wide, and 48.742 meter tall), (2) The Bell Tower, (3) The Great South Gate (Nandai-mon), which is the main gate of Todaiji with 18 giant pillars supporting a 21 meter roof and the structure elevated 25.46 meter above the stone foundation, (4) Nigatsu-do, also known as Second Month Hall because it is used to do rituals every second month on the lunar calendar, (5) Kaidan-in, a baptizing hall having the yard reconstructed during the Edo era, (6) Tegai-mon Gate, a lavish gate with eight pillars and the only 8th century structure left of Todaiji, (7) Daibutsu statue or the Vairocana Buddha, and (8) Nara Park, a public park formerly used as a private park in the base of Mount Wakakusa. This park was erected in 1880.

2.3. Toji Temple Kyoto

The name Toji which etymologically means 'East Temple' is still well preserved until today [11]. The main building was destroyed by earthquakes and fire, but it has been rebuilt to preserve its original layout and architectural style. Many excellent arts are stored in the temple, such as statues, carvings, paintings, and others. Kukai Monk in 816 AC initiated a religious community in Mount Koya as a place to meditate and then Kukai was entrusted by the king to manage the country's temple, which was Toji. He also started the first private school in Japan. Kukai was one of the most prominent religious figures in Japan's history. He was also known as a scholar, calligrapher, and educator. In 823 AC, Emperor Saga gave an award to Monk Kukai and gave him the temple.

Several supporting buildings in Toji temple are: (1) the Five Story Pagoda, which is 54.8 meters tall and is the tallest wooden tower in Jepang, and it is now a city icon, (2) the Kondo (main hall), which has hosted the Yakushi statue since 1903, (3) Miedo, which is the hall that was built as a dedication to Kukai or Kobo Daishi, and, (5) Flea market (Mieku)

3. Discussion

3.1. The Government Policy

Cultural heritage in Indonesia is protected by the 1945 Constitution of the Republic of Indonesia article 32, in which the country is responsible in elevating its people's prosperity and in positioning Indonesia in equally among other nations of the world. The cultural heritage itself according to constitution no 11 year 2010 is defined as something used for improving people's prosperity, while public has responsibility for their preservation. Moreover, the preservation efforts must allow development and utilization for people and their environments, and the government has a role in providing preservation leverage.. Indonesian government manages cultural heritage buildings by registering and giving them national, provincial, and regional ranks.

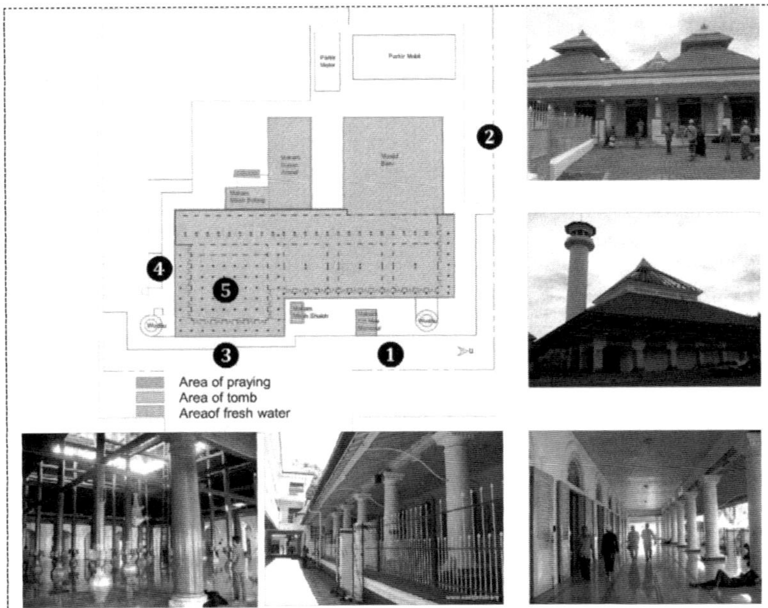

Figure 1. Clockwise: The plan of the site Sunan Ampel, (1) the original mosque relics of Sunan Ampel, (2) the new mosque, (3) the wooden forty-eight doors of original mosque, (4) the row of brick columns supporting inner and outer the original mosque, (5) the teak-wood sixteen poles 17 meters high without any connection supporting the original mosque roof.

Figure 2. Left: The first gate in the shape of split *candi bentar*, newly constructed built by the government.(Image source: http://www.emakmbolang.com/2015/08/menapak-jejak-wali-songo-di-kota-gresik.html) Right: The cupola, with a *joglo* design and low ceiling.

Figure 3. The Great Buddha Hall (Source: field survey, 2015)

Figure 4. The Great Buddha /Daibutsu-den (Source: https://todaiji.or.jp)

Figure 5. Kondo the main Hall (Source: field survey, 2015)

Figure 6. The Five story Pagoda is the highest wooden building in the world. (Source: http://www.toji.or.jp)

In Indonesia, cultural heritage protection is not only about conservation and restoration, but it also about the development of information, education, and tourism. Walisanga's religious sites are the part of the cultural heritage sites protected by The Act No. 11/2011, but the implementation is still weak. For example, in Sunan Giri's site, the restoration done in 2012 has changed the structure of the staircases, from 100 staircases to only 52. This is a carelessness should not be done to a cultural heritage building.

Japanese Government classifies their cultural heritages into two groups: (1) intangible assets, and (2) tangible assets; and both are protected by the acts of cultural resources. The government implements various necessary steps in order to preserve and utilize their cultural resources. Tangible assets are protected through preservation and protection from natural disasters and acquisition. Intangible assets are protected by the available subsidy for training programs for the cultural sustainability Japan's registration system for conserved buildings includes moderate steps for protections. Based on the amendment of the 1975 Act about Cultural Heritage Protection, a system of "Preservation that are for traditional buildings" made to protect historical cities, cities and villages in Japan – including fort cities, postal cities, and cities built around temples and shrines. Additionally, Cultural Affair Office provides monetary support for city projects such as restoration, façade improvements, and disaster mitigation to preserve the area.

There are similarities between Indonesian and Japan governments in treating cultural heritage assets. However, there are several things to be emphasized. Indonesian government gives a quite significant role to the public to participate in the preservation of cultural heritages. It seems to be a good step, but needs clear policies to regulate the rights and responsibilities of both the public and the government. One thing to notice is not to give too much control to the public that it will hinder the government from releasing their policies. Japanese government has more supports from modern technology and funding for activities like the registration of the assets, maintenance and protection, and the utilization of the assets. Law enforcementin the form of rewards and punishments is needed to motivate the public to care about the cultural heritage. The regulation of classifying traditional buildings areas is enforced to assist the operation of cultural preservation.

Todaiji Temple is included in UNESCO's world heritage as "Historic Monuments of Ancient Nara" with 7 other sites in Nara. Todaiji is also one of Japanese National Treasures included in the List of Cultural Properties of Todaiji. Architectural works that are classified as National treasures include Daibutsuden, Nandaimon, Kaizan-do, Shoro, Hokke-do, Nigatsu-do, and Tegamon. Todaiji temple itself is the biggest wooden structure building in the world. This building has witnessed the historical political and cultural changes about religion and life in Japan's capital during the 8th century (710 – 784). During this period and after the capital was moved from Nara, this area was consolidated by the national government. Restoration has been done to these monuments since the late 19th century after the implementation of the Ancient Shrines and Temples Preservation Act in 1897.

3.2. Socio-Cultural Aspects

Social-cultural relations in which the positive aspects of disparate cultures are valued and promoted and there is widespread participation of citizens not only politically but also socially in all areas of urban life environment [12]. According to Jonge (1998), tradition of pilgrimage in Sunan Ampel was conducted with Islamic prayer. This is done by thousands of people every day [13]. People believe that he is a holy figure who can give blessings, fortunes, as well as ranks and position. The same activities are done in Sunan Giri's site and in other *Walisanga* sites. In both Ampel and Giri sites can it be found traditions done by the public such as prayers done by thousands of Muslims every day, congregations every Thursday evening in both of the mosque of Sunan Ampel and Sunan Giri done by hundreds of visitors, Tarawih prayers done in every Ramadhan month, reciting Quran, night meditation, and any other religious activities.

In Todaiji, rituals are done involving the monks and temple officials, but visitors are only observers. Visitors are only welcomed to participate during certain events, such as the *Taimatsu*. Rituals are only done after the sunset from the first night to the last. 6 to 8 meters giant torches long are lit in the balcony of Nugatsudo above the crowd. The cascading sparks and the burnt coal are what the crowd awaits. There is belief which says if there are many sparks; the whole year will be safe and prosperous. If one of the visitors caught the sparks, they will get plenty of harvest and prosperity.

In Ampel, cultural and religious rituals are merged. Visitors are active contributors. In Giri, the people are also the actors in the rituals. There are similarities of purposes of these rituals in Indonesia and Japan namely to get blessings and prosperity. In Ampel and Giri, these rituals are done modestly, showcasing the social cohesion among the actors and the public participation. Here people come because they want to participate in the tradition as a good member of the community. Meanwhile in Todaiji, a lot of the visitors come to watch the show and enjoy celebration.

Physically, the elements found in both of these sites are:

i) Cemetery is one the last resort of the *Wali*, and their friends and families. The original form of Sunan Ampel cemetery is rectangular and has two-tiered *tajug* roof, an adaptation of Hindu's *meru* shape.

ii) The mosque is the almost original mosques built by the *Wali* can be found in Sunan Ampel, Sunan Kudus, Sunan Kalijaga, Sunan Bonang, Sunan Gunung Jati, and others. In the Sultanate period, many mosques were believed to be founded by *Wali* [14]

iii) Gate as a landmark is considered sacred. In Sunan Ampel Mosque there are five split *candi bentar* gates symbolizing five pillars of Islam Meanwhile in Sunan Giri site, a gate in the shape of split *candi bentar* was made of red bricks. This type of temple can be found a lot in the remains of Majapahit kingdom and in the minaret in Kudus, and recently in Hindu temple in Bali.

iv) Statues do not always present, but at the Sunan Giri site are two dragon head statues in front of the gates. During the Hindu and Buddha era, statues were used as a symbol and a media for communication. After Islam came, this culture was faded due to the clash with Islamic teachings. Apparently, Sunan Giri still held influences from the Hindu and Buddha cultures and used the surrounding culture to spread Islam.

v) Wells or water springs are considered as the sources of life and related to holiness. Ampel has a freshwater spring inside the mosque, which is unique since the site is surrounded by brackish water. Water is considered a blessing, as the source of life and the media for ablution. Similar to the *zam-zam* spring in Mecca, this water is sought by the many visitors.

vi) Spaces for economic activity are spread in several strategic locations. In Ampel the businessmen of local people from Arabic and Madurese descendant sell Middle-Eastern praying tools which earn the name "Little Persia" for the place. Meanwhile in Giri, businessmen come from both the local people and the surrounding citizens.. In both sites appear that the creative economy are less creative, it needs to encourage the local potential resources like leisure activities, clothing & culture, indigenous traditions of public life, or intellectual traditions that can be used to express the specialness of a location, although in Giri little more developed thanits counterpart Ampel [15].

Viewing the elements that are compared in both sites (Ampel and Giri), there is the existence of local cultural products which becomes the symbol of transition towards Islam from the previous Hindu and Buddhist cultures. The architecture can be found in the two- tiered *tajug* rood and the gate shaped as a *candi bentar*. *Tajug* is an adapted shape from Hindu/Buddhist culture to Islamic culture. This is the Islamic – Javanese wisdom that teaches tolerance and inherits it to the next generation. The mosque was constructed with woods supported by 16 pillars. These mosques are a symbol of an egalitarian Islam. The existence of the five gates in the Sunan Ampel site is a symbol of the Pillars of Islam and the creative economy activities. The principles of preserving tradition and local culture are strengthening both from the tangible and intangible aspects in which [16] called it the social intensive space, a space in limited area used for many activities frequently done and qualified activities experienced by people [16]. The activities include (1) socio-cultural activities, religious activities, and economic activities. The social intensive space has been found in both Ampel and Giri sites.

Todaiji also uses original natural elements such as trees, rivers, valleys, and fauna (deers) as a unity with the temple that creates wonderful scenery. The temple buildings with their architecture peculiarity of Japan's woodwork are also spectacular, unmatched buildings, even for today. Meanwhile Toji temple uses elements like water, stone, trees, and air in their Japanese garden, temple buildings, and the world's tallest wooden pagoda. Both Todaiji temples and Toji Temples are spectacular works of genius loci. Todaiji temple has maintained its tradition of routine reconstructions. The buildings' shapes, structures, designs, materials, tradition and technicalities are preserved exactly like the original. Japan has proven that the changes in reduced size elements can still respect the materials and technicalities of the original buildings.

A flea market is held in Toji Temple on 21st month to remember the death of Kobo Daishi Monk. This is a traditional bazaar which also serves as a media to socialize. Many craftsmen come from the surrounding cities of Kyoto to participate in this event. This event is also an attraction for the visitors of Toji Temple. Various handmade crafts are sold, such as paintings, kimonos, accessories, gifts, traditional foods, and many others.

Observing the behavior of the thousands visitors performing rituals/prayers both in in the cemetery or in the mosque, according to Pangarsa , shows that there is a rotation of energy in this place which is able to drive people to do something [17]. In this context, this space is able to move visitors to do activities such as praying, using the freshwater in Ampel while praying for blessings, expecting blessings from the religious leaders, doing charity, and others. In other words, humans establish communication with God or with the mystical existence of Sunan Ampel as a holy figure who possesses special blessings with the hope that their wishes will be realized.

Meanwhile the place created in Todaiji is one of the Historic Monuments of Ancient Nara and as the remainder of the 8th century. Todaiji is one of the 8 historical places in Nara's Old Town. Todaiji

Temple itself is Nara's icon, and is a fusion of different wooden buildings with Japanese architecture, open spaces, visitors, and roaming deer.

3.3. Aspect of Balance with Nature
Natural balance emphasizes two things: how to use natural resources and how to exploit it [18]. Over-exploitation will cause natural deterioration, land erosion, aquifer depletion, river swallowing, flood, and all these disasters will be counter productive for the urban development.

The natural elements in Ampel are lowlands around 0.5 meter above sea level, close to the beach. The land around Ampel is alluvial ground from the coast sediment. The city of Surabaya is generally safe from earthquakes or volcanic threats. The site is passed by Kalimas River, originated from Brantas River in the west and Pegirikan River in the east. This has made the region a strategic location since the earliest development period. River is always considered the source of life: drinking water, irrigation, water for rituals, and transportation medium. Moreover, Ampel in Surabaya is part of North Java coast that should concern the rise of sea level globally. This incident is not only due to global circumstances, but also the unbalanced development with the environment, such as the construction of industrial areas, ports, and housings [19]. Ampel is located in Semampir sub district along with 8 other sub district surrounding in the waterside area of Surabaya, with total area of 15,382 ha. Among the nine sub-regencies, Semampir has the highest population density (176,726 in 2013 with 876 ha total area), which is 202 person/ha. The waterside area is almost constantly flooded by the high tide [11]. The tendency of the water to rise constantly is a threat to the industrial complexes, housings, and the waterside tourism. Generally this condition will also harm Surabaya's economy. So Suprijanto (2003) emohazized the importance of its anticipation by rising dams and water pumps in order to conserve the vital activities in this area, especially in Ampel, such as housing and tourism [19].

The whole environmentally friendly development needs to improve the area. On the other side, an effort of environment management needs to be done to the already built environment. Retention pools are needed to control flood, especially in areas with heavy flooding. Creating a green environment also needs to be done, especially in kampong or villages with high population density. Giving pedestrians proper facilities to access nearby housings and the religious sites can help to reduce carbon emission from motor vehicle. Visitors and the public should also be motivated to love the environment by trashing properly, reducing the use of plastics and Styrofoam, and participating actively in maintaining and conserving the environment, because when the public is participative, all efforts to maintain environmental balance will be easier to execute.

3.4. The Aspect of Land Use
Administratively, Ampel is located in Ampel sub district. This site is surrounded by settlements for Arab descents that have lived in the area since the implementation of ethnical centralization during the Dutch invasion (1841). This settlement has the characteristics and patterns of Muslim settlements [19]. The linear pattern of the settlement follows the street networks and the houses are street-oriented. Cul-de-sacs are organized from the labyrinth arrangement and the complicated street networks, which consist of several connected alleys. The mosque is the centre of the orientation. Streets and roads have hierarchy where labyrinth alleys lead to bigger roads. One of the important alleys is Ampel Mosque alley because it has been turned into a well known shopping street named Pasar Guba.

According to the data in Ampel sub district, the area comprises of 29 ha, including 2.5 ha (9%) for settlements, 5 ha for offices (17%), 2 ha for business area or shops (7%), 5 ha for industrial area (17%), 10 ha for public facilities (35%), and others including green space, blue space, and other open spaces amassing 5 ha (17%). Open green space percentage is still below the 30% standard. This will cause the depleting of water absorption area which eventually could damage the ecosystem. If this condition is not treated well, this historical site could be under threats of flooding. The area has 21,873 population in 2014, which means that it is considerably highly dense (754 person/ha). There are few open spaces in Ampel, specifically those which are visible in the site and Kalimas River. Building density is high to provide the increasing businesses and housings inasmuch as Ampel is located in a strategic downtown area. There are also open spaces in the northeast of the site, but these are used as parking areas. Open spaces developments are spontaneous; they are not well planned, not properly managed, and not integrated with the site.

Todaiji temple is located in Nara's modern city, along with the other 78 historical buildings, comprising an area of 617.0 ha which is surrounded by a 1.962 ha area as the buffer zone and a 539.0 ha area as the zone of history and environment harmonization. Todaiji temple is located inside Nara Park which was a previously private public part in the base of Mountain Wakakusa and constructed in 1880. The materials used are cypress, tiles & stones. They avoid the traditional method, demolishing and building the new one, because they aware to conserve the forest. This timber temple was well constructed, so that it can last for long periods (chrispythoughts.wordpress.com). Forest and trees in Japan have significant roles to preserve the nature, so they should be protected and conserved as a heritage [21]. Nara Park is one of the most beautiful places awarded by the Ministry of Education, Culture, Sports, Science and Technology. More than 1,200 Shika deer roam freely in this 502 ha park. The park includes Todaiji, Kofuku-ji, and Kasuga Shrine, where all of them are included as Nara city's natural resources. This park is open to the public, so it can now complement the temples inside it and can be accessed by visitors. Inside the park is also 'Chaya', traditional Japanese building which serves tea and wagashi. The existence of Nara Park as the well managed city's lung is mesmerizing because of its natural resources, botanical resources, as well as the deer and the temples. Deer are believed to be messengers in the Shinto teachings. Nara Park is considered as the symbol of the city and the national heritage, so its existence is protected and preserved by the government of Nara Prefecture. The existence of Nara Park, not only keeping the balance between natural environment and built environment, but also being a buffer zone for Todaiji site and the surrounding sites and to reducing the environmental burden of those sites. Land use and space management by dividing visitors attention to not only focus on the main spaces is urgently needed both sites is order to protect the main spaces and to develop the buffer zone.

Figure 7. Open Space in Ampel is minimal and it is full of supporting activities Clockwise figures: The site plan of Ampel sites; Open space for parking; Open space as the yard of Ampel Mosque; Open space for the blessed freshwater springs; Open space near the Islamic dorms

4. Conclusion

Learning from cases of the best practices in preservation and conservation of historical monuments at Japan to develop *Walisanga* sites in Indonesia, should consider the folloeing factors and facts:

i. Further investigation and research to find the valid boundaries of the area and objects of archaeological both Ampel and Giri sites.
ii. Development of the building and the environment to maintain the original shape, structure, design, materials, traditions and techniques as the original.
iii. Aspects of the legislation in Indonesia are still weak both in the law and its enforcement. Regulation is needed to handle these *Walisanga* sites, and to the base of the operational techniques implementation in conservation and preservation.

iv. Given the role of these extraordinary sites in spreading Islam in Indonesia, it is necessary to promote the *Walisanga* site as world heritage.
v. To keep the management runs well, it needs organizations that manage the site more professionally. These organizations include the combined elements of the central government, local government, and the community.

To get to the harmony between nature and culture is necessary to enhance performance, facilities and infrastructures. Learning from the sites like Todaiji Temple in Nara prefecture and Toji Temple in Kyoto, several efforts can be done as follows:

i. Ampel site, located at the city centre, has been mingled with the surrounding residences, which could threaten the sustainability of the sites. This condition needs restoration . Thus it will save the artifacts, which are difficult to manage because they are spread the population living in the surroundings.
ii. The Giri sites, the balance in environmental conditions that are slightly better, but it is still lack of the balance in the cultural history due to the lack of attention to the historic artefacts.
iii. Both sites have less buffer zone. A buffer zone is needed by both sites to maintain the balance of cultural historic environment and in this case, to keep these sites from the damage done by visitors
iv. From the spatial aspects.
The required additional open spaces / public spaces are the representative to serve as buffer zones or zones of cultural and historical environment balance. It needs activity supports in order to revive the region based on creative economy as long as the placement of activity supports does not cover the sites and does not interfere with the circulation of visitors.

5. Acknowledgments

Authors would like to thank to: (1) Directorate of General of Higher Education of Republic of Indonesia for for the financial support to conduct research in Indonesia, (2) The Sumitomo Foundation for financial support to conduct research in Japan; (3) The National Institute of Technology - Akashi College; (4) Ritsumeikan University; (5) Kyoto Center for Community Collaboration (KCCC); (6) and also Mr. Yasumichi Murakami from Hyogo Prefecture Board of Education; who have kindly shared data and information to us.

6. References

[1] Zahnd, M., 2006. *Urban Design-Model For Dense, Laminar-Based New Quarters For Indonesian Cites*, Inta Conference Proceeding, Yogyakarta : Duta Wacana Cristian Univesity.
[2] De Blij, H., 2008. The Power of Place Geography, Destiny and Globalization's Rough Landscape. New York, Oxford University Press.
[3] Santoso, J., 2013, Understanding Urban Transformation in Asia: Case Study of Jakarta, Journal of Tata Loka, Vol 15, No 2, May 2013.
[4] Samalavicius, A., 2017. Rethinking Modernism and the Built Environment. Cambridge, Cambridge Scholars Publishing.
[5] Tomlinson, J., 1999. Globalization and Culture, Chicago: The University of Chicago Press.
[6] Arefi, M., 1999. Non-Place and Placelessness as Narratives of Loss: Rethinking the Nations of Place. *Journal of Urban Design 4* (2), pp. 179-193.
[7] Serageldin, B., 2001. *The Sacred City*, New York : Routledge.
[8] Rypkema,D., D., 2005. Globalization, Urban Heritage, and The 21th Century Economy, Global Urban Development, Vol 1, Issue 1 May 2005.
[9] Suprapti, A.; Indriastjario; Sardjono B., A., 2016. The Urban Heritage of Masdjid Sunan Ampel Surabaya, Toward the Intelligent Urbanism Development, Procedia Social and Behavioral Sciences 227 (2016): 601-608.
[10] Supriharjo, R., 2004. *Nilai Ruang Pada pada Kawasan Bersejarah Ampel Surabaya*, Disertasi Program Doktor, Jogyakarta, Teknik Arsitektur Universitas Gajahmada
[11] http://www.toji.or.jp
[12] Oktay, D., 2011. Human Sustainable Urbanism: In Pursuit of Echological and Sosio- Cultural Sustainability. Procedia Sosial and Behavioral Sciences 26 (2012) 16-27

[13] Jonge, de, H., 1998. Pilgrimages and Local Islam on Java, Jurnal Studia Islamika Vol V, No 2, Jakarta, IAIN Syarif Hidayatullah.

[14] Budi, S., B., 2005. A Study on the History and Development of the Javanese Mosque Part 2: The Historical Setting and Role of Javanese Mosque under the Sultanates, Journal of Asian architecture and Building Engineering Vol 4 No 1, Japan, J-Stage.

[15] Landry, C. and Bianchini B,. 2000. *The Creative City: A toolkit for Urban Innovators, London UK, Damos*

[16] Suprapti, A.; Budihardjo E.; Pangarsa, W., G,; Kistanto, H., K., 2010. *Ethnography-Architecture in Kampong Kauman Semarang, A Comprehension of Cultural Toward Space*, American Journal of Engineering and Applied Sciences 3 (3).

[17] Pangarsa, W., G., 2006. Merah Putih Arsitektur Nusantara, Yogjakarta: Andi Offset.

[18] Benninger, C., (2001): "Principles of Intelligent Urbanism," in Ekistics, Volume 69, Number 412, pp. 39 –65, Athens.

[19] Suprijanto, I., 2003. Kerentanan Kawasan Tepi Air terhadap Kenaikan Permukaan Air Laut Kasus Kawasan Tepi Air Kota Surabaya, Journal of Architecture & Built Environment 3(1).

[20] Nur, W., K.. 2010. Tipomorfologi dan Kekhasan Tata Ruang Rumah Arab di Ampel Surabaya, Semarang, Jurnal Tesa Arsitektur 8 (1).

[21] Omura, H., 2004. *Trees, Forest and Religion in Japan*, Mountain Research and Development 24(2): 178-182.2004, Washington DC, International Mountain Society.

Principles and concepts in designing tropical-shore settlement in estuary ecosystem, case study: Weriagar District, Bintuni Bay

Firmansyah*, Bintang Nidia Kusuma, Ira Prayuni, Aldo Fernando
Institut Teknologi Bandung, Bandung, Indonesia,

*Corresponding e-mail: firmansyah@ar.itb.ac.id

Abstract. Weriagar District is located in estuary area and is prone to land loss, due to river and coastal erosion. This paper will describe about tropical-shore settlement design and house design in estuary area. The results from analysis phase shows that it's necessary to design a house and settlement that can fulfil the needs of indigenous people, both functionally and aesthetically. Functionally, the house is designed to provide spaces for both private and public needs of the family. It can be used either as a family private space or as a public gathering space between family and their neighbours. Aesthetically, house' architectural form is designed into that identifies the locality of Weriagar District. The houses' design feature highlighted in using local material, rainwater harvesting system, high pitched roof feature as response to hot-humid climate, and elevated-floor feature as response to tidal condition in estuary area. The houses design also considered daily activity pattern and community culture, including appropriate structure, construction, and material availability. The expected result was that the settlement improvement and house design would meet suitable standards and needs of inhabitants in Weriagar District.

Keywords: Weriagar District, Estuary Area, Tropical-Shore, Settlement

1. Introduction

Weriagar District located in Bintuni Bay Regency, West Papua, Indonesia. Bordered by a river in the northern side and the sea in the southern side, Weriagar district is one of the tropical shore settlement in estuary area. The Tropical Shore Settlement Improvement Program was initiated as an effort to rehabilitate physical condition and life quality in Weriagar District. The Rehabilitate activities are to support and improve life quality of communities not only in Weriagar District but also in the Bintuni Bay and beyond by providing a reliable and selected resources. The Tropical Shore Settlement Improvement program scope includes House Design and Settlement Rearrangement, as well as Infrastructure Development. This program phase consists of Site Survey, Site Analysis, Concept Select Matrix, Value Improving Practices, and Detailed Engineering Design. This program was intended to obtain the house design that suits with convenience and functionality, but does not eliminate the architecture value both of function and aesthetics of the building. Infrastructure development program consist of improving and complementing the existing of public infrastructure and social facilities in accordance with the regional priorities of development in Weriagar District. The design and development in this program must be based on the ecology, aesthetic, function, and existing building conditions.

2. Method

Some problem issues found during site survey were soil erosion and abrasion that risk the settlements and livelihood. Existing soil condition (peat, mud, and tidal) affect the material structure design and construction methods for the settlement development. Besides that, Weriagar District is located in a remote area on the bank of the river and foreshore caused some problem, including difficulties in finding clean water, strong wind gusts, frequent occurrence of tidal, so that the house design must be accordance with the weather conditions in the area. The Tropical Shore Settlement Improvement Program was intended to find the most prospective solution to answer the issue through design. The method used in the program was site observation, site survey, data collection through primer and secondary sources, site analysis, concept selection process, and detailed engineering design.

3. Resilient Settlement Design

There are several factors that can affect the settlement development [1], namely: Geographical Location, Population, Facility and Infrastructure, Economic and affordability of purchasing power, Social and Cultural, Science and Technology, Institutional, and Community Participation. In this Tropical Shore Settlement Improvement program, resilient design is one of the important criteria. Resilient design is needed to protect natural features that provide ecosystem services [2]. The first step in resilient design for inland flooding is to identifying and mapping areas of any existing natural features and resources that provide ecosystem services in absorbing rainfall to develop a plan that protects or restores these features [3]. These include Swales, depressions, and flow pathways; Wetlands, headwater streams, and stream systems; Vegetation; Geology, soils, and slopes; Geotechnical information for building construction, such as soil borings.

There are some criteria for designing in riparian area [3]; (a) Rainwater Harvesting as an extension of cisterns, additional measures may include designs to recover and reuse all water that falls on-site through integrated rainwater harvesting; (b) Native Planting once established, native planting requires little or no maintenance and is a cost- effective measure by which to reduce municipal costs and improve environmental benefits by replacing traditional lawn planting and care. They are adapted to local conditions and resistant to local insects; (c) Riparian Buffers are areas established by setbacks from streams, rivers, and wetlands. A riparian buffer preserves the stream's natural characteristics. It filters sediments, nutrients, and chemicals from surface run off and shallow groundwater; (d) Wetland Protection and Restoration. A wetland is an exquisitely complex biological system that cleans water and air and provides a natural sponge for varying water flows [4]. Where some detention is required for large storm events, newly created wetland areas will improve water quality and create habitat while providing detention

At the building scale, there are some building elements criteria for flood design [3]. Conditions on waterfront sites are often unsuitable for supporting shallow foundations. Foundations need to be constructed at sufficient depth to prevent undermining by eroding soils and scour. Foundation for waterfront structure require substantial bracing to resist hydrodynamic wave and wind forces. Pile foundations typically need bracing bolted to the pilings. Structures should be securely anchored to the pilings with corrosion resistance bolts. For Structure, it should be Wind Resistant Design, the porosity of construction and openings either left open or created during the storm determine the extent of wind penetration to the interior and contribute additive (or subtract from) external pressure. In coastal environments exposed to salt air, corrosion resistant structural connectors are required. Salt is very corrosive to metals, including concrete reinforcement that is exposed over time by cracking or spalling of the concrete. For the material, it should consider the use of flood damage resistant material, The International Building Code (IBC) defines flood damage resistant materials as "capable of withstanding direct and prolonged contact with floodwaters without sustaining any significant damage that requires more than cosmetic repair".

4. Discussion

4.1 Weriagar District Existing Condition

Site condition were collected on the Site Survey on May 2014.

Figure 1. Site Map.

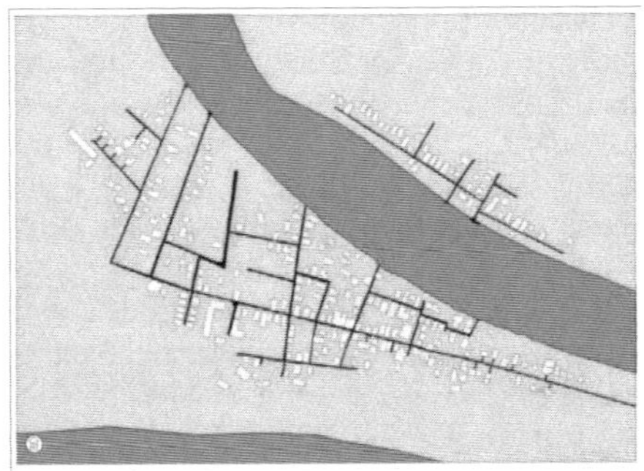

Figure 2. Green Area in Weriagar District.

4.1.1. Water Availability and Sanitation System.

The water with high salinity enter into the bay and mix with low salinity water so it causes mixing between the fresh water masses and bay water stuck in the head of bay. This factor was the one that make it difficult in finding clean water. Based on the information gathered from the site survey, the water available is mostly infected with E. Coli bacteria and is overly salty. That condition of water is unsuitable to be used for daily purposes for the villagers in the Weriagar District Area. The infected clean water in this area is caused by contamination from waste water of the existing houses. At the moment, there are no waste water sanitation system available.

4.1.2. Green area and vegetation.

The riverbank areas and seaside that have the heaviest process of erosion and abrasion are those where the natural cover has been disturbed and that have become an open land. Nature can no longer be a supporting and inhibiting element of the rate of erosion due to the fact that local plants such as Nypa palm and mangrove forests have lost their media to be able to grow back naturally. Built environment especially in the dense area of the village have experience the most serious degradation in its environment and spatial quality with the decrease of green open spaces. These green open spaces are supposed to be inhibiting the rate of the tide that erodes the fertile soil layer on the wet land.

4.1.3 Building Materials.

The use of local materials becomes dominant because of limited availability of materials from the outside area and lack of supporting accessibility to send the material into the site. The local materials have been processed beforehand to ensure their strength, durability and to facilitate their maintenance. Local materials are used as structural elements of the building and as well as cover the walls of the building. Wood is the dominant material used in the existing buildings found in the Weriagar District Area. However, the fact that materials used in residential buildings are mainly unprocessed, local materials or raw materials indicate that the construction quality does not meet appropriateness standards especially those related to durability against weather condition.

4.2 Development of Design Criteria

4.2.1. Site plan design criteria

Some principles were used as consideration in arranging Weriagar District site plan [5]:

a. Geographical conditions: such as water, soil and climate affect the engineering, design, and construction on the development area. To prevent soil erosion and abrasion in Weriagar District that often happened, it is necessary to make structural protection. The tidal conditions also have to be considered in the design of the building and walkway in Weriagar District.
b. Image of Region: a visual image of character which captured when seeing a region to be maintained and created in the Weriagar District is natural and comfortable coastal village.
c. Accessibility: Existing access in this area should be supported by improving its infrastructure. Weriagar district can only be accessed via the water lane, especially the river. Therefore, the jetty

dock and boat dock is the important basic infrastructure. While in the area there is a walkway as the main circulation infrastructures.

d. Administrative Boundaries: The boundaries of the land ownership/community rights, village boundaries that should not change, and the minimum number of resident remain the same.
e. The building in this region is dominated by houses and public infrastructure. The building is generally a house on stilts made of wood material. Tidal condition affecting the floor level of the house and social facilites, as well as the surface level of walkway as public infrastructure. The highest tidal water level measured is approximately 1 m, therefore the floor level of the house and social facilities were designed above the tide level, approximately 1.5 m from the existing ground level. Meanwhile, the surface level of the walkway is 1.3 m from the existing ground. The distance between the new houses on either side is at least 6 m, and the distance to the front porch of the main walkway is at least 3 m.
f. Landscape Arrangement is indispensable especially in the border areas of the river and coast area. The potential of erosion and abrasion is very high.
g. Completeness infrastructure of region: Weriagar district as one administrative area in Bintuni Bay Regency, need to have regional infrastructure. Public infrastructure consists of a walkway, drainage, boat dock, and the Social Facilities were Sanggar Pemuda, Posyandu and a football field.
h. Social and Cultural: Culture and habits of the local people should not be overlooked in settlement improvement program. There are many social and historical values that are embedded in their living that made a principle of life for the next generations, such as ceremonies when moving house, sacred house, do routine activity in the edge of the river and many other things.
i. Regulation: The estuary area has a variety of restrictions or rules in size and complexity. Thus, in the arrangement Weriagar District need to consider regulation of river border, coastal border, and emphasizes the protection of the environment and restore natural function as an area that is always experiencing tidal conditions.

4.2.2. House design criteria has some of point that must be suits with the site condition, such as:

1. Climate

The design creates a new identity that responds to its climate and context. The main purpose of building is to give its occupants comfort inside the building. The comfort parameters include [6]: (a) Natural Daylight and Heat Control; Natural lighting is what the design seeks the most. The building has to design not too wide so that the sun can reach into innermost room. Points to control are, the form of building mass; the pattern, position, and shape of building openings; the system of building's cover and shade control of building. (b) Air Circulation and Humidity Control; Good air circulation, especially inside the building, makes the occupants able to do the activity well. The more opening in the lower side of the building allows air to flow inside, while the opening in the upper side pulls the air below and get it out from the building. This air movement is the optimal condition of the building.

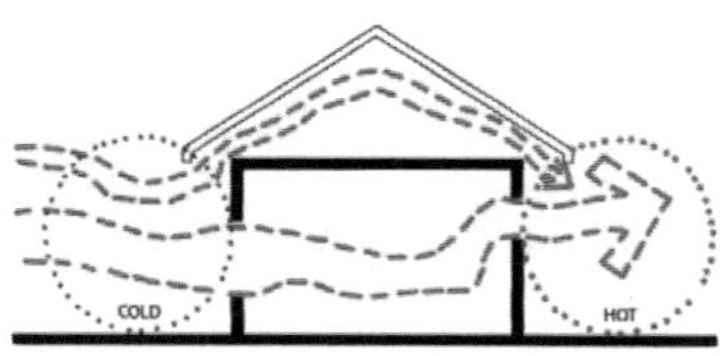

Figure 3. Air Circulation Diagram

2. Functionality

Building design with modular system to accommodate typical functions is applied in every room of the house. Modular system is a way to keep the buildings in unity. The modular system leads to easiness in design and construction process. Open layout system allows function flexibility, adaptive to function, interior layout change, and easiness for development and expansion of building module and mass.

3. Design Character and Façade
   Design character of the buildings gives the most impression about the character of the house itself. Vernacular character is contextual with local and cultural condition of Weriagar District. The house design has to represent the local richness by applying it in the design elements.
4. Constructability
   Another aspect that should be considered is the easiness in the building construction process. The building material has to be fixed in early design process. It will decide how large a space can be or how tall the building will be. The material should be surveyed to make sure its availability in the location. With large number of availability, the construction can be carried out using the same material to avoid social gap. The structural modular system is benefit guideline in the initial design process as well as each development.

4.3 Design Concept

4.3.1. Design Concept of Weriagar District Settlement Plan

a. Housing Arrangement Concept

Based on the analysis of neighborhood space usage, community activities, and the layout of the existing building, as well as the structuring principle of settlement in estuary there are several concepts used in the arrangement of the house. Building Boundary Line Determination is based on security considerations, health, comfort, and harmony with the environment as well as the building height.

b. Public Infrastructure Arrangement Concept

The public infrastructure items in Weriagar District are Walkway, Drainage and Boat Dock. The arrangement of the new Walkway path follow the existing walkway path. Walkway path aims to connect each village in the Weriagar Disrict, ease for the people to access boat dock and jetty as the main transportation, and to anticipate tidal conditions.

Drainage system arrangement aimed to speed up the flow of the tides and the flow of surface water to get out from the landside. Drainage was placed on one side of walkway to ease inspection. In drainage system, water flow that coming out from the site will be heading to the river. Arrangement of boat dock that serves as transportation infrastructure for small boat belonging to people who used daily both as a mode of transportation.

Figure 4. Arrangement Pattern of House

Figure 5. Arrangement Pattern of Public Infrastructure

c. Social Facilities Arrangement Concept

Social facilities are facilities, equipment, or installations that constructed and needed for the proper function of society system especially in social service system. The purpose is to equate the distribution of social facilities in each village and to realize people aspirations of the people of community needs.

Figure 6. Arrangement Pattern of Social Facilities

Figure 7. Arrangement Pattern of River Border and Coastal Border

d. River Boundary Line and Coastal Boundary Line Arrangement Concept
River and coastal in Weriagar District planned to be built an embankment as protection from erosion and abrasion, then the boundary line rules of river and embankment coastal. Therefore, the river boundary line in Weriagar District at least 5 meters from the outer side of the embankment. Based on consideration to natural protection from abrasion and geobag protection from community activities, the river boundary line space will add 5 meters to 10 meters to plant Nypa palm. Based on national regulation that states the maximum building height is 15 meters and coastal boundary line is at least 100 meters, the site plan must be accommodated the 100-meter border and then planted with mangroves. Mangroves planting is intended as one of the efforts to protect the beach from erosion [7].

4.3.2. Building Concept

a. General Concept

Based on the analysis result of space usage preferences, inhabitant's activities, and existing house layout, the following recommendations on house layout are made, house layout should follow daily activity pattern, should follow room usage and function pattern that different from public buildings, should consider the building growth to anticipate additional room needs which have not been fulfilled in the initial house construction stage. The house shape and character should still be maintained while integrating original and additional parts [8].

b. Building Mass and Form

The concept of the building mass is to express the traditional aspect of Weriagar District local house. The building is adapted from the existing traditional house found in Weriagar District. One of the main identity of a Weriagar House is the elevated system of the house. The house floor is elevated at the maximum height of the water from the river located near the site, as a response to the swampy and wet soil condition. The elements which is adapted from the Weriagar District traditional house is the high-pitched roof. Not only to create the traditional expression of the house, but also to use natural ventilation system [9].

Figure 8. Building Mass

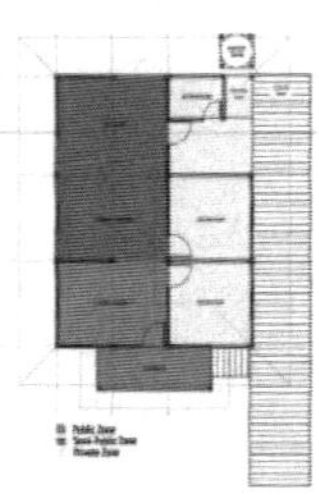

Figure 9. Zoning Diagram

The roofing system was modified from the old traditional roofing system by adding vents at both sides of the roof. The purpose is to allow the trapped hot/warm air in the roof to flow through (crossing ventilation). The circulation of air in the roof will decrease the number of heat that received inside the house.

c. Zoning
Zoning in the house designed based on function of space to anticipate the presence of space that has not been fulfilled at the early construction so the house character can still be maintained. The zones in the house divided into three zones, they are the public zone, the semi-public zone, and the private zone. The positioning of the zones is shown in the diagram above.

d. Air Circulation
One design approach for the indigenous people in Weriagar District Area is the natural ventilation to provide good indoor environment quality. It maintains a healthy, comfortable, and productive indoor climate. Natural ventilation systems rely on pressure differences to move fresh air through buildings. The amount of ventilation will depend on the placement of openings in the building.

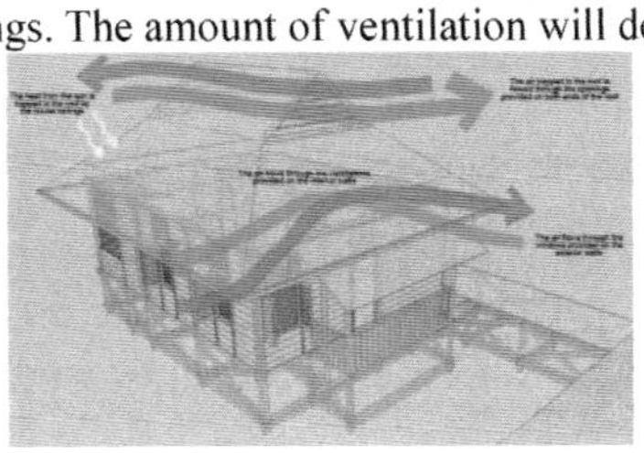

Figure 10. Air Circulation Concept

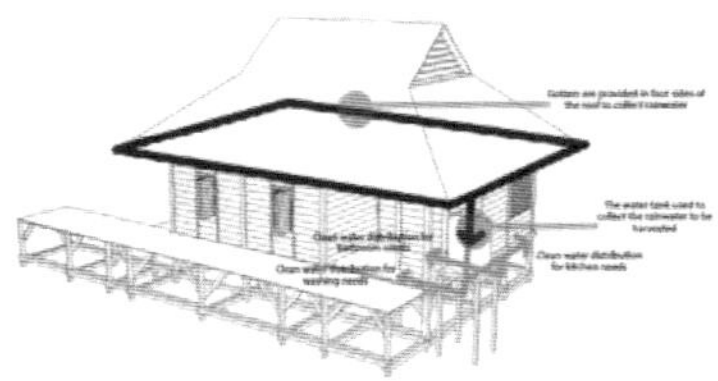

Figure 11. Water Treatment Concept

e. Water Treatment
Due to the site condition where the clean water supply system is still insufficient to the needs of the people, rain water harvesting is one of the solution applied in this housing design. The roof of the house was designed to tilt in four directions and completed with gutters in all four sides. The gutters will collect the rainwater which then be channelled into the water tank provided at the back of the house. The water tank is equipped with a filter system using a sheet of gauze placed on the openings of the water tank to ensure that the water that enters the water tank is free from dirt. By using the filter, the rain water held in the tank can directly be used for daily needs in the house.

f. Natural Lighting
Lighting is one of the most critical qualities of an indoor space. Exposure to natural light can be the solution to create a comfortable space inside the house. In Weriagar District Area, electricity is still limited. The indigenous people still relies on solar machine (genset) as the energy source. With that condition, natural lighting is one of the important concept to be used in this design. The diagram shows that the house design enables sunlight to enter the house easily as natural lighting. But the roof is designed wide enough to prevent the heat from the sun to enter in the middle of the day. This is intended to prevent the house users to feel overly heated during the day.

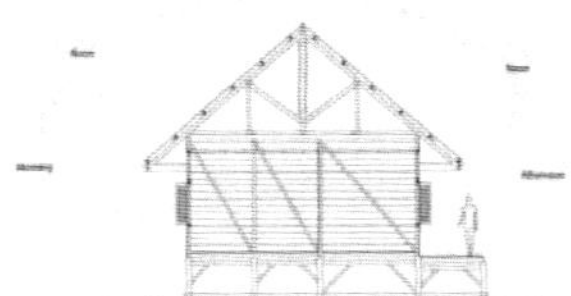

Figure 12. Natural Lighting Concept

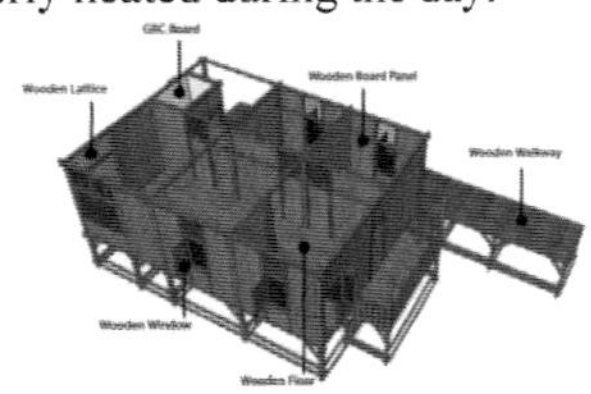

Figure 13. Building material Concept

g. Building Material
Weriagar District is located in the remote area with difficult access, so the usage of materials both of component and systems structure using local material that are easily found on the location of the design will minimize the cost of shipping materials. The material must be processed beforehand so

the quality of construction can fulfil the building standard of comfort and can be adapted with weather conditions in the region, and the material chosen is wood. Wood is the dominant material used in the existing buildings found in the Weriagar District Area.

h. Modular Structure System
A modular structure is used as an approach to reduce cost (due to less customization and shorter learning time) and to offer flexibility in the house design. The module used in this housing design varies in the multiple of 1,50 m, 2,00 m, and 3,00 m. The smallest module used has the length of 1,50 m x 2,00 m. While the biggest module used has the length of 3,00 m x 3,00 m. The length of the module used is based on the availability of the material in the area.

5. Conclusion

Weriagar District is included into the category of settlement in the estuary area by its location in the coastal area that flanked by river and sea. By developing criteria and concept design based on resilient design principle and criteria, The Tropical Shore Settlement Improvement Program attempt to upgrading physical condition of Weriagar District and to support and improve quality of life of communities in the Bintuni Bay and beyond by providing a reliable and selected resources. Some of the design criteria for the settlements in the estuary area are divided into two concepts. Site plan concept that describes the fulfilment of criteria design used as consideration in structuring the site plan Weriagar District and the concept building based on analysis result of space usage preferences, inhabitant's activities, existing house layout and design criteria for buildings in estuary areas.

References:

[1] Supriharyono. 2007. *Konservasi Ekosistem Sumber Daya Hayati*. Pustaka Pelajar. Yogyakarta.
[2] Bengen, D.G. 2004. *Menuju Pengelolaan Wilayah Pesisir Terpadu Berbasis Daerah Aliran Sungai (DAS)*, dalam *Interaksi daratan dan Lautan*. Jakarta: Kedeputian Ilmu Pengetahuan Kebumian, Lembaga Ilmu Pengetahuan Indonesia.
[3] Watson Donald, Adams Michele. 2011. *Design for Flooding Architecture, Landscape, and Urban Design for Resilience to Flooding and Climate Change.*
[4] Wijanarka. 2008. *Desain Tepi Sungai*. Yogyakarta
[5] Brenn, Ann and Rigby, Dick. 1994. *Waterfronts: Cities Reclaim Their Edge.*
[6] Kay. R, Alder J. 1999. *Coastal Planning and Management*
[7] Peraturan Pemerintah Republik Indonesia no. 47 Tahun 1997 tentang Tata Ruang Wilayah dan Peraturan Menteri Pekerjaan Umum No.63/PRT/1993 tentang Garis Sempadan Sungai, Daerah Manfaat Sungai, daerah Penguasaan Sungai dan Bekas Sungai.
[8] Departemen PU. Dirjen Cipta Karya (1999). Petunjuk Teknis Pembangunan Perumahan Nelayan. Dep. PU. Jakarta.
[9] Undang-undang No. 28 Th 2002 mengenai Bangunan Gedung dan Peraturan Menteri Pekerjaan Umum Nomor: 29/PRT/M/2006 Tentang Pedoman Persyaratan Teknis Bangunan Gedung.

Towards the new urban agenda of safe cities: urban crimes in four Indonesian cities

B. Setiawan

Urban and Regional Planning, Gadjah Mada University, Yogyakarta Indonesia

Corresponding e-mail: bobi.setiawan@yahoo.com

Abstract. As has stated in the Habitat III agreement, named as The New Urban Agenda/NUA, signed in Quito, Ecuador, October 2016, safe cities is one among important global goals that should be achieved in the next twenty years - safe city is defined generally by city that is equated with reduced violence on the one hand, and on the other, with reduced uncertainty. The fact that under a rapid urbanization and urban transformation in Indonesia, urban crimes tend to increase in many Indonesian cities questions whether Indonesian cities are ready to ensure the idea of safe city. This paper aims to document the incidents of urban crimes, spatial dimensions, the socio-factors and how local community responses to such crimes. It utilised a multiple case studies of four cities in Indonesia which are: Yogyakarta, Surabaya, Makasar, and Manado, Data and information for the paper were primarily collected through secondary data, but some field observations and interviews with stakeholders related to urban crimes in four cities were also conducted. The paper could be considered as a preliminary study to understand the urban crimes phenomena in Indonesia and finds that the incidents of urban crimes are fluctuated in four cities explored. It argues that there is a significant correlation between urban crimes incidents and spatial characteristics of the city.

Key words: safe city, urban crimes, Indonesia

1. Introduction

1.1. Background: urbanization, urban growth, and urban challenge's in Indonesia

With rapid level of urbanization and urban growth, the sustainability and the quality lives of the majority of the Indonesian population will depend on the quality of the urban environment. Recently, more than half the residents of Indonesia (53,3%) reside in urban areas. In 2025, with the total population of around 273 million, approximately 68 percent of them will be living in urban areas. Such rapid urbanization rate, which is then manifested in the rapid growth of the city, should be anticipated with good planning and appropriate management in order to guarantee its sustainability.

This current status and future projection of urbanization in Indonesia brings two related aspects. First, urbanization and urban growth will facilitate the improvement of the quality of life and the environment of a large part of Indonesia society – something that we all hope. Second, however, this rapid urbanization phenomenon could increasingly perpetuate many problems that are already exist, and even leads to the worsening quality of the urban environment and communities.

One among serious problems faced by cities and towns in Indonesia is the increasing rate of urban crimes, which is relate to a broader issue of urban safety. As has been circulated through mass media, almost everyday, incidents of urban crimes, in various types from pity pick pocket, thief, social conflicts among neighbourhoods, robbery, kids abused, rape, to murder and even terrorism are happening in many cities in Indonesia. Such phenomena are worrying as they threatening the urban safety specifically and even urban sustainability in general.

1.2. The New Urban Agenda, Habitat III

The United Nations Conference on Housing and Sustainable Urban Development has wrapped up in Quito, Ecuador, in October 2016, with delegations adopting the New Urban Agenda – a new framework that lays out how cities should be planned and managed to best promote sustainable urbanization (http://www.worldurbancampaign.org/sites/default/files/newsletters/nov2016/index.html) It resulted in the adoption of what famous as the New Urban Agenda/NUA, which sets a new global standard for sustainable urban development, and will help us rethink how we plan, manage and live in cities. The NUA is roadmap for building cities that can serve as engines of prosperity and centers of cultural and social well-being while protecting the environment. It also provides guidance for achieving the Sustainable Development Goals and provides the underpinning for actions to address climate change. It will guide the efforts around urbanization of a wide range of actors — nation states, city and regional leaders, international development funders, United Nations programmed and civil society — for the next 20 years (http://un.orgh/sustainabledevelopment/blog/2016/10/newurbanagenda)

For safe cities, the NUA calls for the integration of crime prevention strategies into all urban planning efforts. It commits signatory states and cities to 'integrate measures for urban safety and violence, and crime prevention into all urban planning efforts, including in informal areas, and pay particular attention to vulnerability and cultural factors in the development of public security policies, including by eliminating the stigmatization of certain groups as security threats'.

To stakeholders in the safe-cities space, this commitment is important because: (1) it lays emphasis on data ("measures") that is vitally important in understanding the aggregate and disaggregated picture; (2) it draws attention to the 863 million slum dwellers that live in informal settlements; (3) it calls for cultural sensitivity in developing local solutions; and (4) it calls for the breaking down of the notion that 'criminals', 'offenders' and 'perpetrators' are static categories predetermined by an individual's or group's identity.

1.3. Studies on urban crime and safety

Crime is defined by the Merriam-Webster dictionary as "an act or the commission of an act that is forbidden or the omission of a duty that is commanded by a public law and that makes the offender liable to punishment by that law; especially : a gross violation of law." Meanwhile, the general legal definition of crime would be: "a violation of a law in which there is injury to the public or a member of the public and a term in jail or prison, and/or a fine as possible penalties. In the literature, there is some sentiment for excluding from the "crime" category crimes without victims, such as consensual acts, or violations in which only the perpetrator is hurt or is involved in something such as the personal use of illegal drugs." [2]

The studies on urban crime are generally of two types. The first type of studies are comparative studies, seeking to understand why some have higher crime rates than others. The second type are studies that focus on explaining variations in crime levels within cities [6]. However, both types of studies use similar theories and focus on the same social forces to understand their observations. The primary theories used to study urban crime are social disorganization, subculture, and conflict theories. While the spatial and environmental factors tend to be seen as the secondary ones [7].

Recently, urban crime has become a global phenomenon; criminal activities are becoming increasingly proliferated in many urban areas neither homes nor are markets and schools places safe because of frequent occurrence of crime incidents. This strengthens the proposition that population increase (urbanization) with it attended problems might be a factor for high level of crimes in urban

areas and it is also hypothesized that both property and violent crimes will crop up in densely populated urban areas [5].
Within the same city, criminal activities tend to be higher in the city canters than in the suburbs [5]. This may be connected with the intensities of activities and potentialities for offenders to commit crime in the city centre. In other words, there is a tendency where crime and criminal activities are constantly performing and growing in a sporadic faced.

1.4. The objectives of the paper
This paper has a general objective to document and evaluate the evidences of urban crime in Indonesian city. Such general goal is elaborated into three specific objectives as follows: 1) to document the incidents of urban crimes in four cities; 2) to analyze the spatial patterns of urban crimes and the socio-characteristics of the cities; and 3) to discuss local community responses to urban crimes.

2. Methods
The research is a multiple case studies on four cities in Indonesia which are: Yogyakarta, Surabaya, Makasar, and Manado. It is descriptive exploratory type of study, using primarily secondary data and information. Some field observations and interviews with stakeholders related to urban crimes, particularly local community leaders, in four cities were also conducted, particularly on issues related to local community response to urban crimes. Analysis was done qualitatively, while simple quantitative data was used to support discussions. The four cities were selected to represent big and medium size city in Indonesia.

The research is part of the research road map related to the issues raised in the New Urban Agenda Habitat III, particularly four key issues: inclusiveness, safety, resilient, and sustainability of the city. It is, however, has started several years before the Habitat III in the Graduate Program in Urban and Regional Planning, Department of Architecture and Planning, Gadjah Mada University, involving master students doing research related to urban crimes issues.

3. Findings and Discussions

3.1. Brief description of four cities
As has discussed previously, urban crimes are closely related to a much broader socio, economy, and political context of a given setting, in this case the urban setting. It is therefore important to understand urban crime with a better understanding of the socio, economy, political, and also physical and environmental setting of the city. The following table summarizes the general descriptions of the four cities explored in this paper.

Table 1. General Discriptions of Four Cities

No.	Aspects	Yogyakarta	Surabaya	Makasar	Manado
1	Area (Km 2)	32.5	350.54	175.77	157.26
2	Population	554,320	2,853,661	1,130,384	423,257
3	Population density (Pop/Km2)	17,056	8,140	6,431	2,691
4	Percentage of Green Public Space	17 %	21%	7 %	18%
5	Poverty Rate	8,67%	4,00%	5,36%	4,8%
6	Number of Unemployment (2014)	34,460	83,000	20,000	17,344
7	Annual Local Revenue, in billion	471	3,300	900	1,120

	rupiah (2014)					
7	Characteristicsof the cities	Inland city; Historic/traditional city; Center for education; Tourism city; famous for its harmony and social capital	Trade city; coastal city; historic city, strong social capital;	Coastal city; melting pot for eastren Indonesia; trade city	Coastal city, trade and service city.	

Sources: BPS of four cities (2015)

3.2. Incidents of urban crimes in four cities

Because urban crime is considered a recent or contemporary phenomena, studies on urban crime is limited in Indonesia. Data and information related to urban crime is therefore also limited. The only data recorded annual rate of urban crime can be found in the Police office in each city. In general, all policy office provides quite good data on urban crime, both types and spatial distribution. Data on urban crime incidents in this study were collected from Police office in every city. However, since there is no standard definition of urban crime incidents in Indonesia, data and information in this paper should be critically viewed and interpreted. As can be seen in the table below, data on urban crimes in four cities is recorded in term number in the past five year. The table also provides the level of urban crime incidents per 100 persons in the year 2014.

Table 2. Incidents of urban crimes in four cities

No.	City	2010	2011	2012	2013	2014	Incident of crime per 1000 pop.
1	Yogyakarta	421	435	400	369	422	0.70
2	Surabaya	5532	4563	4232	4232	4125	1.44
3	Makasar	2940	2905	3492	3493	3296	2.91
4	Manado	1280	1203	1008	1008	997	2.35

Note: For a compilation, in Jakarta, the crime rate per capita during 2008 was about 3.1 cases per 1000 people

As the table above indicates, the urban crime fluctuate in the past five years in Yogyakarta and Makasar, while it tend to consistently went slowdown in Surabaya and Manado. In Yogyakarta, although the incidents of urban crime is considered the lowest among that of other cities, the incidents of urban crime tends to be the same in the past five years. In Makasar, the incidents of urban crime tend to increase in the past two years (2013 and 2014).

3.3. The relation between cities characteristics, spatial patterns, and crime levels

As shown in the table 1 above, some characteristics are important to understand the context of urban crime in each city. Several aspects could be learned as follows. First, is that these four cities have different level of population density. Yogyakarta is considered the ones that has a very high population density, but with less crimes' incidents. Manado, on the other hands, has the lower population density, but with relatively higher incidents of urban crimes. Makasar, which has the highest urban crimes incidents of four cities explored has less population density compare to Yogyakarta. Surabaya has the similar population density compare to Makasar, but with only a half of urban crimes compare to Makasar. This figure is important as references from other cities in different countries indicate that there is a significant correlation between population density and urban crime rate.

Second, as has argued in several literatures, poverty rate in a given city has a direct correlation with urban crime incidents. The case of Indonesia cities, however, show a quite different phenomena. As shown in the table above, the poverty rate as well as the level of unemployment in the city of

IOP Conf. Series: Earth and Environmental Science **99** (2017) 012037 doi:10.1088/1755-1315/99/1/012037

Yogyakarta is considered high, but it does not correlate directly to the level of crime incidents in the city. Such phenomena is interesting and should be further explored as it possibly relates to another factor of urban crime which is crime prevention.

Lastly and correspond to the second point, the level of local revenue of the city, which is could be seen as indicator of the economic or financial capacity of the city, again, does not always relate to the urban crime incidents. Among four cities, again, Yogyakarta is considered the ones that has the lower local revenue, but it does not means that the level of urban crime in this city is the highest. While Makasar, which has the highest urban crimes among four cities studied, relatively has a lower local revenue compare to Manado.

As this study is not comprehensive in terms of study coverage, there is not yet conclusive understanding on the relation between urban crime and the spatial pattern of the city. It however, indicates that there is an indication that types of land use of the city has a significant correlation to the level of urban crimes. In Makasar city, for example, several types of land use such as sea port, commercial, and recreational activities tend to be the area with high intensity of urban crimes. While the housing area tends to be an area with low or moderate level of urban crimes.

In addition, there is also indication that in some city, particularly Makasar and Surabaya, urban crimes tend to happened most in urban fringe area rather in urban center. Such phenomena may relate to the fact that in the fringe area, many area are used for industrial area with no local community organized local respond to urban crimes.

In brief, the spatial, economy, and social condition of a city should be explored in trying to understand the urban crime. Theories explaining urban crimes from Western context, however could not always used to portray what happened in Indonesia. There is a need to further explore the socio, cultural context of a society to be able to comprehensively understand urban crime.

3.4. The livability index and people perceptions on urban crimes and safety

In the literature, urban crimes tend to correlate with people's perception towards the livability of the city. In four cities explored in this studies, it is quite interesting that people's perception on urban livability has somehow correlate with the incidents of urban crimes. As shown in the table below, Yogyakarta is considered as the city that has the most livability index according the survey by AIP, Association of Indonesian Planner and at the same time also has the lowest urban crimes incidents. In the case of Makasar and Manado there is a quite significant correlation between the incidents of crime and the level of livability perceived by people. In Surabaya, which has relatively low of urban crimes incidents, rank as the 6st of livability index.

These two data and information, however, should be critically interpreted as these two are taken in different years and also did not design purposively to see the relation between the two. It would be very interesting and important to further explore and study the link between these two, as they are very important to understand better about the whole concept of urban crime, safety, and livability.

Table 4. Level of livability of ten cities in Indonesia

No.	City	Livebility rank	Incidents of crime per 1000 population
1	Yogyakarta	1	0,7
2	Bandung	2	1,1
3	Semarang	3	1,6
4	Solo	4	0,8
5	Bogor	5	1,01
6	Surabaya	6	1,44
7	Malang	7	0,95
8	Manado	8	2,35
9	Denpasar	9	0,7
10	Makasar	10	2,91

Source: Indonesian Association of Planner/IAP (2014)

3.5. Local Responses To Urban Crimes

Recently many countries had implemented crime prevention as one of the most effective action to reduce the future risk of crime. It involves the police, the justice system, surveillance systems and guards, local cooperation between schools, social authorities and agencies and technical crime prevention such as; locks, alarms on doors and windows, video cameras and monitor ethic, to careful urban planning. Many services and programs contribute to create a safer communities. Government, non government organizations and the police all have strategies and programs to reduce crime. From the experienced of many developed countries, besides increasing the partnership and the relationship between the local communities, police and the government, the role of government in supporting local projects, funding, developing capacity, fostering analysis, setting guidelines and raising public and professional awareness are also important factors to promotes effective crime prevention.

In developing countries, however, where the capacity of the local governments tend not to be strong, the role of local community responses to crime are then become significant. Such local community responses to crimes are usually already developed as part of the community organization and management to maintain social harmony and solidarity. In Indonesia, for example, local community has a long history of organization and management as a means of achieving community harmony and solidarity. Both in rural and urban area, community is organized into a very strong sub-neighborhood and neighborhood groups. As argued by Sullivan and Guiness [6], [7], the sub-neighborhood organization called *Rukun Tetangga/RT* and the neighborhood organization called *Rukun Warga/RW*, which originally borrowed from rural community practices were then effectively also adopted in urban setting.

As community crimes and safety is part of the *RT* and *RW* organizations mandate, communities' responses toward crimes are already practices in many urban communities. Their effectiveness, however, depends on many factors. Studies in four cites in Indonesia indicate several factors contribute to the effectiveness of community responses towards urban crimes such as: 1) community solidarity; 2) community leaderships; 3) technical physical/design innovation; and 4) relations with external community including local government. Table below summaries the result of communities responses toward urban crimes in four cities, based on interviews with stakeholders involved in local community responses to urban crimes in four cities explored in this study.

Table 5. Examples of communities responses toward urban crimes

No.	Cities	Effectiveness of community responses	Factors	Notes/ Sources
1	Yogyakarta	Effective	Community solidarity, leadership, relations with local government	[11]
2	Surabaya	Effective	Community solidarity, leadership, relations with local government, technical/physical innovation	[12]
3	Makasar	Less effective	Community solidarity, technical innovation	[10]
4	Manado	Less effective	Community solidarity	[13]

Sources: Qualitative analysis based on interviews with stakeholders in four cities.

As can be seen from the table above, Yogyakarta and Surabaya, which has relatively low urban crime incidents, have more effective community responses toward urban crimes. While in Makasar and Manado have less effective local community responses. Another important finding is that in Yogyakarta and Surabaya, a more advanced technology and urban design is crucial to support the already exist, but conventional community responses. The uses of more advanced technologies such as CCTV, innovative urban design, physical barriers (i.e. by blocking entrance or by using portal and security gates, good lighting) tend to increasingly used in Yogyakarta and Surabaya, but less in Makasar and Manado.

In brief, an effective crime prevention can be achieved by implementing both social and technology crimes prevention means. Such approaches are already recognized in the literature as Crime Prevention Through Social Development (CPTSD) which is a non physical crime prevention system, and Crime Prevention Through Environmental Design(CPTED) that is physical crime prevention system.[8,9]

4. Conclusion

This paper finds three important points. The first is that urban crimes tend to fluctuate in some Indonesian cities, some are increase while some are decrease. Both in big cities such as Surabaya, Makasar, and also medium size cities like Yogyakarta and Manado, urban crimes are fluctuated. Secondly, this study indicates that in terms of spatial distribution, urban crimes tends to be happened in urban fringe areas. This finding is contrary to the finding commonly found in urban areas in Western countries where urban crimes tend to be found in urban centres. The reasons for such phenomena may relate to the spatial patterns of urban growth in Indonesia whereas urban fringes are characterized by new urban uses including university campuses, industrial zones, and others functions including residential areas, a very strategic area, but with no local communities having long tradition of community responds towards crimes.

Third, the research find that socio-factor contributed to urban crimes in urban fringes include: social gap, local government responds, and community responses. Among these factors, community responds seems to be the more crucial ones. The stronger community develop crimes prevention, the area would have a less crimes rate/incidents. It however, should be supported by both good and effective relations with local governments and more innovative technology and physical interventions.

This paper recommends that there is a need for developing a comprehensive and systematic policy and program to achieve safe city, as integral parts of the Habitat III goals in achieving inclusive, safe, resilient, and sustainable city. Such policy and program needs to be integrated with the SDGs goals, since many factors contributed to urban crimes and urban safety relate directly and indirectly with the seventeen's goals within SDGs.

Further, it is curial for the local government to develop their capacity in developing a better urban crimes prevention system. It is important for the local/urban governments to give more priority to urban crimes and urban safety as they are important element in achieving just and sustainable city and community. More effective coordination needs to be developed, particularly between local government and local community in developing urban crimes prevention systems. At the same time, it is crucial to strengthen local community to be able to develop a more effective and efficient local crimes prevention system. The long-tradition model of crime prevention through *gotong royong* system needs to be adjust into a more efficient and modern system, using appropriate technology such as CCTV and others means.

5. Acknowledgments

Author would like to thanks Canadian Council for Science which support the author to start doing research on the issue of urban inclusiveness in the early 2010s, which then become one among research agenda of the author. In addition, author also wishing to acknowledge students working on research agenda related to urban crimes issues under author's supervision. Those students have provided important data and important information for this paper – names of the students and their topic thesis are provided in the references below.

References

[1] BPS – National Bureau of Statistics Indonesia. 2014.
[2] Rank.J. 2007. Urban Crime - Explaining Urban Crime - Violence, Social, Subculture, and Values - http://law.jrank.org/pages/2223/Urban-Crime-Explaining-urban-crime.html#ixzz4iLsYr11F.

[3] Baur, Jenelle, 2007, "Fear of Crime: The Impact of Age , Victimisation, PerceivedVulnerability to Victimisation and Neighborhood Characteristic". ACPR ISSUES, Australia, http://www.acpr.gov.au./pdf/ACPR_16pdf.

[4] Bayley, David H., "Community Policing in Australia an Appraisal Working Paper", National Police Research Unit, Payneham, 1986, http://www.acpr.gov.au/pdf/ACPR35.pdf.

[5] Claude, Fisher, 1995, "The Subcultural Theory of Urbanism: A twentieth Year Assessment." American Journal of Sociology: http://everything2.com/title/subcultural theory.

[6] Sullivan, John, 1980. Back alley neighbourhood: Kampung as urban community in Yogyakarta. Center of Southeast Asian Studies, Monash University;

[7] Guiness, Patrick, 1986, Harmony and Hierarchy in A Javanese Kampung. Oxford University Press.

[8] Crowe, Timothy D.,Crime Prevention Through Environmental Design, Woburn: Butterworth-Heinemann, 1991, http://law.jrank.org/Prevention Police Role.html. Delinquent and Criminal Subcultures, http://law.jrank.org/pages/952 .

[9] Friedmann, Robert R., Community Policing, Comparative Perspectives and Prospect http://books.google .com.

[10] Afandy, Alfian. 2006. Aspek Spasial Kejahatan Perkotaan Studi Kasus Kota Makasar (The Spatial Dimension of Urban Crimes, Case Study of Makasar City. . Unpublished Thesis, Master Program in Urban and Regional Planning, Gadjah Mada University.

[11] Urasatri, Diah. 2009. Local Community Crime Prevention System to Urban Crimes in Indonesia, Case study of Yogyakarta City. . Unpublished Thesis, Master Program in Urban and Regional Planning, Gadjah Mada University.

[12] Pradinie, Karina. 2011. Pola Ruang Kriminalitas Kota, Studi Kasus: Surabaya (The Spatial Pattern of Urban Crimes: The Case of Surabaya. Unpublished Thesis, Master Program in Urban and Regional Planning, Gadjah Mada University.

[13] Permana, Ade. 2015. Environmental Design Affecting Crimes Case Study Bandung City. . Unpublished Thesis, Master Program in Urban and Regional Planning, Gadjah Mada University.